PRAISE FOR MICHIO KAKU'S

THE FUTURE OF THE MIND

"Elegant and thought-provoking. . . . Will cause you to wonder if in the not-too-distant future, the power of our own minds will change our lives in ways we can barely imagine. . . . Kaku's image-filled descriptions of the biology of the brain and brain scanning and measurement tools are wonderfully informative and clear."

—*Washington Independent Review of Books*

"A mind-bending study of the possibilities of the brain. . . . A clear and readable guide to what is going on at a time of astonishingly rapid change."

—*The Telegraph* (London)

"An endlessly fascinating read, truly eye-opening in terms of what could be possible." —*The Big Issue*

"Intriguing. . . . Extraordinary findings." —*Nature*

"Science fiction fans might experience a sort of breathless thrill when reading the book—this stuff is happening! It's really happening!—and for general readers who have never really thought of the brain in all its glorious complexity and potential, the book could be a seriously mind-opening experience." —*Booklist*

"[An] expansive, illuminating journey through the mind. . . . These new mental frontiers make for captivating reading."
—*Publishers Weekly*

"[Author of] *Physics of the Impossible* and *Physics of the Future*, Kaku . . . turns his attention to the human mind with equally satisfying results."
—*Kirkus Reviews*

MICHIO KAKU

THE FUTURE OF THE MIND

Michio Kaku is a professor of physics at the City University of New York, cofounder of string field theory, and the author of several widely acclaimed science books, including *Hyperspace*, *Beyond Einstein*, *Physics of the Impossible*, and *Physics of the Future*; and host of numerous television specials and a national science radio show.

www.mkaku.org

ALSO BY MICHIO KAKU

Physics of the Future
Physics of the Impossible
Parallel Worlds
Hyperspace
Visions
Einstein's Cosmos
Beyond Einstein

DR. MICHIO KAKU

PROFESSOR OF THEORETICAL PHYSICS

CITY UNIVERSITY OF NEW YORK

THE FUTURE OF THE MIND

THE SCIENTIFIC QUEST TO UNDERSTAND,
ENHANCE, AND EMPOWER THE MIND

ANCHOR BOOKS

A DIVISION OF RANDOM HOUSE LLC

NEW YORK

FIRST ANCHOR BOOKS EDITION, FEBRUARY 2015

 Published in the United States by Anchor Books, a division of Random House LLC, New York, and distributed in Canada by Random House of Canada Limited, Toronto, Penguin Random House companies. Originally published in hardcover in the United States by Doubleday, a division of Random House LLC, New York, in 2014.

The Library of Congress has cataloged the Doubleday edition as follows:
Kaku, Michio.
The future of the mind : the scientific quest to understand, enhance, and empower the mind / Dr. Michio Kaku, professor of Theoretical Physics, City University of New York. — First edition.
pages cm
Includes bibliographical references.
1. Neuropsychology. 2. Mind and body—Research.
3. Brain—Mathematical models. 4. Cognitive neuroscience.
5. Brain-computer interfaces. I. Title.
QP360.K325 2014
612.8—dc23 2013017338

Anchor Books Trade Paperback ISBN: 978-0-307-47334-9
eBook ISBN: 978-0-385-53083-5

Book design by Pei Loi Koay
Illustrations by Jeffrey L. Ward

www.anchorbooks.com

Printed in the United States of America
10 9

This book is dedicated to my loving wife, Shizue,

and my daughters, Michelle and Alyson

CONTENTS

BOOK III: ALTERED CONSCIOUSNESS

ACKNOWLEDGMENTS

It has been my great pleasure to have interviewed and interacted with the following prominent scientists, all of them leaders in their fields. I would like to thank them for graciously giving up their time for interviews and discussions about the future of science. They have given me guidance and inspiration, as well as a firm foundation in their respective fields.

I would like to thank these pioneers and trailblazers, especially those who have agreed to appear on my TV specials for the BBC, Discovery, and Science TV channels, and also on my national radio shows, *Science Fantastic* and *Explorations.*

Peter Doherty, Nobel laureate, St. Jude Children's Research Hospital
Gerald Edelman, Nobel laureate, Scripps Research Institute
Leon Lederman, Nobel laureate, Illinois Institute of Technology
Murray Gell-Mann, Nobel laureate, Santa Fe Institute and Cal Tech
the late Henry Kendall, Nobel laureate, MIT
Walter Gilbert, Nobel laureate, Harvard University
David Gross, Nobel laureate, Kavli Institute for Theoretical Physics
Joseph Rotblat, Nobel laureate, St. Bartholomew's Hospital
Yoichiro Nambu, Nobel laureate, University of Chicago
Steven Weinberg, Nobel laureate, University of Texas at Austin
Frank Wilczek, Nobel laureate, MIT

. . .

Amir Aczel, author of *Uranium Wars*
Buzz Aldrin, NASA astronaut, second man to walk on the moon
Geoff Andersen, U.S. Air Force Academy, author of *The Telescope*
Jay Barbree, author of *Moon Shot*
John Barrow, physicist, Cambridge University, author of *Impossibility*
Marcia Bartusiak, author of *Einstein's Unfinished Symphony*
Jim Bell, Cornell University astronomer
Jeffrey Bennet, author of *Beyond UFOs*
Bob Berman, astronomer, author *The Secrets of the Night Sky*
Leslie Biesecker, National Institutes of Health
Piers Bizony, author of *How to Build Your Own Starship*
Michael Blaese, National Institutes of Health
Alex Boese, founder of Museum of Hoaxes
Nick Bostrom, transhumanist, Oxford University
Lt. Col. Robert Bowman, Institute for Space and Security Studies
Cynthia Breazeal, artificial intelligence, MIT Media Lab
Lawrence Brody, National Institutes of Health
Rodney Brooks, director of the MIT Artificial Intelligence Laboratory
Lester Brown, Earth Policy Institute
Michael Brown, astronomer, Cal Tech
James Canton, author of *The Extreme Future*
Arthur Caplan, director of the Center for Bioethics at the University of Pennsylvania
Fritjof Capra, author of *The Science of Leonardo*
Sean Carroll, cosmologist, Cal Tech
Andrew Chaikin, author of *A Man on the Moon*
Leroy Chiao, NASA astronaut
Eric Chivian, International Physicians for the Prevention of Nuclear War
Deepak Chopra, author of *Super Brain*
George Church, director of Harvard's Center for Computational Genetics
Thomas Cochran, physicist, Natural Resources Defense Council
Christopher Cokinos, astronomer, author of *Fallen Sky*

Francis Collins, National Institutes of Health
Vicki Colvin, nanotechnologist, University of Texas
Neal Comins, author of *Hazards of Space Travel*
Steve Cook, NASA spokesperson
Christine Cosgrove, author of *Normal at Any Cost*
Steve Cousins, CEO of Willow Garage Personal Robots Program
Phillip Coyle, former assistant secretary of defense for the U.S. Defense Department
Daniel Crevier, AI, CEO of Coreco
Ken Croswell, astronomer, author of *Magnificent Universe*
Steven Cummer, computer science, Duke University
Mark Cutkowsky, mechanical engineering, Stanford University
Paul Davies, physicist, author of *Superforce*
Daniel Dennet, philosopher, Tufts University
the late Michael Dertouzos, computer science, MIT
Jared Diamond, Pulitzer Prize winner, UCLA
Marriot DiChristina, *Scientific American*
Peter Dilworth, MIT AI Lab
John Donoghue, creator of Braingate, Brown University
Ann Druyan, widow of Carl Sagan, Cosmos Studios
Freeman Dyson, Institute for Advanced Study, Princeton University
David Eagleman, neuroscientist, Baylor College of Medicine
John Ellis, CERN physicist
Paul Erlich, environmentalist, Stanford University
Daniel Fairbanks, author of *Relics of Eden*
Timothy Ferris, University of California, author of *Coming of Age in the Milky Way Galaxy*
Maria Finitzo, stem cell expert, Peabody Award winner
Robert Finkelstein, AI expert
Christopher Flavin, World Watch Institute
Louis Friedman, cofounder of the Planetary Society
Jack Gallant, neuroscientist, University of California at Berkeley
James Garwin, NASA chief scientist
Evelyn Gates, author of *Einstein's Telescope*
Michael Gazzaniga, neurologist, University of California at Santa Barbara
Jack Geiger, cofounder, Physicians for Social Responsibility

David Gelertner, computer scientist, Yale University, University of California
Neal Gershenfeld, MIT Media Lab
Daniel Gilbert, psychologist, Harvard University
Paul Gilster, author of *Centauri Dreams*
Rebecca Goldberg, Environmental Defense Fund
Don Goldsmith, astronomer, author of *Runaway Universe*
David Goodstein, assistant provost of Cal Tech
J. Richard Gott III, Princeton University, author of *Time Travel in Einstein's Universe*
Late Stephen Jay Gould, biologist, Harvard University
Ambassador Thomas Graham, spy satellites and intelligence gathering
John Grant, author of *Corrupted Science*
Eric Green, National Institutes of Health
Ronald Green, author of *Babies by Design*
Brian Greene, Columbia University, author of *The Elegant Universe*
Alan Guth, physicist, MIT, author of *The Inflationary Universe*
William Hanson, author of *The Edge of Medicine*
Leonard Hayflick, University of California at San Francisco Medical School
Donald Hillebrand, Argonne National Labs, future of the car
Frank N. von Hippel, physicist, Princeton University
Allan Hobson, psychiatrist, Harvard University
Jeffrey Hoffman, NASA astronaut, MIT
Douglas Hofstadter, Pulitzer Prize winner, Indiana University, author of *Gödel, Escher, Bach*
John Horgan, Stevens Institute of Technology, author of *The End of Science*
Jamie Hyneman, host of *MythBusters*
Chris Impey, astronomer, author of *The Living Cosmos*
Robert Irie, AI Lab, MIT
P. J. Jacobowitz, *PC* magazine
Jay Jaroslav, MIT AI Lab
Donald Johanson, anthropologist, discoverer of Lucy
George Johnson, *New York Times* science journalist
Tom Jones, NASA astronaut

Steve Kates, astronomer
Jack Kessler, stem cell expert, Peabody Award winner
Robert Kirshner, astronomer, Harvard University
Kris Koenig, astronomer
Lawrence Krauss, Arizona State University, author of *Physics of Star Trek*
Lawrence Kuhn, filmmaker and philosopher, *Closer to Truth*
Ray Kurzweil, inventor, author of *The Age of Spiritual Machines*
Robert Lanza, biotechnology, Advanced Cell Technologies
Roger Launius, author of *Robots in Space*
Stan Lee, creator of Marvel Comics and Spider-Man
Michael Lemonick, senior science editor of *Time*
Arthur Lerner-Lam, geologist, volcanist
Simon LeVay, author of *When Science Goes Wrong*
John Lewis, astronomer, University of Arizona
Alan Lightman, MIT, author of *Einstein's Dreams*
George Linehan, author of *Space One*
Seth Lloyd, MIT, author of *Programming the Universe*
Werner R. Loewenstein, former director of Cell Physics Laboratory, Columbia University
Joseph Lykken, physicist, Fermi National Laboratory
Pattie Maes, MIT Media Lab
Robert Mann, author of *Forensic Detective*
Michael Paul Mason, author of *Head Cases: Stories of Brain Injury and Its Aftermath*
Patrick McCray, author of *Keep Watching the Skies*
Glenn McGee, author of *The Perfect Baby*
James McLurkin, MIT, AI Lab
Paul McMillan, director of Space Watch
Fulvia Melia, astronomer, University of Arizona
William Meller, author of *Evolution Rx*
Paul Meltzer, National Institutes of Health
Marvin Minsky, MIT, author of *The Society of Minds*
Hans Moravec, author of *Robot*
Late Phillip Morrison, physicist, MIT
Richard Muller, astrophysicist, University of California at Berkeley
David Nahamoo, IBM Human Language Technology

Christina Neal, volcanist
Miguel Nicolelis, neuroscientist, Duke University
Shinji Nishimoto, neurologist, University of California at Berkeley
Michael Novacek, American Museum of Natural History
Michael Oppenheimer, environmentalist, Princeton University
Dean Ornish, cancer and heart disease specialist
Peter Palese, virologist, Mount Sinai School of Medicine
Charles Pellerin, NASA official
Sidney Perkowitz, author of *Hollywood Science*
John Pike, GlobalSecurity.org
Jena Pincott, author of *Do Gentlemen Really Prefer Blondes?*
Steven Pinker, psychologist, Harvard University
Thomas Poggio, MIT, artificial intelligence
Correy Powell, editor of *Discover* magazine
John Powell, founder of JP Aerospace
Richard Preston, author of *Hot Zone* and *Demon in the Freezer*
Raman Prinja, astronomer, University College London
David Quammen, evolutionary biologist, author of *The Reluctant Mr. Darwin*
Katherine Ramsland, forensic scientist
Lisa Randall, Harvard University, author of *Warped Passages*
Sir Martin Rees, Royal Astronomer of Great Britain, Cambridge University, author of *Before the Beginning*
Jeremy Rifkin, Foundation for Economic Trends
David Riquier, MIT Media Lab
Jane Rissler, Union of Concerned Scientists
Steven Rosenberg, National Institutes of Health
Oliver Sacks, neurologist, Columbia University
Paul Saffo, futurist, Institute of the Future
Late Carl Sagan, Cornell University, author of *Cosmos*
Nick Sagan, coauthor of *You Call This the Future?*
Michael H. Salamon, NASA's Beyond Einstein program
Adam Savage, host of *MythBusters*
Peter Schwartz, futurist, founder of Global Business Network
Michael Shermer, founder of Skeptic Society and *Skeptic* magazine
Donna Shirley, NASA Mars program

Seth Shostak, SETI Institute
Neil Shubin, author of *Your Inner Fish*
Paul Shurch, SETI League
Peter Singer, author of *Wired for War*
Simon Singh, author of *The Big Bang*
Gary Small, author of *iBrain*
Paul Spudis, author of *Odyssey Moon Limited*
Stephen Squyres, astronomer, Cornell University
Paul Steinhardt, Princeton University, author of *Endless Universe*
Jack Stern, stem cell surgeon
Gregory Stock, UCLA, author of *Redesigning Humans*
Richard Stone, author of *NEOs* and *Tunguska*
Brian Sullivan, Hayden Planetarium
Leonard Susskind, physicist, Stanford University
Daniel Tammet, author of *Born on a Blue Day*
Geoffrey Taylor, physicist, University of Melbourne
Late Ted Taylor, designer of U.S. nuclear warheads
Max Tegmark, cosmologist, MIT
Alvin Toffler, author of *The Third Wave*
Patrick Tucker, World Future Society
Chris Turney, University of Wollongong, author of *Ice, Mud and Blood*
Neil de Grasse Tyson, director of Hayden Planetarium
Sesh Velamoor, Foundation for the Future
Robert Wallace, author of *Spycraft*
Kevin Warwick, human cyborgs, University of Reading, UK
Fred Watson, astronomer, author of *Stargazer*
Late Mark Weiser, Xerox PARC
Alan Weisman, author of *The World Without Us*
Daniel Wertheimer, SETI at Home, University of California at Berkeley
Mike Wessler, MIT AI Lab
Roger Wiens, astronomer, Los Alamos National Laboratory
Author Wiggins, author of *The Joy of Physics*
Anthony Wynshaw-Boris, National Institutes of Health
Carl Zimmer, biologist, author of *Evolution*
Robert Zimmerman, author of *Leaving Earth*
Robert Zubrin, founder of Mars Society

I would also like to thank my agent, Stuart Krichevsky, who has been at my side all these years and has given me helpful advice about my books. I have always benefited from his sound judgment. In addition, I would like to thank my editors, Edward Kastenmeier and Melissa Danaczko, who have guided my book and provided invaluable editorial advice. And I would like to thank Dr. Michelle Kaku, my daughter and a neurology resident at Mount Sinai Hospital in New York, for stimulating, thoughtful, and fruitful discussions about the future of neurology. Her careful and thorough reading of the manuscript has greatly enhanced the presentation and content of this book.

THE FUTURE OF THE MIND

INTRODUCTION

The two greatest mysteries in all of nature are the mind and the universe. With our vast technology, we have been able to photograph galaxies billions of light-years away, manipulate the genes that control life, and probe the inner sanctum of the atom, but the mind and the universe still elude and tantalize us. They are the most mysterious and fascinating frontiers known to science.

If you want to appreciate the majesty of the universe, just turn your gaze to the heavens at night, ablaze with billions of stars. Ever since our ancestors first gasped at the splendor of the starry sky, we have puzzled over these eternal questions: Where did it all come from? What does it all mean?

To witness the mystery of our mind, all we have to do is stare at ourselves in the mirror and wonder, What lurks behind our eyes? This raises haunting questions like: Do we have a soul? What happens to us after we die? Who am "I" anyway? And most important, this brings us to the ultimate question: Where do we fit into this great cosmic scheme? As the great Victorian biologist Thomas Huxley once said, "The question of all questions for humanity, the problem which lies behind all others and is more interesting than any of them, is that of the determination of man's place in Nature and his relation to the Cosmos."

There are 100 billion stars in the Milky Way galaxy, roughly the same as the number of neurons in our brain. You may have to travel twenty-four trillion miles, to the first star outside our solar system, to find an object as complex as what is sitting on your shoulders. The mind and the universe pose the greatest scientific challenge of all, but they also share a curious relationship. On one hand they are polar opposites. One is concerned with the vastness of outer space, where we encounter strange denizens like black holes, exploding stars, and colliding galaxies. The other is concerned with inner space, where we find our most intimate and private hopes and desires. The mind is no farther than our next thought, yet we are often clueless when asked to articulate and explain it.

But although they may be opposites in this respect, they also have a common history and narrative. Both were shrouded in superstition and magic since time immemorial. Astrologers and phrenologists claimed to find the meaning of the universe in every constellation of the zodiac and in every bump on your head. Meanwhile, mind readers and seers have been alternately celebrated and vilified over the years.

The universe and the mind continue to intersect in a variety of ways, thanks in no small part to some of the eye-opening ideas we often encounter in science fiction. Reading these books as a child, I would daydream about being a member of the Slan, a race of telepaths created by A. E. van Vogt. I marveled at how a mutant called the Mule could unleash his vast telepathic powers and nearly seize control of the Galactic Empire in Isaac Asimov's *Foundation Trilogy.* And in the movie *Forbidden Planet,* I wondered how an advanced civilization millions of years beyond ours could channel its enormous telekinetic powers to reshape reality to its whims and wishes.

Then when I was about ten, "The Amazing Dunninger" appeared on TV. He would dazzle his audience with his spectacular magic tricks. His motto was "For those who believe, no explanation is necessary; for those who do not believe, no explanation will suffice." One day, he declared that he would send his thoughts to millions of people throughout the country. He closed his eyes and began to concentrate, stating that he was beaming the name of a president of the United States. He asked people to write down the name that popped into their heads on a postcard and mail it in. The next week, he announced triumphantly that thousands of postcards had come pouring in with the name "Roosevelt," the very same name he was "beaming" across the United States.

I wasn't impressed. Back then, the legacy of Roosevelt was strong among those who had lived through the Depression and World War II, so this came as no surprise. (I thought to myself that it would have been truly amazing if he had been thinking of President Millard Fillmore.)

Still, it stoked my imagination, and I couldn't resist experimenting with telepathy on my own, trying to read other people's minds by concentrating as hard as I could. Closing my eyes and focusing intently, I would attempt to "listen" to other people's thoughts and telekinetically move objects around my room.

I failed.

Maybe somewhere telepaths walked the Earth, but I wasn't one of them. In the process, I began to realize that the wondrous exploits of telepaths were probably impossible—at least without outside assistance. But in the years that followed, I also slowly learned another lesson: to fathom the greatest secrets in the universe, one did not need telepathic or superhuman abilities. One just had to have an open, determined, and curious mind. In particular, in order to understand whether the fantastic devices of science fiction are possible, you have to immerse yourself in advanced physics. To understand the precise point when the possible becomes the impossible, you have to appreciate and understand the laws of physics.

These two passions have fired up my imagination all these years: to understand the fundamental laws of physics, and to see how science will shape the future of our lives. To illustrate this and to share my excitement in probing the ultimate laws of physics, I have written the books *Hyperspace, Beyond Einstein,* and *Parallel Worlds.* And to express my fascination with the future, I have written *Visions, Physics of the Impossible,* and *Physics of the Future.* Over the course of writing and researching these books, I was continually reminded that the human mind is still one of the greatest and most mysterious forces in the world.

Indeed, we've been at a loss to understand what it is or how it works for most of history. The ancient Egyptians, for all their glorious accomplishments in the arts and sciences, believed the brain to be a useless organ and threw it away when embalming their pharaohs. Aristotle was convinced that the soul resided in the heart, not the brain, whose only function was to cool down the cardiovascular system. Others, like Descartes, thought that the soul entered the body through the tiny pineal gland of the brain. But in the absence of any solid evidence, none of these theories could be proven.

This "dark age" persisted for thousands of years, and with good reason. The brain weighs only three pounds, yet it is the most complex object in the solar system. Although it occupies only 2 percent of the body's weight, the brain has a ravenous appetite, consuming fully 20 percent of our total energy (in newborns, the brain consumes an astonishing 65 percent of the baby's energy), while fully 80 percent of our genes are coded for the brain. There are an estimated 100 billion neurons residing inside the skull with an exponential amount of neural connections and pathways.

Back in 1977, when the astronomer Carl Sagan wrote his Pulitzer Prize–winning book, *The Dragons of Eden*, he broadly summarized what was known about the brain up to that time. His book was beautifully written and tried to represent the state of the art in neuroscience, which at that time relied heavily on three main sources. The first was comparing our brains with those of other species. This was tedious and difficult because it involved dissecting the brains of thousands of animals. The second method was equally indirect: analyzing victims of strokes and disease, who often exhibit bizarre behavior because of their illness. Only an autopsy performed after their death could reveal which part of the brain was malfunctioning. Third, scientists could use electrodes to probe the brain and slowly and painfully piece together which part of the brain influenced which behavior.

But the basic tools of neuroscience did not provide a systematic way of analyzing the brain. You could not simply requisition a stroke victim with damage in the specific area you wanted to study. Since the brain is a living, dynamic system, autopsies often did not uncover the most interesting features, such as how the parts of the brain interact, let alone how they produced such diverse thoughts as love, hate, jealousy, and curiosity.

TWIN REVOLUTIONS

Four hundred years ago, the telescope was invented, and almost overnight, this new, miraculous instrument peered into the heart of the celestial bodies. It was one of the most revolutionary (and seditious) instruments of all time. All of a sudden, with your own two eyes, you could see the myths and dogma of the past evaporate like the morning mist. Instead of being perfect examples of divine wisdom, the moon had jagged craters, the sun had black spots, Jupiter had moons, Venus had phases, and Saturn had rings. More was learned about the universe in the fifteen years after the invention of the telescope than in all human history put together.

Like the invention of the telescope, the introduction of MRI machines and a variety of advanced brain scans in the mid-1990s and 2000s has transformed neuroscience. We have learned more about the brain in the last fifteen years than in all prior human history, and the mind, once considered out of reach, is finally assuming center stage.

Nobel laureate Eric R. Kandel of the Max Planck Institute in Tübingen, Germany, writes, "The most valuable insights into the human mind to emerge during this period did not come from the disciplines traditionally concerned with the mind—philosophy, psychology, or psycho-analysis. Instead they came from a merger of these disciplines with the biology of the brain. . . ."

Physicists have played a pivotal role in this endeavor, providing a flood of new tools with acronyms like MRI, EEG, PET, CAT, TCM, TES, and DBS that have dramatically changed the study of the brain. Suddenly with these machines we could see thoughts moving within the living, thinking brain. As neurologist V. S. Ramachandran of the University of California, San Diego, says, "All of these questions that philosophers have been studying for millennia, we scientists can begin to explore by doing brain imaging and by studying patients and asking the right questions."

Looking back, some of my initial forays into the world of physics intersected with the very technologies that are now opening up the mind for science. In high school, for instance, I became aware of a new form of matter, called antimatter, and decided to conduct a science project on the topic. As it is one of the most exotic substances on Earth, I had to appeal to the old Atomic Energy Commission just to obtain a tiny quantity of sodium-22, a substance that naturally emits a positive electron (anti-electron, or positron). With my small sample in hand, I was able to build a cloud chamber and powerful magnetic field that allowed me to photograph the trails of vapor left by antimatter particles. I didn't know it at the time, but sodium-22 would soon become instrumental in a new technology, called PET (positron emission tomography), which has since given us startling new insights into the thinking brain.

Yet another technology I experimented with in high school was magnetic resonance. I attended a lecture by Felix Bloch of Stanford University, who shared the 1952 Nobel Prize for Physics with Edward Purcell for the discovery of nuclear magnetic resonance. Dr. Bloch explained to us high school kids that if you had a powerful magnetic field, the atoms would align vertically in that field like compass needles. Then if you applied a radio pulse to these atoms at a precise resonant frequency, you could make them flip over. When they eventually flipped back, they would emit another pulse, like an echo, which would allow you to determine the identity of these atoms. (Later, I used the principle of magnetic resonance to build a 2.3-million-electron-volt particle accelerator in my mom's garage.)

Just a couple of years later, as a freshman at Harvard University, it was an honor to have Dr. Purcell teach me electrodynamics. Around that same time, I also had a summer job and got a chance to work with Dr. Richard Ernst, who was trying to generalize the work of Bloch and Purcell on magnetic resonance. He succeeded spectacularly and would eventually win the Nobel Prize for Physics in 1991 for laying the foundation for the modern MRI (magnetic resonance imaging) machine. The MRI machine, in turn, has given us detailed photographs of the living brain in even finer detail than PET scans.

EMPOWERING THE MIND

Eventually I became a professor of theoretical physics, but my fascination with the mind remained. It is thrilling to see that, just within the last decade, advances in physics have made possible some of the feats of mentalism that excited me when I was a child. Using MRI scans, scientists can now read thoughts circulating in our brains. Scientists can also insert a chip into the brain of a patient who is totally paralyzed and connect it to a computer, so that through thought alone that patient can surf the web, read and write e-mails, play video games, control their wheelchair, operate household appliances, and manipulate mechanical arms. In fact, such patients can do anything a normal person can do via a computer.

Scientists are now going even further, by connecting the brain directly to an exoskeleton that these patients can wear around their paralyzed limbs. Quadriplegics may one day lead near-normal lives. Such exoskeletons may also give us superpowers enabling us to handle deadly emergencies. One day, our astronauts may even explore the planets by mentally controlling mechanical surrogates from the comfort of their living rooms.

As in the movie *The Matrix*, we might one day be able to download memories and skills using computers. In animal studies, scientists have already been able to insert memories into the brain. Perhaps it's only a matter of time before we, too, can insert artificial memories into our brains to learn new subjects, vacation in new places, and master new hobbies. And if technical skills can be downloaded into the minds of workers and scientists, this may even affect the world economy. We might even be able to share these memories as well. One day, scientists might construct an "Internet of the mind," or a brain-net, where thoughts and emotions are sent electronically around the world. Even dreams will be videotaped and then "brain-mailed" across the Internet.

Technology may also give us the power to enhance our intelligence. Progress has been made in understanding the extraordinary powers of "savants" whose mental, artistic, and mathematical abilities are truly astonishing. Furthermore, the genes that separate us from the apes are now being sequenced, giving us an unparalleled glimpse into the evolutionary origins of the brain. Genes have already been isolated in animals that can increase their memory and mental performance.

The excitement and promise generated by these eye-opening advances are so enormous that they have also caught the attention of the politicians. In fact, brain science has suddenly become the source of a transatlantic rivalry between the greatest economic powers on the planet. In January 2013, both President Barack Obama and the European Union announced what could eventually become multibillion-dollar funding for two independent projects that would reverse engineer the brain. Deciphering the intricate neural circuitry of the brain, once considered hopelessly beyond the scope of modern science, is now the focus of two crash projects that, like the Human Genome Project, will change the scientific and medical landscape. Not only will this give us unparalleled insight into the mind, it will also generate new industries, spur economic activity, and open up new vistas for neuroscience.

Once the neural pathways of the brain are finally decoded, one can envision understanding the precise origins of mental illness, perhaps leading to a cure for this ancient affliction. This decoding also makes it possible to create a copy of the brain, which raises philosophical and ethical questions. Who are we, if our consciousness can be uploaded into a computer? We can also toy with the concept of immortality. Our bodies may eventually decay and die, but can our consciousness live forever?

And beyond that, perhaps one day in the distant future the mind will be freed of its bodily constraints and roam among the stars, as several scientists have speculated. Centuries from now, one can imagine placing our entire neural blueprint on laser beams, which will then be sent into deep space, perhaps the most convenient way for our consciousness to explore the stars.

A brilliant new scientific landscape that will reshape human destiny is now truly opening up. We are now entering a new golden age of neuroscience.

In making these predictions, I have had the invaluable assistance of scientists who graciously allowed me to interview them, broadcast their ideas on national radio, and even take a TV crew into their laboratories. These are the scientists who are laying the foundation for the future of the mind. For their ideas to be incorporated into this book, I made only two requirements: (1) their predictions must rigorously obey the laws of physics; and (2) prototypes must exist to show proof-of-principle for these far-reaching ideas.

TOUCHED BY MENTAL ILLNESS

I once wrote a biography of Albert Einstein, called *Einstein's Cosmos,* and had to delve into the minute details of his private life. I had known that Einstein's youngest son was afflicted with schizophrenia, but did not realize the enormous emotional toll that it had taken on the great scientist's life. Einstein was also touched by mental illness in another way; one of his closest colleagues was the physicist Paul Ehrenfest, who helped Einstein create the theory of general relativity. After suffering bouts of depression, Ehrenfest tragically killed his own son, who had Down's syndrome, and then committed suicide. Over the years, I have found that many of my colleagues and friends have struggled to manage mental illness in their families.

Mental illness has also deeply touched my own life. Several years ago, my mother died after a long battle with Alzheimer's disease. It was heartbreaking to see her gradually lose her memories of her loved ones, to gaze into her eyes and realize that she did not know who I was. I could see the glimmer of humanity slowly being extinguished. She had spent a lifetime struggling to raise a family, and instead of enjoying her golden years, she was robbed of all the memories she held dear.

As the baby boomers age, the sad experience that I and many others have had will be repeated across the world. My wish is that rapid advances in neuroscience will one day alleviate the suffering felt by those afflicted with mental illness and dementia.

WHAT IS DRIVING THIS REVOLUTION?

The data pouring in from brain scans are now being decoded, and the progress is stunning. Several times a year, headlines herald a fresh breakthrough. It took 350 years, since the invention of the telescope, to enter the space age, but it has taken only fifteen years since the introduction of the MRI and advanced brain scans to actively connect the brain to the outside world. *Why so quickly, and how much is there to come?*

Part of this rapid progress has occurred because physicists today have a good understanding of electromagnetism, which governs the electrical signals racing through our neurons. The mathematical equations of James Clerk Maxwell, which are used to calculate the physics of antennas, radar, radio receivers, and microwave towers, form the very cornerstone of MRI technology. It took centuries to finally solve the secret of electromagnetism, but neuroscience can enjoy the fruits of this grand endeavor. In Book I, I will survey the history of the brain and explain how a galaxy of new instruments has left the physics labs and given us glorious color pictures of the mechanics of thought. Because consciousness plays so central a role in any discussion of the mind, I also give a physicist's perspective, offering a definition of consciousness that includes the animal kingdom as well. In fact, I provide a ranking of consciousness, showing how it is possible to assign a number to various types of consciousness.

But to fully answer the question of how this technology will advance, we also have to look at Moore's law, which states that computer power doubles every two years. I often surprise people with the simple fact that your cell phone today has more computer power than *all* of NASA when it put two men on the moon in 1969. Computers are now powerful enough to record the electrical signals emanating from the brain and partially decode them into a familiar digital language. This makes it possible for the brain to directly interface with computers to control any object around it. The fast-growing field is called BMI (brain-machine interface), and the key technology is the computer. In Book II, I'll explore this new technology, which has made recording memories, mind reading, videotaping our dreams, and telekinesis possible.

In Book III, I'll investigate alternate forms of consciousness, from dreams, drugs, and mental illness to robots and even aliens from outer space. Here we'll also learn about the potential to control and manipulate the brain to manage diseases such as depression, Parkinson's, Alzheimer's, and many more. I will also elaborate on the Brain Research Through Advancing Innovative Neurotechnologies (or BRAIN) project announced by President Obama, and the Human Brain Project of the European Union, which will potentially allocate billions of dollars to decode the pathways of the brain, all the way down to the neural level. These two crash programs will undoubtedly open up entirely new research areas, giving us new ways to treat mental illness and also revealing the deepest secrets of consciousness.

After we have given a definition of consciousness, we can use it to explore nonhuman consciousness as well (i.e., the consciousness of robots). How advanced can robots become? Can they have emotions? Will they pose a threat? And we can also explore the consciousness of aliens, who may have goals totally different from ours.

In the Appendix, I will discuss perhaps the strangest idea in all of science, the concept from quantum physics that consciousness may be the fundamental basis for reality.

There is no shortage of proposals for this exploding field. Only time will tell which ones are mere pipe dreams created by the overheated imagination of science-fiction writers and which ones represent solid avenues for future scientific research. Progress in neuroscience has been astronomical, and in many ways the key has been modern physics, which uses the full power of the electromagnetic and nuclear forces to probe the great secrets hidden within our minds.

I should stress that I am not a neuroscientist. I am a theoretical physicist with an enduring interest in the mind. I hope that the vantage point of a physicist can help further enrich our knowledge and give a fresh new understanding of the most familiar and alien object in the universe: our mind.

But given the dizzying pace with which radically new perspectives are being developed, it is important that we have a firm grasp on how the brain is put together.

So let us first discuss the origins of modern neuroscience, which some historians believe began when an iron spike sailed through the brain of a certain Phineas Gage. This seminal event set off a chain reaction that helped open the brain to serious scientific investigation. Although it was an unfortunate event for Mr. Gage, it paved the way for modern science.

BOOK I THE MIND AND CONSCIOUSNESS

My fundamental premise about the brain is that its workings—what we sometimes call "mind"—are a consequence of its anatomy and physiology, and nothing more.

—CARL SAGAN

1 UNLOCKING THE MIND

In 1848, Phineas Gage was working as a railroad foreman in Vermont, when dynamite accidentally went off, propelling a three-foot, seven-inch spike straight into his face, through the front part of his brain, and out the top of his skull, eventually landing eighty feet away. His fellow workers, shocked to see part of their foreman's brain blown off, immediately called for a doctor. To the workers' (and even the doctor's) amazement, Mr. Gage did not die on-site.

He was semiconscious for weeks, but eventually made what seemed like a full recovery. (A rare photograph of Gage surfaced in 2009, showing a handsome, confident man, with an injury to his head and left eye, holding the iron rod.) But after this incident, his coworkers began to notice a sharp change in his personality. A normally cheerful, helpful foreman, Gage became abusive, hostile, and selfish. Ladies were warned to stay clear of him. Dr. John Harlow, the doctor who treated him, observed that Gage was "capricious and vacillating, devising many plans of future operations, which are no sooner arranged than they are abandoned in turn for others appearing more feasible. A child in his intellectual capacity and manifestations, yet with the animal passions of a strong man." Dr. Harlow noted that he was "radically changed" and that his fellow workers said that "he was no longer Gage." After Gage's death in 1860, Dr. Harlow preserved both his skull and the rod that had smashed into it. Detailed X-ray scans of the skull have since confirmed that the iron rod caused massive destruction in the area of the brain behind the forehead known as the frontal lobe, in both the left and right cerebral hemispheres.

This incredible accident would not only change the life of Phineas Gage, it would alter the course of science as well. Previously, the dominant thinking was that the brain and the soul were two separate entities, a philosophy called dualism. But it became increasingly clear that damage to the frontal lobe of his brain had caused abrupt changes in Gage's personality. This, in turn, created a paradigm shift in scientific thinking: perhaps specific areas of the brain could be traced to certain behaviors.

BROCA'S BRAIN

In 1861, just a year after Gage's death, this view was further cemented through the work of Pierre Paul Broca, a physician in Paris who documented a patient who appeared normal except that he had a severe speech deficit. The patient could understand and comprehend speech perfectly, but he could utter only one sound, the word "tan." After the patient died, Dr. Broca confirmed during the autopsy that the patient suffered from a lesion in his left temporal lobe, a region of the brain near his left ear. Dr. Broca would later confirm twelve similar cases of patients with damage to this specific area of the brain. Today patients who have damage to the temporal lobe, usually in the left hemisphere, are said to suffer from Broca's aphasia. (In general, patients with this disorder can understand speech but cannot say anything, or else they drop many words when speaking.)

Soon afterward, in 1874, German physician Carl Wernicke described patients who suffered from the opposite problem. They could articulate clearly, but they could not understand written or spoken speech. Often these patients could speak fluently with correct grammar and syntax, but with nonsensical words and meaningless jargon. Sadly, these patients often didn't know they were spouting gibberish. Wernicke confirmed after performing autopsies that these patients had suffered damage to a slightly different area of the left temporal lobe.

The works of Broca and Wernicke were landmark studies in neuroscience, establishing a clear link between behavioral problems, such as speech and language impairment, and damage to specific regions of the brain.

Another breakthrough took place amid the chaos of war. Throughout history, there were many religious taboos prohibiting the dissection of the human body, which severely restricted progress in medicine. In warfare, however, with tens of thousands of bleeding soldiers dying on the battlefield, it became an urgent mission for doctors to develop any medical treatment that worked. During the Prusso-Danish War in 1864, German doctor Gustav Fritsch treated many soldiers with gaping wounds to the brain and happened to notice that when he touched one hemisphere of the brain, the opposite side of the body often twitched. Later Fritsch systematically showed that, when he electrically stimulated the brain, the left hemisphere controlled the right side of the body, and vice versa. This was a stunning discovery, demonstrating that the brain was basically electrical in nature and that a particular region of the brain controlled a part on the other side of the body. (Curiously, the use of electrical probes on the brain was first recorded a couple of thousand years earlier by the Romans. In the year A.D. 43, records show that the court doctor to the emperor Claudius used electrically charged torpedo fish, which were applied to the head of a patient suffering from severe headaches.)

The realization that there were electrical pathways connecting the brain to the body wasn't systematically analyzed until the 1930s, when Dr. Wilder Penfield began working with epilepsy patients, who often suffered from debilitating convulsions and seizures that were potentially life-threatening. For them, the last option was to have brain surgery, which involved removing parts of the skull and exposing the brain. (Since the brain has no pain sensors, a person can be conscious during this entire procedure, so Dr. Penfield used only a local anesthetic during the operation.)

Dr. Penfield noticed that when he stimulated certain parts of the cortex with an electrode, different parts of the body would respond. He suddenly realized that he could draw a rough one-to-one correspondence between specific regions of the cortex and the human body. His drawings were so accurate that they are still used today in almost unaltered form. They had an immediate impact on both the scientific community and the general public. In one diagram, you could see which region of the brain roughly controlled which function, and how important each function was. For example, because our hands and mouth are so vital for survival, a considerable amount of brain power is devoted to controlling them, while the sensors in our back hardly register at all.

Furthermore, Penfield found that by stimulating parts of the temporal lobe, his patients suddenly relived long-forgotten memories in a crystal-clear fashion. He was shocked when a patient, in the middle of brain surgery, suddenly blurted out, "It was like . . . standing in the doorway at [my] high school. . . . I heard my mother talking on the phone, telling my aunt to come over that night." Penfield realized that he was tapping into memories buried deep inside the brain. When he published his results in 1951, they created another transformation in our understanding of the brain.

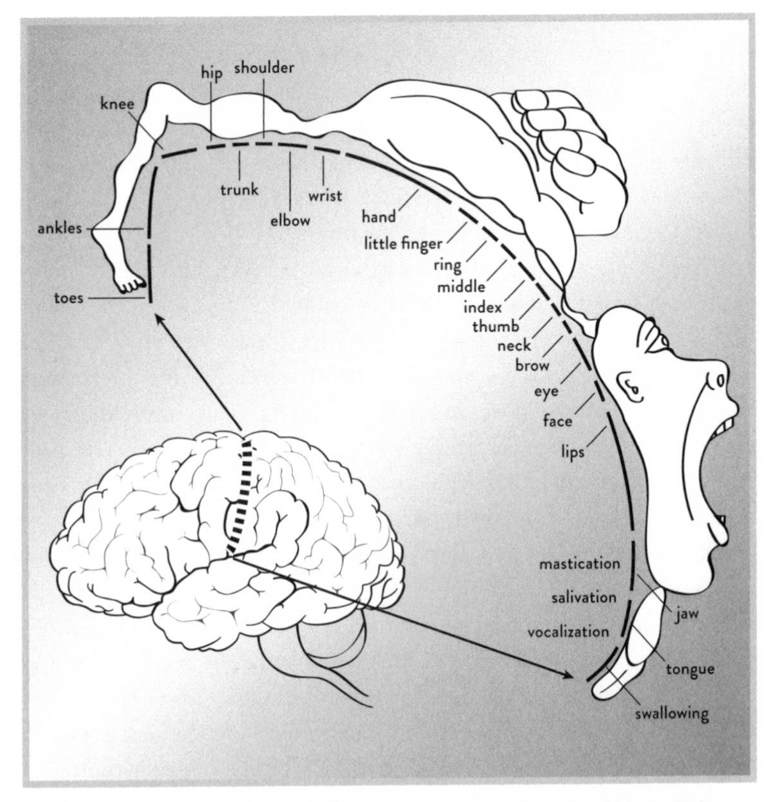

Figure 1. This is the map of the motor cortex that was created by Dr. Wilder Penfield, showing which region of the brain controls which part of the body.

A MAP OF THE BRAIN

By the 1950s and '60s, it was possible to create a crude map of the brain, locating different regions and even identifying the functions of a few of them.

In Figure 2, we see the neocortex, which is the outer layer of the brain, divided into four lobes. It is highly developed in humans. All the lobes of the brain are devoted to processing signals from our senses, except for one: the frontal lobe, located behind the forehead. The prefrontal cortex, the foremost part of the frontal lobe, is where most rational thought is processed. The information you are reading right now is being processed in your prefrontal cortex. Damage to this area can impair your ability to plan or con-

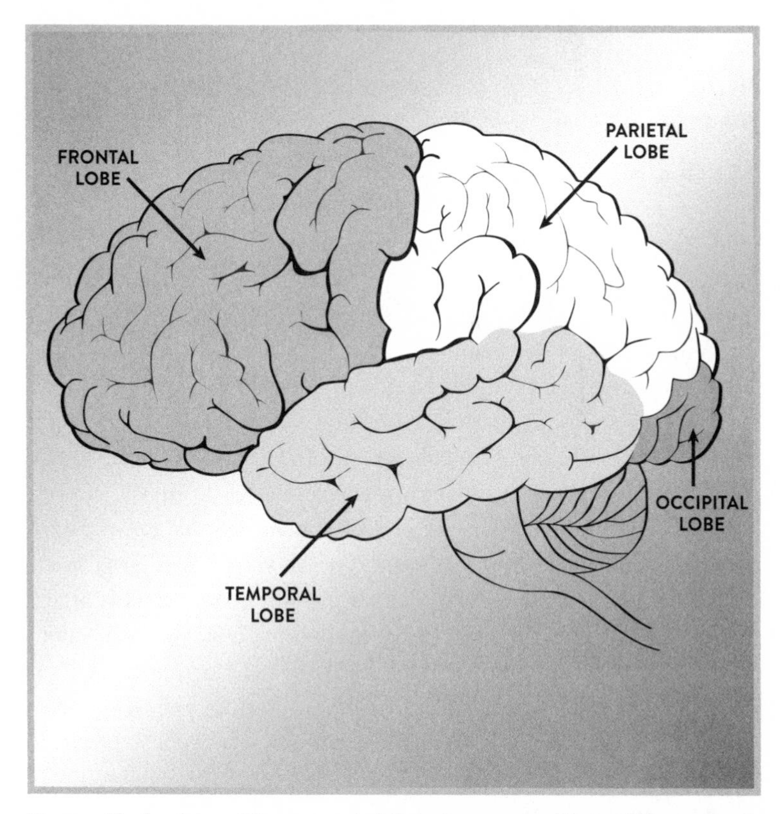

Figure 2. The four lobes of the neocortex of the brain are responsible for different, though related, functions.

template the future, as in the case of Phineas Gage. This is the region where information from our senses is evaluated and a future course of action is carried out.

The parietal lobe is located at the top of our brains. The right hemisphere controls sensory attention and body image; the left hemisphere controls skilled movements and some aspects of language. Damage to this area can cause many problems, such as difficulty in locating parts of your own body.

The occipital lobe is located at the very back of the brain and processes visual information from the eyes. Damage to this area can cause blindness and visual impairment.

The temporal lobe controls language (on the left side only), as well as the visual recognition of faces and certain emotional feelings. Damage to this lobe can leave us speechless or without the ability to recognize familiar faces.

THE EVOLVING BRAIN

When you look at other organs of the body, such as our muscles, bones, and lungs, there seems to be an obvious rhyme and reason to them that we can immediately see. But the structure of the brain might seem slapped together in a rather chaotic fashion. In fact, trying to map the brain has often been called "cartography for fools."

To make sense of the seemingly random structure of the brain, in 1967 Dr. Paul MacLean of the National Institute of Mental Health applied Charles Darwin's theory of evolution to the brain. He divided the brain into three parts. (Since then, studies have shown that there are refinements to this model, but we will use it as a rough organizing principle to explain the overall structure of the brain.) First, he noticed that the back and center part of our brains, containing the brain stem, cerebellum, and basal ganglia, are almost identical to the brains of reptiles. Known as the "reptilian brain," these are the oldest structures of the brain, governing basic animal functions such as balance, breathing, digestion, heartbeat, and blood pressure. They also control behaviors such as fighting, hunting, mating, and territoriality, which are necessary for survival and reproduction. The reptilian brain can be traced back about 500 million years. (See Figure 3.)

But as we evolved from reptiles to mammals, the brain also became more complex, evolving outward and creating entirely new structures. Here we

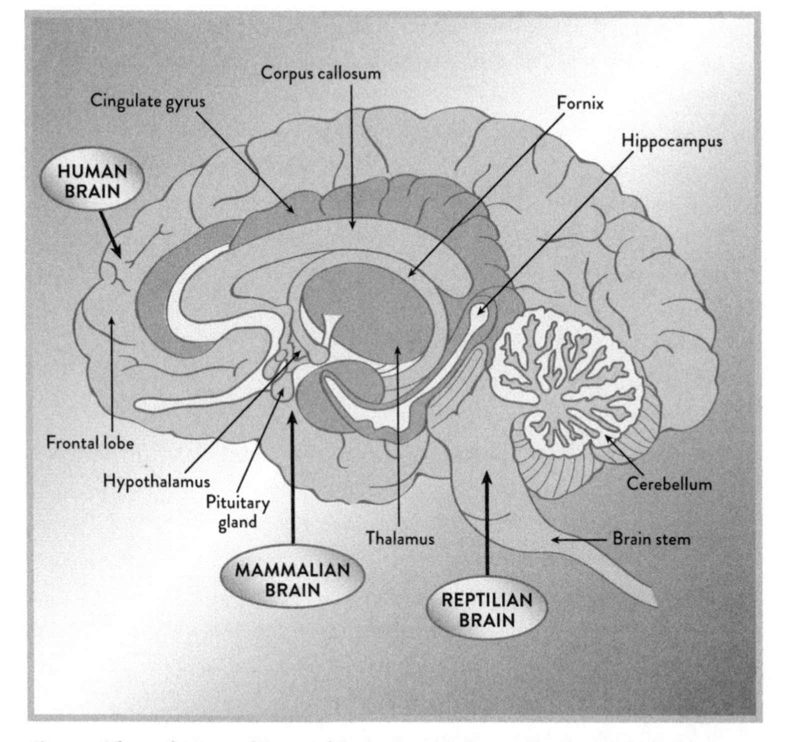

Figure 3. The evolutionary history of the brain, with the reptilian brain, the limbic system (the mammalian brain), and the neocortex (the human brain). Roughly speaking, one can argue that the path of our brain's evolution passed from the reptilian brain to the mammalian brain to the human brain.

encounter the "mammalian brain," or the limbic system, which is located near the center of the brain, surrounding parts of the reptilian brain. The limbic system is prominent among animals living in social groups, such as the apes. It also contains structures that are involved in emotions. Since the dynamics of social groups can be quite complex, the limbic system is essential in sorting out potential enemies, allies, and rivals.

The different parts of the limbic system that control behaviors crucial for social animals are:

- The hippocampus. This is the gateway to memory, where short-term memories are processed into long-term memories. Its name means "seahorse," which describes its strange shape. Damage here will destroy the ability to make new long-term memories. You are left a prisoner of the present.

- The amygdala. This is the seat of emotions, especially fear, where emotions are first registered and generated. Its name means "almond."
- The thalamus. This is like a relay station, gathering sensory signals from the brain stem and then sending them out to the various cortices. Its name means "inner chamber."
- The hypothalamus. This regulates body temperature, our circadian rhythm, hunger, thirst, and aspects of reproduction and pleasure. It lies below the thalamus—hence its name.

Finally, we have the third and most recent region of the mammalian brain, the cerebral cortex, which is the outer layer of the brain. The latest evolutionary structure within the cerebral cortex is the neocortex (meaning "new bark"), which governs higher cognitive behavior. It is most highly developed in humans: it makes up 80 percent of our brain's mass, yet is only as thick as a napkin. In rats the neocortex is smooth, but it is highly convoluted in humans, which allows a large amount of surface area to be crammed into the human skull.

In some sense, the human brain is like a museum containing remnants of all the previous stages in our evolution over millions of years, exploding outward and forward in size and function. (This is also roughly the path taken when an infant is born. The infant brain expands outward and toward the front, perhaps mimicking the stages of our evolution.)

Although the neocortex seems unassuming, looks are deceiving. Under a microscope you can appreciate the intricate architecture of the brain. The gray matter of the brain consists of billions of tiny brain cells called neurons. Like a gigantic telephone network, they receive messages from other neurons via dendrites, which are like tendrils sprouting from one end of the neuron. At the other end of the neuron, there is a long fiber called the axon. Eventually the axon connects to as many as ten thousand other neurons via their dendrites. At the juncture between the two, there is a tiny gap called the synapse. These synapses act like gates, regulating the flow of information within the brain. Special chemicals called neurotransmitters can enter the synapse and alter the flow of signals. Because neurotransmitters like dopamine, serotonin, and noradrenaline help control the stream of information moving across the myriad pathways of the brain, they exert a powerful effect on our moods, emotions, thoughts, and state of mind. (See Figure 4.)

This description of the brain roughly represented the state of knowledge through the 1980s. In the 1990s, however, with the introduction of new technologies from the field of physics, the mechanics of thought began to be revealed in exquisite detail, unleashing the current explosion of scientific discovery. One of the workhorses of this revolution has been the MRI machine.

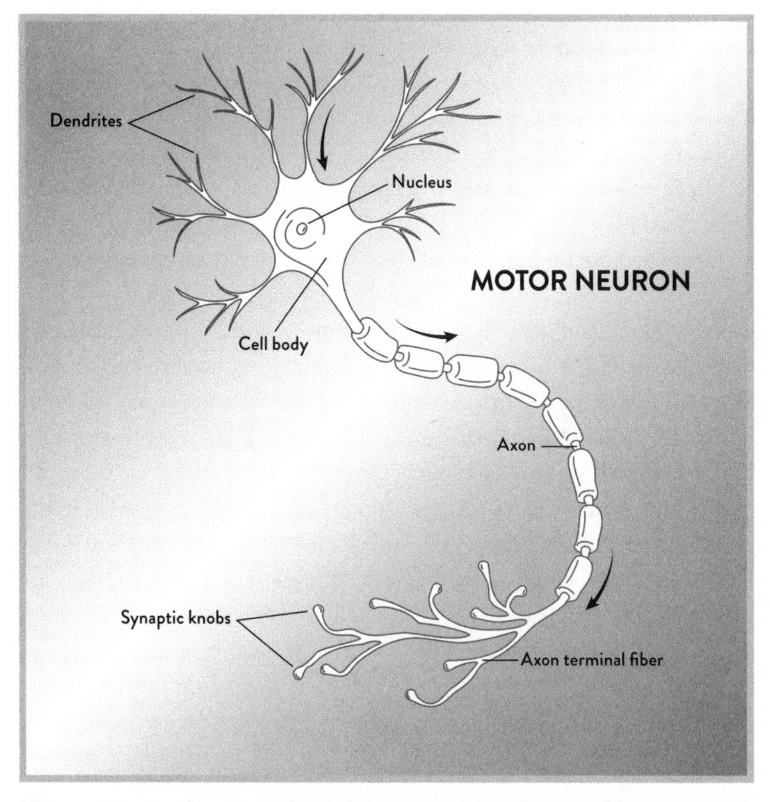

Figure 4. Diagram of a neuron. Electrical signals travel along the axon of the neuron until they hit the synapse. Neurotransmitters can regulate the flow of electrical signals past the synapse.

THE MRI: WINDOW INTO THE BRAIN

To understand the reason why this radical new technology has helped decode the thinking brain, we have to turn our attention to some basic principles of physics.

Radio waves, a type of electromagnetic radiation, can pass right through tissue without doing damage. MRI machines take advantage of this fact, allowing electromagnetic waves to freely penetrate the skull. In the process, this technology has given us glorious photographs of something once thought to be impossible to capture: the inner workings of the brain as it experiences sensations and emotions. Watching the dance of lights flickering in a MRI machine, one can trace out the thoughts moving within the brain. It's like being able to see the inside of a clock as it ticks.

The first thing you notice about an MRI machine is the huge, cylindrical magnetic coils, which can produce a magnetic field twenty to sixty thousand times greater than the strength of Earth's. The giant magnet is one of the principal reasons why an MRI machine can weigh a ton, fill up an entire room, and cost several million dollars. (MRI machines are safer than X-ray machines because they don't create harmful ions. CT scans, which can also create 3-D pictures, flood the body with many times the dosage from an ordinary X-ray, and hence have to be carefully regulated. By contrast, MRI machines are safe when used properly. One problem, however, is the carelessness of workers. The magnetic field is powerful enough to send tools hurling through the air at high velocity when turned on at the wrong time. People have been injured and even killed in this way.)

MRI machines work as follows: Patients lie flat and are inserted into a cylinder containing two large coils, which create the magnetic field. When the magnetic field is turned on, the nuclei of the atoms inside your body act very much like a compass needle: they align horizontally along the direction of the field. Then a small pulse of radio energy is generated, which causes some of the nuclei in our body to flip upside down. When the nuclei later revert back to their normal position, they emit a secondary pulse of radio energy, which is then analyzed by the MRI machine. By analyzing these tiny "echoes," one can then reconstruct the location and nature of these atoms. Like a bat, which uses echoes to determine the position of objects in its path, the echoes created by the MRI machine allow scientists to re-create a remarkable image of the inside of the brain. Computers then reconstruct the position of the atoms, giving us beautiful diagrams in three dimensions.

When MRIs were originally introduced, they were able to show the static structure of the brain and its various regions. However, in the mid-1990s, a new type of MRI was invented, called "functional" MRI, or fMRI, which detected the presence of oxygen in the blood in the brain. (For different types of MRI machines, scientists sometimes put a lowercase letter in front of "MRI," but we will use the abbreviation MRI to denote all the various types of MRI machines.) MRI scans cannot directly detect the flow of electricity in the neurons, but since oxygen is necessary to provide the energy for the neurons, oxygenated blood can indirectly trace the flow of electrical energy in the neurons and show how various regions of the brain interact with one another.

Already these MRI scans have definitively disproven the idea that thinking is concentrated in a single center. Instead, one can see electrical energy circulating across different parts of the brain as it thinks. By tracing the path taken by our thoughts, MRI scans have shed new light into the nature of Alzheimer's, Parkinson's, schizophrenia, and a host of other mental diseases.

The great advantage of MRI machines is their exquisite ability to locate minute parts of the brain, down to a fraction of a millimeter in size. An MRI scan will create not just dots on a two-dimensional screen, called pixels, but dots in three-dimensional space, called "voxels," yielding a bright collection of tens of thousands of colored dots in 3-D, in the shape of a brain.

Since different chemical elements respond to different frequencies of radio, you can change the frequency of the radio pulse and therefore identify different elements of the body. As noted, fMRI machines zero in on the oxygen atom contained within blood in order to measure blood flow, but MRI machines can also be tuned to identify other atoms. In just the last decade, a new form of MRI was introduced called "diffusion tensor imaging" MRI, which detects the flow of water in the brain. Since water follows the neural pathways of the brain, DTI yields beautiful pictures that resemble networks of vines growing in a garden. Scientists can now instantly determine how certain parts of the brain are hooked up with other parts.

There are a couple of drawbacks to MRI technology, however. Although they are unparalleled in spatial resolution, locating voxels down to the size of a pinpoint in three dimensions, MRIs are not that good in temporal resolution. It takes almost a full second to follow the path of blood in the brain, which may not sound like a lot, but remember that electrical signals travel almost instantly throughout the brain, and hence MRI scans can miss some of the intricate details of thought patterns.

Another snag is the cost, which runs in the millions of dollars, so doctors often have to share the machines. But like most technology, developments should bring down the cost over time.

In the meantime, exorbitant costs haven't stalled the hunt for commercial applications. One idea is to use MRI scans as lie detectors, which, according to some studies, can identify lies with 95 percent accuracy or higher. The level of accuracy is still controversial, but the basic idea is that when a person tells a lie, he simultaneously has to know the truth, concoct the lie, and rapidly analyze the consistency of this lie with previously known facts. Today some companies are claiming that MRI technology shows that the prefrontal and parietal lobes light up when someone tells a lie. More specifically, the "orbitofrontal cortex" (which can serve, among other functions, as the brain's "fact-checker" to warn us when something is wrong) becomes active. This area is located right behind the orbits of our eyes, and hence the name. The theory goes that the orbitofrontal cortex understands the difference between the truth and a lie and kicks into overdrive as a result. (Other areas of the brain also light up when someone tells a lie, such as the superiormedial and inferolateral prefrontal cortices, which are involved in cognition.)

Already there are several commercial firms offering MRI machines as lie detectors, and cases involving these machines are entering the court system. But it's important to note that these MRI scans indicate increased brain activity only in certain areas. While DNA results can sometimes have an accuracy of one part in 10 billion or better, MRI scans cannot, because it takes many areas of the brain to concoct a lie, and these same areas of the brain are responsible for processing other kinds of thoughts as well.

EEG SCANS

Another useful tool to probe deep inside the brain is the EEG, the electroencephalogram. The EEG was introduced all the way back in 1924, but only recently has it been possible to employ computers to make sense out of all the data pouring in from each electrode.

To use the EEG machine, the patient usually puts on a futuristic-looking helmet with scores of electrodes on the surface. (More advanced versions place a hairnet over the head containing a series of tiny electrodes.) These electrodes detect the tiny electrical signals that are circulating in the brain.

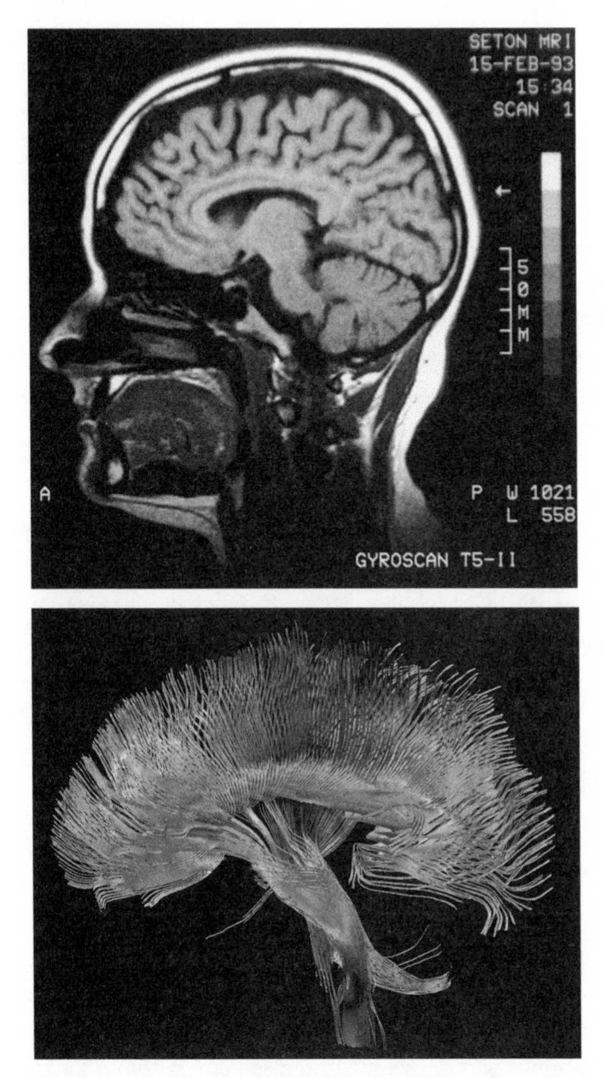

Figure 5. At the top, we see an image taken by a functional MRI machine, showing regions of high mental activity. In the bottom image, we see the flowerlike pattern created by a diffusion MRI machine, which can follow the neural pathways and connections of the brain.

An EEG scan differs from an MRI scan in several crucial ways. The MRI scan, as we have seen, shoots radio pulses into the brain and then analyzes the "echoes" that come back. This means you can vary the radio pulse to select different atoms for analysis, making it quite versatile. The EEG machine, however, is strictly passive; that is, it analyzes the tiny electromagnetic signals the brain naturally emits. The EEG excels at recording the broad electromagnetic signals that surge across the entire brain, which allows scientists to measure the overall activity of the brain as it sleeps, concentrates, relaxes, dreams, etc. Different states of consciousness vibrate at different frequencies. For example, deep sleep corresponds to delta waves, which vibrate at .1 to 4 cycles per second. Active mental states, such as problem solving, correspond to beta waves, vibrating from 12 to 30 cycles per second. These vibrations allow various parts of the brain to share information and communicate with one another, even if they are located on opposite sides of the brain. And while MRI scans measuring blood flow can be taken only several times a second, EEG scans measure electrical activity instantly.

The greatest advantage of the EEG scan, though, is its convenience and cost. Even high school students have done experiments in their living rooms with EEG sensors placed over their heads.

However, the main drawback to the EEG, which has held up its development for decades, is its very poor spatial resolution. The EEG picks up electrical signals that have already been diffused after passing through the skull, making it difficult to detect abnormal activity when it originates deep in the brain. Looking at the output of the muddled EEG signals, it is almost impossible to say for sure which part of the brain created it. Furthermore, slight motions, like moving a finger, can distort the signal, sometimes rendering it useless.

PET SCANS

Yet another useful tool from the world of physics is the positron emission topography (PET) scan, which calculates the flow of energy in the brain by locating the presence of glucose, the sugar molecule that fuels cells. Like the cloud chamber I made as a high school student, PET scans make use of the subatomic particles emitted from sodium-22 within the glucose. To start the PET scan, a special solution containing slightly radioactive sugar is injected into the patient. The sodium atoms inside the sugar molecules have been replaced by radioactive sodium-22 atoms. Every time a sodium atom decays, it emits a positive electron, or positron, which is easily detected by sensors. By following the path of the radioactive sodium atoms in sugar, one can then trace out the energy flow within the living brain.

The PET scan shares many of the same advantages of MRI scans but does not have the fine spatial resolution of an MRI photo. However, instead of measuring blood flow, which is only an indirect indicator of energy consumption in the body, PET scans measure energy consumption, so it is more closely related to neural activity.

There is another drawback to PET scans, however. Unlike MRI and EEG scans, PET scans are slightly radioactive, so patients cannot continually take them. In general, a person is not allowed to have a PET scan more than once a year because of the risk from radiation.

MAGNETISM IN THE BRAIN

Within the last decade, many new high-tech devices have entered the tool kit of neuroscientists, including the transcranial electromagnetic scanner (TES), magnetoencephalography (MEG), near-infrared spectroscopy (NIRS), and optogenetics, among others.

In particular, magnetism has been used to systematically shut down specific parts of the brain without cutting it open. The basic physics behind these new tools is that a rapidly changing electric field can create a magnetic field, and vice versa. MEGs passively measure the magnetic fields produced by the changing electric fields of the brain. These magnetic fields are weak and extremely tiny, only a billionth of Earth's magnetic field. Like the EEG, the MEG is extremely good at time resolution, down to a thousandth of a second. Its spatial resolution, however, is only a cubic centimeter.

Unlike the passive measurement of the MEG, the TES generates a large pulse of electricity, which in turn creates a burst of magnetic energy. The TES is placed next to the brain, so the magnetic pulse penetrates the skull and creates yet another electric pulse inside the brain. This secondary electrical pulse, in turn, is sufficient to turn off or dampen the activity of selected areas of the brain.

Historically, scientists had to rely on strokes or tumors to silence certain parts of the brain and hence determine what they do. But with the TES, one can harmlessly turn off or dampen parts of the brain at will. By shooting magnetic energy at a particular spot in the brain, one can determine its function by simply watching how a person's behavior has changed. (For example, by shooting magnetic pulses into the left temporal lobe, one can see that this adversely affects our ability to talk.)

One potential drawback of the TES is that these magnetic fields do not penetrate very far into the interior of the brain (because magnetic fields decrease much faster than the usual inverse square law for electricity). TES is quite useful in turning off parts of the brain near the skull, but the mag-

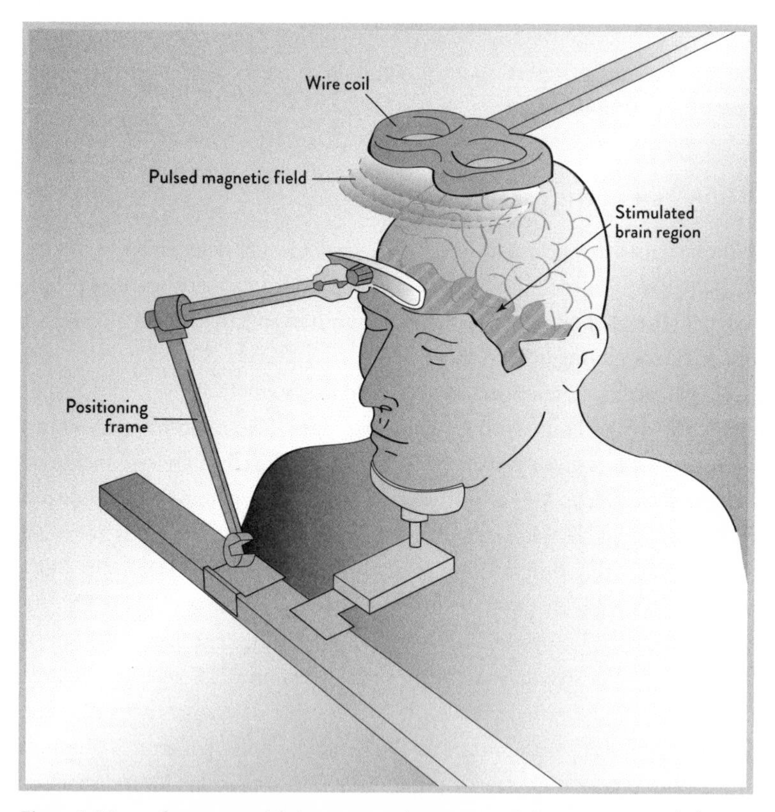

Figure 6. We see the transcranial electromagnetic scanner and the magnetoencephalograph, which uses magnetism rather than radio waves to penetrate the skull and determine the nature of thoughts within the brain. Magnetism can temporarily silence parts of the brain, allowing scientists to safely determine how these regions perform without relying on stroke victims.

netic field cannot reach important centers located deep in the brain, such as the limbic system. But future generations of TES devices may overcome this technical problem by increasing the intensity and precision of the magnetic field.

DEEP BRAIN STIMULATION

Yet another tool that has proven vital to neurologists is deep brain stimulation (DBS). The probes originally used by Dr. Penfield were relatively crude. Today these electrodes can be hairlike and reach specific areas of the brain deep within its interior. Not only has DBS allowed scientists to locate the function of various parts of the brain, it can also be used to treat mental disorders. DBS has already proven its worth with Parkinson's disease, in which certain regions of the brain are overactive and often create uncontrollable shaking of the hands.

More recently, these electrodes have targeted a new area of the brain (called Brodmann's area number 25) that is often overactive in depressed patients who do not respond to psychotherapy or drugs. Deep brain stimulation has given almost miraculous relief after decades of torment and agony for these long-suffering patients.

Every year, new uses for deep brain stimulation are being found. In fact, nearly all the major disorders of the brain are being reexamined in light of this and other new brain-scanning technologies. This promises to be an exciting new area for diagnosing and even treating illnesses.

OPTOGENETICS—LIGHTING UP THE BRAIN

But perhaps the newest and most exciting instrument in the neurologist's tool kit is optogenetics, which was once considered science fiction. Like a magic wand, it allows you to activate certain pathways controlling behavior by shining a light beam on the brain.

Incredibly, a light-sensitive gene that causes a cell to fire can be inserted, with surgical precision, directly into a neuron. Then, by turning on a light beam, the neuron is activated. More importantly, this allows scientists to excite these pathways, so that you can turn on and off certain behaviors by flicking a switch.

Although this technology is only a decade old, optogenetics has already proven successful in controlling certain animal behaviors. By turning on a light switch, it is possible to make fruit flies suddenly fly off, worms stop wiggling, and mice run around madly in circles. Monkey trials are now beginning, and even human trials are in discussion. There is great hope that this technology will have a direct application in treating disorders like Parkinson's and depression.

THE TRANSPARENT BRAIN

Like optogenetics, another spectacular new development is making the brain fully transparent so that its neural pathways are exposed to the naked eye. In 2013, scientists at Stanford University announced that they had successfully made the entire brain of a mouse transparent, as well as parts of a human brain. The announcement was so stunning that it made the front page of the *New York Times,* with the headline "Brain as Clear as Jell-O for Scientists to Explore."

At the cellular level, cells seen individually are transparent, with all their microscopic components fully exposed. However, once billions of cells come together to form organs like the brain, the addition of lipids (fats, oils, waxes, and chemicals not soluble in water) helps make the organ opaque. The key to the new technique is to remove the lipids while keeping the neurons intact. The scientists at Stanford did this by placing the brain in hydrogel (a gel-like substance mainly made of water), which binds to all the brain's molecules except the lipids. By placing the brain in a soapy solution with an electric field, the solution can be flushed out of the brain, carrying along the lipids, leaving the brain transparent. The addition of dyes can then make the neural pathways visible. This will help to identify and map the many neural pathways of the brain.

Making tissue transparent is not new, but getting precisely the right conditions necessary to make the entire brain transparent took a lot of ingenuity. "I burned and melted more than a hundred brains," confessed Dr. Kwanghun Chung, one of the lead scientists in the study. The new technique, called Clarity, can also be applied to other organs (and even organs preserved years ago in chemicals like formalin). He has already created transparent livers, lungs, and hearts. This new technique has startling applications across all of medicine. In particular, it will accelerate locating the neural pathways of the brain, which is the focus of intense research and funding.

FOUR FUNDAMENTAL FORCES

The success of this first generation of brain scans has been nothing less than spectacular. Before their introduction, only about thirty or so regions of the brain were known with any certainty. Now the MRI machine alone can identify two to three hundred regions of the brain, opening up entirely new frontiers for brain science. With so many new scanning technologies being introduced from physics just within the last fifteen years, one might wonder: Are there more? The answer is yes, but they will be variations and refinements of the previous ones, not radically new technologies. This is because there are only four fundamental forces—gravitational, electromagnetic, weak nuclear, and strong nuclear—that rule the universe. (Physicists have tried to find evidence for a fifth force, but so far all such attempts have failed.)

The electromagnetic force, which lights up our cities and represents the energy of electricity and magnetism, is the source of almost all the new scanning technologies (with the exception of the PET scan, which is governed by the weak nuclear force). Because physicists have had over 150 years of experience working with the electromagnetic force, there is no mystery in creating new electric and magnetic fields, so any new brain-scanning technology will most likely be a novel modification of existing technologies, rather than being something entirely new. As with most technology, the size and cost of these machines will drop, vastly increasing the widespread use of these sophisticated instruments. Already physicists are doing the basic calculations necessary to make an MRI machine fit into a cell phone. At the same time, the fundamental challenge facing these brain scans is resolution, both spatial and temporal. The spatial resolution of MRI scans will increase as the magnetic field becomes more uniform and as the electronics become more sensitive. At present, MRI scans can see only dots or voxels within a fraction of a millimeter. But each dot may contain hundreds of thousands of neurons. New scanning technology should reduce this even further. The holy grail of this approach would be to create an MRI-like machine that could identify individual neurons and their connections.

The temporal resolution of MRI machines is also limited because they analyze the flow of oxygenated blood in the brain. The machine itself has very good temporal resolution, but tracing the flow of blood slows it down. In the future, other MRI machines will be able to locate different substances that are more directly connected to the firing of neurons, thereby allowing real-time analysis of mental processes. No matter how spectacular the successes of the past fifteen years, then, they were just a taste of the future.

NEW MODELS OF THE BRAIN

Historically, with each new scientific discovery, a new model of the brain has emerged. One of the earliest models of the brain was the "homunculus," a little man who lived inside the brain and made all the decisions. This picture was not very helpful, since it did not explain what was happening in the brain of the homunculus. Perhaps there was a homunculus hiding inside the homunculus.

With the arrival of simple mechanical devices, another model of the brain was proposed: that of a machine, such as a clock, with mechanical wheels and gears. This analogy was useful for scientists and inventors like Leonardo da Vinci, who actually designed a mechanical man.

During the late 1800s, when steam power was carving out new empires, another analogy emerged, that of a steam engine, with flows of energy competing with one another. This hydraulic model, historians have conjectured, affected Sigmund Freud's picture of the brain, in which there was a continual struggle between three forces: the ego (representing the self and rational thought), the id (representing repressed desires), and the superego (representing our conscience). In this model, if too much pressure built up because of a conflict among these three, there could be a regression or general breakdown of the entire system. This model was ingenious, but as even Freud himself admitted, it required detailed studies of the brain at the neuronal level, which would take another century.

Early in the last century, with the rise of the telephone, another analogy surfaced—that of a giant switchboard. The brain was a mesh of telephone lines connected into a vast network. Consciousness was a long row of telephone operators sitting in front of a large panel of switches, constantly plugging and unplugging wires. Unfortunately, this model said nothing about how these messages were wired together to form the brain.

With the rise of the transistor, yet another model became fashionable: the computer. The old-fashioned switching stations were replaced by microchips containing hundreds of millions of transistors. Perhaps the "mind" was just a software program running on "wetware" (i.e., brain tissue rather than transistors). This model is an enduring one, even today, but it has limitations. The transistor model cannot explain how the brain performs computations that would require a computer the size of New York City. Plus the brain has no programming, no Windows operating system or Pentium chip. (Also, a PC with a Pentium chip is extremely fast, but it has a bottleneck. All calculations must pass through this single processor. The brain is the opposite. The firing of each neuron is relatively slow, but it more than makes up for this by having 100 billion neurons processing data simultaneously. Therefore a slow parallel processor can trump a very fast single processor.)

The most recent analogy is that of the Internet, which lashes together billions of computers. Consciousness, in this picture, is an "emergent" phenomenon, miraculously arising out of the collective action of billions of neurons. (The problem with this picture is that it says absolutely nothing about how this miracle occurs. It brushes all the complexity of the brain under the rug of chaos theory.)

No doubt each of these analogies has kernels of truth, but none of them truly captures the complexity of the brain. However, one analogy for the brain that I have found useful (albeit still imperfect) is that of a large corporation. In this analogy, there is a huge bureaucracy and lines of authority, with vast flows of information channeled between different offices. But the important information eventually winds up at the command center with the CEO. There the final decisions are made.

If this analogy of the brain to a large corporation is valid, then it should be able to explain certain peculiar features of the brain:

- **Most information is "subconscious"**—that is, the CEO is blissfully unaware of the vast, complex information that is constantly flowing inside the bureaucracy. In fact, only a tiny amount of information finally reaches the desk of the CEO, who can be compared to the prefrontal cortex. The CEO just has to know information important enough to get his attention; otherwise, he would be paralyzed by an avalanche of extraneous information.

This arrangement is probably a by-product of evolution, since our ancestors would have been overwhelmed with superfluous, subconscious information flooding their brains when facing an emergency. We are all mercifully unaware of the trillions of calculations being processed in our brains. Upon encountering a tiger in the forest, one does not have to be bothered with the status of our stomach, toes, hair, etc. All one has to know is how to run.

- **"Emotions" are rapid decisions made independently at a lower level.** Since rational thought takes many seconds, this means that it is often impossible to make a reasoned response to an emergency; hence lower-level brain regions must rapidly assess the situation and make a decision, an emotion, without permission from the top.

  So emotions (fear, anger, horror, etc.) are instantaneous red flags made at a lower level, generated by evolution, to warn the command center of possibly dangerous or serious situations. We have little conscious control over emotions. For example, no matter how much we practice giving a speech to a large audience, we still feel nervous.

  Rita Carter, author of *Mapping the Mind,* writes, "Emotions are not feelings at all but a set of body-rooted survival mechanisms that have evolved to turn us away from danger and propel us forward to things that may be of benefit."

- **There is a constant clamoring for the attention of the CEO.** There is no single homunculus, CPU, or Pentium chip making decisions; instead, the various subcenters within the command center are in constant competition with one another, vying for the attention of the CEO. So there is no smooth, steady continuity of thought, but the cacophony of different feedback loops competing with one another. The concept of "I," as a single, unified whole making all decisions continuously, is an illusion created by our own subconscious minds.

  Mentally we feel that our mind is a single entity, continuously and smoothly processing information, totally in charge of our decisions. But the picture emerging from brain scans is quite different from the perception we have of our own mind.

  MIT professor Marvin Minsky, one of the founding fathers of artificial intelligence, told me that the mind is more like a "society of minds," with different submodules, each trying to compete with the others.

When I interviewed Steven Pinker, a psychologist at Harvard University, I asked him how consciousness emerges out of this mess. He said that consciousness was like a storm raging in our brain. He elaborated on this when he wrote that "the intuitive feeling we have that there's an executive 'I' that sits in a control room of our brain, scanning the screens of the senses and pushing the buttons of our muscles, is an illusion. Consciousness turns out to consist of a maelstrom of events distributed across the brain. These events compete for attention, and as one process outshouts the others, the brain rationalizes the outcome after the fact and concocts the impression that a single self was in charge all along."

- **Final decisions are made by the CEO in the command center.** Almost all the bureaucracy is devoted to accumulating and assembling information for the CEO, who meets only with the directors of each division. The CEO tries to mediate all the conflicting information pouring into the command center. The buck stops here. The CEO, located in the prefrontal cortex, has to make the final decision. While most decisions are made by instinct in animals, humans make higher-level decisions after sifting through different bodies of information from our senses.
- **Information flows are hierarchical.** Because of the vast amount of information that must flow upward toward the CEO's office, or downward to the support staff, information must be arranged in complex arrays of nested networks, with many branches. Think of a pine tree, with the command center on top and a pyramid of branches flowing downward, branching out into many subcenters.

There are, of course, differences between a bureaucracy and the structure of thought. The first rule of any bureaucracy is that "it expands to fill the space allotted to it." But wasting energy is a luxury the brain cannot afford. The brain consumes only about twenty watts of power (the power of a dim lightbulb), but that is probably the maximum energy it can consume before the body becomes dysfunctional. If it generates more heat, it will cause tissue damage. Therefore the brain is constantly using shortcuts to conserve energy. We will see throughout this book the clever and ingenious devices that evolution has crafted, without our knowledge, to cut corners.

IS "REALITY" REALLY REAL?

Everyone knows the expression "seeing is believing." Yet much of what we see is actually an illusion. For example, when we see a typical landscape, it seems like a smooth, movielike panorama. In reality, there is a gaping hole in our field of vision, corresponding to the location of the optic nerve in the retina. We should see this large ugly black spot wherever we look. But our brains fill in that hole by papering it over, by averaging it out. This means that part of our vision is actually fake, generated by our subconscious minds to deceive us.

Also, we see only the center of our field of vision, called the fovea, with clarity. The peripheral part is blurry, in order to save energy. But the fovea is very small. To capture as much information as possible with the tiny fovea, the eye darts around constantly. This rapid, jiggling motion of our eyes is called saccades. All this is done subconsciously, giving us the false impression that our field of vision is clear and focused.

When I was a child and first saw a diagram showing the electromagnetic spectrum in its true glory, I was shocked. I had been totally unaware that huge parts of the EM spectrum (e.g., infrared light, UV light, X-rays, gamma rays) were totally invisible to us. I began to realize that what I saw with my eyes was only a tiny, crude approximation of reality. (There is an old saying: "If appearance and essence were the same thing, there would be no need for science.") We have sensors in the retina that can detect only red, green, and blue. This means that we've never actually seen yellow, brown, orange, and a host of other colors. These colors do exist, but our brain can approximate each of them only by mixing different amounts of red, green, and blue. (You can see this if you look at an old color-TV screen very carefully. You see only a collection of red, green, and blue dots. Color TV is actually an illusion.)

Our eyes also fool us into thinking we can see depth. The retinas of our eyes are two-dimensional, but because we have two eyes separated by a few inches, the left and right brain merge these two images, giving us the false sense of a third dimension. For more distant objects, we can judge how far an object is by observing how they move when we move our head. This is called parallax.

(This parallax explains the fact that children sometimes complain that "the moon is following me." Because the brain has difficulty comprehending the parallax of an object as distant as the moon, it appears as if the moon is always a fixed distance "behind" them, but it's just an illusion caused by the brain taking a shortcut.)

THE SPLIT-BRAIN PARADOX

One way in which this picture, based on the corporate hierarchy of a company, deviates from the actual structure of the brain can be seen in the curious case of split-brain patients. One unusual feature of the brain is that it has two nearly identical halves, or hemispheres, the left and right. Scientists have long wondered why the brain has this unnecessary redundancy, since the brain can operate even if one entire hemisphere is completely removed. No normal corporate hierarchy has this strange feature. Furthermore, if each hemisphere has consciousness, does this mean that we have two separate centers of consciousness inside one skull?

Dr. Roger W. Sperry of the California Institute of Technology won the Nobel Prize in 1981 for showing that the two hemispheres of the brain are not exact carbon copies of each other, but actually perform different duties. This result created a sensation in neurology (and also spawned a cottage industry of dubious self-help books that claim to apply the left-brain, right-brain dichotomy to your life).

Dr. Sperry was treating epileptics, who sometimes suffer from grand mal seizures often caused by feedback loops between the two hemispheres that go out of control. Like a microphone screeching in our ears because of a feedback loop, these seizures can become life-threatening. Dr. Sperry began by severing the corpus callosum, which connects the two hemispheres of the brain, so that they no longer communicated and shared information between the left and right side of the body. This usually stopped the feedback loop and the seizures.

At first, these split-brain patients seemed perfectly normal. They were alert and could carry on a natural conversation as if nothing had happened. But a careful analysis of these individuals showed that something was very different about them.

Normally the hemispheres complement each other as thoughts move back and forth between the two. The left brain is more analytical and logical. It is where verbal skills are found, while the right brain is more holistic and artistic. But the left brain is the dominant one and makes the final decisions. Commands pass from the left brain to the right brain via the corpus callosum. But if that connection is cut, it means that the right brain is now free from the dictatorship of the left brain. Perhaps the right brain can have a will of its own, contradicting the wishes of the dominant left brain.

In short, there could be two wills acting within one skull, sometimes struggling for control of the body. This creates the bizarre situation where the left hand (controlled by the right brain) starts to behave independently of your wishes, as if it were an alien appendage.

There is one documented case in which a man was about to hug his wife with one hand, only to find that the other hand had an entirely different agenda. It delivered a right hook to her face. Another woman reported that she would pick out a dress with one hand, only to see her other hand grab an entirely different outfit. Meanwhile, one man had difficulty sleeping at night thinking that his other rebellious hand might strangle him.

At times, split-brain people think they are living in a cartoon, where one hand struggles to control the other. Physicians sometimes call this the Dr. Strangelove syndrome, because of a scene in the movie in which one hand has to fight against the other hand.

Dr. Sperry, after detailed studies of split-brain patients, finally concluded that there could be two distinct minds operating in a single brain. He wrote that each hemisphere is "indeed a conscious system in its own right, perceiving, thinking, remembering, reasoning, willing, and emoting, all at a characteristically human level, and . . . both the left and right hemisphere may be conscious simultaneously in different, even in mutually conflicting, mental experiences that run along in parallel."

When I interviewed Dr. Michael Gazzaniga of the University of California, Santa Barbara, an authority on split-brain patients, I asked him how experiments can be done to test this theory. There are a variety of ways to communicate separately to each hemisphere without the knowledge of the other hemisphere. One can, for example, have the subject wear special glasses on which questions can be shown to each eye separately, so that directing questions to each hemisphere is easy. The hard part is trying to get an answer from each hemisphere. Since the right brain cannot speak (the speech centers are located only in the left brain), it is difficult to get answers from the right brain. Dr. Gazzaniga told me that to find out what the right brain was thinking, he created an experiment in which the (mute) right brain could "talk" by using Scrabble letters.

He began by asking the patient's left brain what he would do after graduation. The patient replied that he wanted to become a draftsman. But things got interesting when the (mute) right brain was asked the same question. The right brain spelled out the words: "automobile racer." Unknown to the dominant left brain, the right brain secretly had a completely different agenda for the future. The right brain literally had a mind of its own.

Rita Carter writes, "The possible implications of this are mind-boggling. It suggests that we might all be carrying around in our skulls a mute prisoner with a personality, ambition, and self-awareness quite different from the day-to-day entity we believe ourselves to be."

Perhaps there is truth to the oft-heard statement that "inside him, there is someone yearning to be free." This means that the two hemispheres may even have different beliefs. For example, the neurologist V. S. Ramanchandran describes one split-brain patient who, when asked if he was a believer or not, said he was an atheist, but his right brain declared he was a believer. Apparently, it is possible to have two opposing religious beliefs residing in the same brain. Ramachandran continues: "If that person dies, what happens? Does one hemisphere go to heaven and the other go to hell? I don't know the answer to that."

(It is conceivable, therefore, that a person with a split-brain personality might be both Republican and Democrat at the same time. If you ask him whom he will vote for, he will give you the candidate of the left brain, since the right brain cannot speak. But you can imagine the chaos in the voting booth when he has to pull the lever with one hand.)

WHO IS IN CHARGE?

One person who has spent considerable time and done much research to understand the problem of the subconscious mind is Dr. David Eagleman, a neuroscientist at the Baylor College of Medicine. When I interviewed him, I asked him, If most of our mental processes are subconscious, then why are we ignorant of this important fact? He gave an example of a young king who inherits the throne and takes credit for everything in the kingdom, but hasn't the slightest clue about the thousands of staff, soldiers, and peasants necessary to maintain the throne.

Our choice of politicians, marriage partners, friends, and future occupations are all influenced by things that we are not conscious of. (For example, it is an odd result, he says, that "people named Denise or Dennis are disproportionately likely to become dentists, while people named Laura or Lawrence are more likely to become lawyers, and people with names like George or Georgina to become geologists.") This also means that what we consider to be "reality" is only an approximation that the brain makes to fill in the gaps. Each of us sees reality in a slightly different way. For example, he pointed out, "at least 15 percent of human females possess a genetic mutation that gives them an extra (fourth) type of color photoreceptor—and this allows them to discriminate between colors that look identical to the majority of us with a mere three types of color photoreceptors."

Clearly, the more we understand the mechanics of thought, the more questions arise. Precisely what happens in the command center of the mind when confronted with a rebellious shadow command center? What do we mean by "consciousness" anyway, if it can be split in half? And what is the relationship between consciousness and "self" and "self-awareness"?

If we can answer these difficult questions, then perhaps it will pave the way for understanding nonhuman consciousness, the consciousness of robots and aliens from outer space, for example, which may be entirely different from ours.

So let us now propose a clear answer to this deceptively complex question: What is consciousness?

The mind of man is capable of anything . . . because everything is in it, all the past as well as all the future.

—JOSEPH CONRAD

Consciousness can reduce even the most fastidious thinker to blabbering incoherence.

—COLIN MCGINN

2 CONSCIOUSNESS—A PHYSICIST'S VIEWPOINT

The idea of consciousness has intrigued philosophers for centuries, but it has resisted a simple definition, even to this day. The philosopher David Chalmers has cataloged more than twenty thousand papers written on the subject; nowhere in science have so many devoted so much to create so little consensus. The seventeenth-century thinker Gottfried Leibniz once wrote, "If you could blow the brain up to the size of a mill and walk about inside, you would not find consciousness."

Some philosophers doubt that a theory of consciousness is even possible. They claim that consciousness can never be explained since an object can never understand itself, so we don't even have the mental firepower to solve this perplexing question. Harvard psychologist Steven Pinker writes, "We cannot see ultraviolet light. We cannot mentally rotate an object in the fourth dimension. And perhaps we cannot solve conundrums like free will and sentience."

In fact, for most of the twentieth century, one of the dominant theories of psychology, behaviorism, denied the importance of consciousness entirely. Behaviorism is based on the idea that only the objective behavior of animals and people is worthy of study, not the subjective, internal states of the mind.

Others have given up trying to define consciousness, and try simply to describe it. Psychiatrist Giulio Tononi has said, "Everybody knows what consciousness is: it is what abandons you every night when you fall into dreamless sleep and returns the next morning when you wake up."

Although the nature of consciousness has been debated for centuries, there has been little resolution. Given that physicists created many of the inventions that have made the explosive advancements in brain science possible, perhaps it will be useful to follow an example from physics in reexamining this ancient question.

HOW PHYSICISTS UNDERSTAND THE UNIVERSE

When a physicist tries to understand something, first he collects data and then he proposes a "model," a simplified version of the object he is studying that captures its essential features. In physics, the model is described by a series of parameters (e.g., temperature, energy, time). Then the physicist uses the model to predict its future evolution by simulating its motions. In fact, some of the world's largest supercomputers are used to simulate the evolution of models, which can describe protons, nuclear explosions, weather patterns, the big bang, and the center of black holes. Then you create a better model, using more sophisticated parameters, and simulate it in time as well.

For example, when Isaac Newton was puzzling over the motion of the moon, he created a simple model that would eventually change the course of human history: he envisioned throwing an apple in the air. The faster you threw the apple, he reasoned, the farther it would travel. If you threw it fast enough, in fact, it would encircle the Earth entirely, and might even return to its original point. Then, Newton claimed, this model represented the path of the moon, so the forces that guided the motion of the apple circling the Earth were identical to the forces guiding the moon.

But the model, by itself, was still useless. The key breakthrough came when Newton was able to use his new theory to simulate the future, to calculate the future position of moving objects. This was a difficult problem, requiring him to create an entirely new branch of mathematics, called calculus. Using this new mathematics, Newton was then able to predict the trajectory of not just the moon, but also Halley's Comet and the planets.

Since then, scientists have used Newton's laws to simulate the future path of moving objects, from cannonballs, machines, automobiles, and rockets to asteroids and meteors, and even stars and galaxies.

The success or failure of a model depends on how faithfully it reproduces the basic parameters of the original. In this case, the basic parameter was the location of the apple and the moon in space and time. By allowing this parameter to evolve (i.e., letting time move forward), Newton unlocked, for the first time in history, the action of moving bodies, which is one of the most important discoveries in science.

Models are useful, until they are replaced by even more accurate models described by better parameters. Einstein replaced Newton's picture of forces acting on apples and moons with a new model based on a new parameter, the curvature of space and time. An apple moved not because the Earth exerted a force on it, but because the fabric of space and time was stretched by the Earth, so the apple was simply moving along the surface of a curved space-time. From this, Einstein could then simulate the future of the entire universe. Now, with computers, we can run simulations of this model into the future and create gorgeous pictures presenting the collisions of black holes.

Let us now incorporate this basic strategy into a new theory of consciousness.

DEFINITION OF CONSCIOUSNESS

I've taken bits and pieces from previous descriptions of consciousness in the fields of neurology and biology in order to define consciousness as follows:

> **Consciousness is the process of creating a model of the world using multiple feedback loops in various parameters (e.g., in temperature, space, time, and in relation to others), in order to accomplish a goal (e.g., find mates, food, shelter).**

I call this the "space-time theory of consciousness," because it emphasizes the idea that animals create a model of the world mainly in relation to space, and to one another, while humans go beyond and create a model of the world in relation to time, both forward and backward.

For example, the lowest level of consciousness is Level 0, where an organism is stationary or has limited mobility and creates a model of its place using feedback loops in a few parameters (e.g., temperature). For example, the simplest level of consciousness is a thermostat. It automatically turns on an air conditioner or heater to adjust the temperature in a room, without any help. The key is a feedback loop that turns on a switch if the temperature gets too hot or cold. (For example, metals expand when heated, so a thermostat can turn on a switch if a metal strip expands beyond a certain point.)

Each feedback loop registers "one unit of consciousness," so a thermostat would have a single unit of Level 0 consciousness, that is, Level 0:1.

In this way, we can rank consciousness numerically, on the basis of the number and complexity of the feedback loops used to create a model of the world. Consciousness is then no longer a vague collection of undefined, circular concepts, but a system of hierarchies that can be ranked numerically. For example, a bacterium or a flower has many more feedback loops, so they would have a higher level of Level 0 consciousness. A flower with ten feedback loops (which measure temperature, moisture, sunlight, gravity, etc.), would have a Level 0:10 consciousness.

Organisms that are mobile and have a central nervous system have Level I consciousness, which includes a new set of parameters to measure their changing location. One example of Level I consciousness would be reptiles. They have so many feedback loops that they developed a central nervous system to handle them. The reptilian brain would have perhaps one hundred or more feedback loops (governing their sense of smell, balance, touch, sound, sight, blood pressure, etc., and each of these contains more feedback loops). For example, eyesight alone involves a large number of feedback loops, since the eye can recognize color, movement, shapes, light intensity, and shadows. Similarly, the reptile's other senses, such as hearing and taste, require additional feedback loops. The totality of these numerous feedback loops creates a mental picture of where the reptile is located in the world, and where other animals (e.g., prey) are located as well. Level I consciousness, in turn, is governed mainly by the reptilian brain, located in the back and center of the human head.

Next we have Level II consciousness, where organisms create a model of their place not only in space but also with respect to others (i.e., they are social animals with emotions). The number of feedback loops for Level II consciousness explodes exponentially, so it is useful to introduce a new numerical ranking for this type of consciousness. Forming allies, detecting enemies, serving the alpha male, etc., are all very complex behaviors requiring a vastly expanded brain, so Level II consciousness coincides with the formation of new structures of the brain in the form of the limbic system. As noted earlier, the limbic system includes the hippocampus (for memories), amygdala (for emotions), and the thalamus (for sensory information), all of which provide new parameters for creating models in relation to others. The number and type of feedback loops therefore change.

We define the degree of Level II consciousness as the total number of distinct feedback loops required for an animal to interact socially with members of its grouping. Unfortunately, studies of animal consciousness are extremely limited, so little work has been done to catalog all the ways in which animals communicate socially with one another. But to a crude first approximation, we can estimate Level II consciousness by counting the number of fellow animals in its pack or tribe and then listing the total number of ways in which the animal interacts emotionally with each one. This would include recognizing rivals and friends, forming bonds with others, reciprocating favors, building coalitions, understanding your status and the social ranking of others, respecting the status of your superiors, displaying your power over your inferiors, plotting to rise on the social ladder, etc. (We exclude insects from Level II, because although they have social relations with members of their hive or group, they have no emotions as far as we can tell.)

Despite the lack of empirical studies of animal behaviors, we can give a very rough numerical rank to Level II consciousness by listing the total number of distinct emotions and social behaviors that the animal can exhibit. For example, if a wolf pack consists of ten wolves, and each wolf interacts with all the others with fifteen different emotions and gestures, then its level of consciousness, to a first approximation, is given by the product of the two, or 150, so it would have Level II:150 consciousness. This number takes into account both the number of other animals it has to interact with as well as the number of ways it can communicate with each one. This number only approximates the total number of social interactions that the animal can display, and will undoubtedly change as we learn more about its behavior.

(Of course, because evolution is never clean and precise, there are caveats that we have to explain, such as the level of consciousness of social animals that are solitary hunters. We will do so in the notes.)

LEVEL III CONSCIOUSNESS: SIMULATING THE FUTURE

With this framework for consciousness, we see that humans are not unique, and that there is a continuum of consciousness. As Charles Darwin once commented, "The difference between man and the higher animals, great as it is, is certainly one of degree and not of kind." But what separates human consciousness from the consciousness of animals? Humans are alone in the animal kingdom in understanding the concept of tomorrow. Unlike animals, we constantly ask ourselves "What if?" weeks, months, and even years into the future, so I believe that Level III consciousness creates a model of its place in the world and then simulates it into the future, by making rough predictions. We can summarize this as follows:

> **Human consciousness is a specific form of consciousness that creates a model of the world and then simulates it in time, by evaluating the past to simulate the future. This requires mediating and evaluating many feedback loops in order to make a decision to achieve a goal.**

By the time we reach Level III consciousness, there are so many feedback loops that we need a CEO to sift through them in order to simulate the future and make a final decision. Accordingly, our brains differ from those of other animals, especially in the expanded prefrontal cortex, located just behind the forehead, which allows us to "see" into the future.

Dr. Daniel Gilbert, a Harvard psychologist, has written, "The greatest achievement of the human brain is its ability to imagine objects and episodes that do not exist in the realm of the real, and it is this ability that allows us to think about the future. As one philosopher noted, the human brain is an 'anticipation machine,' and 'making the future' is the most important thing it does."

Using brain scans, we can even propose a candidate for the precise area of the brain where simulation of the future takes place. Neurologist Michael Gazzaniga notes that "area 10 (the internal granular layer IV), in the lateral prefrontal cortex, is almost twice as large in humans as in apes. Area 10 is involved with memory and planning, cognitive flexibility, abstract thinking, initiating appropriate behavior, and inhibiting inappropriate behavior, learning rules, and picking out relevant information from what is perceived through the senses." (For this book, we will refer to this area, in which decision making is concentrated, as the dorsolateral prefrontal cortex, although there is some overlap with other areas of the brain.)

Although animals may have a well-defined understanding of their place in space and some have a degree of awareness of others, it is not clear if they systematically plan for the future and have an understanding of "tomorrow." Most animals, even social animals with well-developed limbic systems, react to situations (e.g., the presence of predators or potential mates) by relying mainly on instinct, rather than systematically planning into the future.

For instance, mammals do not plan for the winter by preparing to hibernate, but largely follow instinct as the temperature drops. There is a feedback loop that regulates their hibernation. Their consciousness is dominated by messages coming in from their senses. There is no evidence that they systematically sift through various plans and schemes as they prepare to hibernate. Predators, when they use cunning and disguise to stalk an unsuspecting prey, do anticipate future events, but this planning is limited only to instinct and the duration of the hunt. Primates are adept at devising short-term plans (e.g., finding food), but there is no indication that they plan more than a few hours ahead.

Humans are different. Although we do rely on instinct and emotions in many situations, we also constantly analyze and evaluate information from many feedback loops. We do this by running simulations sometimes even beyond our own life span and even thousands of years into the future. The point of running simulations is to evaluate various possibilities to make the best decision to fulfill a goal. This occurs in the prefrontal cortex, which allows us to simulate the future and evaluate the possibilities in order to chart the best course of action.

This ability evolved for several reasons. First, having the ability to peer into the future has enormous evolutionary benefits, such as evading predators and finding food and mates. Second, it allows us to choose among several different outcomes and to select the best one.

Third, the number of feedback loops explodes exponentially as we go from Level 0 to Level I to Level II, so we need a "CEO" to evaluate all these conflicting, competing messages. Instinct is no longer enough. There has to be a central body that evaluates each of these feedback loops. This distinguishes human consciousness from that of the animals. These feedback loops are evaluated, in turn, by simulating them into the future to obtain the best outcome. If we didn't have a CEO, chaos would ensue and we would have sensory overload.

A simple experiment can demonstrate this. David Eagleman describes how you can take a male stickleback fish and have a female fish trespass on its territory. The male gets confused, because it wants to mate with the female, but it also wants to defend its territory. As a result, the male stickleback fish will simultaneously attack the female while initiating courtship behavior. The male is driven into a frenzy, trying to woo and kill the female at the same time.

This works for mice as well. Put an electrode in front of a piece of cheese. If the mouse gets too close, the electrode will shock it. One feedback loop tells the mouse to eat the cheese, but another one tells the mouse to stay away and avoid being shocked. By adjusting the location of the electrode, you can get the mouse to oscillate, torn between two conflicting feedback loops. While a human has a CEO in its brain to evaluate the pros and cons of the situation, the mouse, governed by two conflicting feedback loops, goes back and forth. (This is like the proverb about the donkey that starves to death because it is placed between two equal bales of hay.)

Precisely how does the brain simulate the future? The human brain is flooded by a large amount of sensory and emotional data. But the key is to simulate the future by making causal links between events—that is, if A happens, then B happens. But if B happens, then C and D might result. This sets off a chain reaction of events, eventually creating a tree of possible cascading futures with many branches. The CEO in the prefrontal cortex evaluates the results of these causal trees in order to make the ultimate decision.

Let's say you want to rob a bank. How many realistic simulations of this event can you make? To do this, you have to think of the various causal links involving the police, bystanders, alarm systems, relations with fellow criminals, traffic conditions, the DA's office, etc. For a successful simulation of the robbery, hundreds of causal links may have to be evaluated.

It is also possible to measure this level of consciousness numerically. Let's say that a person is given a series of different situations like the one above and is asked to simulate the future of each. The sum total number of causal links that the person can make for all these situations can be tabulated. (One complication is that there are an unlimited number of causal links that a person might make for a variety of conceivable situations. To get around this complication, we divide this number by the average number of causal links obtained from a large control group. Like the IQ exam, one may multiply this number by 100. So a person's level of consciousness, for example, might be Level III:100, meaning that the person can simulate future events just like the average person.)

We summarize these levels of consciousness in the following diagram:

LEVELS OF CONSCIOUSNESS FOR DIFFERENT SPECIES

LEVEL	SPECIES	PARAMETER	BRAIN STRUCTURE
0	Plants	Temperature, sunshine	None
I	Reptiles	Space	Brain stem
II	Mammals	Social relations	Limbic system
III	Humans	Time (esp. future)	Prefrontal cortex

Space-time theory of consciousness. We define consciousness as the process of creating a model of the world using multiple feedback loops in various parameters (e.g., in space, time, and in relation to others), in order to accomplish a goal. Human consciousness is a particular type that involves mediating between these feedback loops by simulating the future and evaluating the past.

(Notice that these categories correspond to the rough evolutionary levels we find in nature—e.g., reptiles, mammals, and humans. However, there are also gray areas, such as animals that might possess tiny aspects of different levels of consciousness, animals that do some rudimentary planning, or even single cells that communicate with one another. This chart is meant only to give you the larger, global picture of how consciousness is organized across the animal kingdom.)

WHAT IS HUMOR? WHY DO WE HAVE EMOTIONS?

All theories have to be falsifiable. The challenge for the space-time theory of consciousness is to explain *all* aspects of human consciousness in this framework. It can be falsified if there are patterns of thought that cannot be brought into this theory. A critic might say that surely our sense of humor is so quixotic and ephemeral that it is beyond explanation. We spend a great deal of time laughing with our friends or at comedians, yet it seems that humor has nothing to do with our simulations of the future. But consider this. Much of humor, such as telling a joke, depends on the punch line.

When hearing a joke, we can't help but simulate the future and complete the story ourselves (even if we're unaware that we're doing so). We know enough about the physical and social world that we can anticipate the ending, so we burst out with laughter when the punch line gives us a totally unexpected conclusion. The essence of humor is when our simulation of the future is suddenly disrupted in surprising ways. (This was historically important for our evolution since success depends, in part, on our ability to simulate future events. Since life in the jungle is full of unanticipated events, anyone who can foresee unexpected outcomes has a better chance at survival. In this way, having a well-developed sense of humor is actually one indication of our Level III consciousness and intelligence; that is, the ability to simulate the future.)

For example, W. C. Fields was once asked a question about social activities for youth. He was asked, "Do you believe in clubs for young people?" He replied, "Only when kindness fails."

The joke has a punch line only because we mentally simulate a future in which children have social clubs, while W. C. Fields simulates a different future involving clubs as a weapon. (Of course, if a joke is deconstructed, it loses its power, since we have already simulated various possible futures in our minds.)

This also explains what every comedian knows: timing is the key to humor. If the punch line is delivered too quickly, then the brain hasn't had time to simulate the future, so there is no experience of the unanticipated. If the punch line is delivered too late, the brain has already had time to simulate various possible futures, so again the punch line loses the element of surprise.

(Humor has other functions, of course, such as bonding with fellow members of our tribe. In fact, we use our sense of humor as a way to size up the character of others. This, in turn, is essential to determine our status within society. So in addition, laughter helps define our position in the social world, i.e., Level II consciousness.)

WHY DO WE GOSSIP AND PLAY?

Even seemingly trivial activities, such as engaging in idle gossip or horsing around with our friends, must be explained in this framework. (If a Martian were to visit a supermarket checkout line and view the huge display of gossip magazines, it might conclude that gossip is the main activity of humans. This observation would not be far off.)

Gossiping is essential for survival because the complex mechanics of social interactions are constantly changing, so we have to make sense of this ever-shifting social terrain. This is Level II consciousness at work. But once we hear a piece of gossip, we immediately run simulations to determine how this will affect our own standing in the community, which moves us to Level III consciousness. Thousands of years ago, in fact, gossip was the *only* way to obtain vital information about the tribe. One's very life often depended on knowing the latest gossip.

Something as superfluous as "play" is also an essential feature of consciousness. If you ask children why they like to play, they will say, "Because it's fun." But that invites the next question: What is fun? Actually, when children play, they are often trying to reenact complex human interactions in simplified form. Human society is extremely sophisticated, much too involved for the developing brains of young children, so children run simplified simulations of adult society, playing games such as doctor, cops and robber, and school. Each game is a model that allows children to experiment with a small segment of adult behavior and then run simulations into the future. (Similarly, when adults engage in play, such as a game of poker, the brain constantly creates a model of what cards the various players possess, and then projects that model into the future, using previous data about people's personality, ability to bluff, etc. The key to games like chess, cards, and gambling is the ability to simulate the future. Animals, which live largely in the present, are not as good at games as humans are, especially if they involve planning. Infant mammals do engage in a form of play, but this is more for exercise, testing one another, practicing future battles, and establishing the coming social pecking order rather than simulating the future.)

My space-time theory of consciousness might also shed light on another controversial topic: intelligence. Although IQ exams claim to measure "intelligence," IQ exams actually give no definition of intelligence in the first place.

In fact, a cynic may claim, with some justification, that IQ is a measure of "how well you do on IQ exams," which is circular. In addition, IQ exams have been criticized for being too culturally biased. In this new framework, however, intelligence may be viewed as the complexity of our simulations of the future. Hence, a master criminal, who may be a dropout and functionally illiterate and score dismally low on an IQ exam, may also far outstrip the ability of the police. Outwitting the cops may entail simply being able to run more sophisticated simulations of the future.

LEVEL I: STREAM OF CONSCIOUSNESS

Humans are probably alone on this planet in being able to operate on all levels of consciousness. Using MRI scans, we can break down the different structures involved in each level of consciousness.

For us, Level I stream of consciousness is largely the interplay between the prefrontal cortex and the thalamus. When taking a leisurely stroll in the park, we are aware of the smells of the plants, the sensation of a gentle breeze, the visual stimuli from the sun, and so on. Our senses send signals to the spinal cord, the brain stem, and then to the thalamus, which operates like a relay station, sorting out the stimuli and sending them on to the various cortices of the brain. The images of the park, for example, are sent to the occipital cortex in the back of the brain, while the sense of touch from the wind is sent to the parietal lobe. The signals are processed in appropriate cortices, and then sent to the prefrontal cortex, where we finally become conscious of all these sensations.

This is illustrated in Figure 7.

LEVEL II: FINDING OUR PLACE IN SOCIETY

While Level I consciousness uses sensations to create a model of our physical location in space, Level II consciousness creates a model of our place in society.

Let's say we are going to an important cocktail party, in which people essential to our job will be present. As we scan the room, trying to identify people from our workplace, there is an intense interplay between the hippocampus (which processes memories), the amygdala (which processes emotions), and the prefrontal cortex (which puts all this information together).

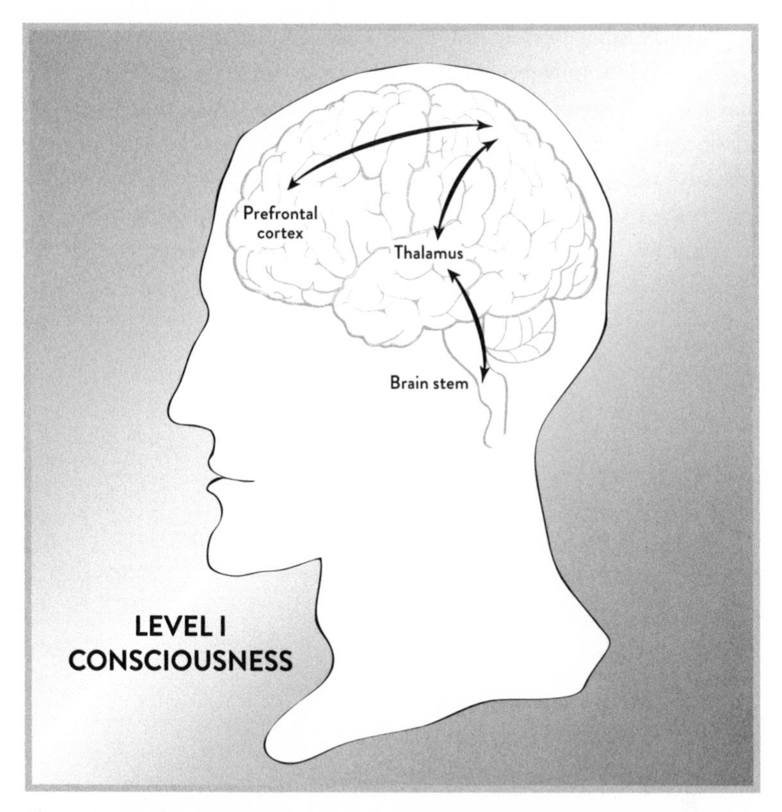

Figure 7. In Level I consciousness, sensory information travels through the brain stem, past the thalamus, onto the various cortices of the brain, and finally to the prefrontal cortex. Thus this stream of Level I consciousness is created by the flow of information from the thalamus to the prefrontal cortex.

With each image, the brain automatically attaches an emotion, such as happiness, fear, anger, or jealousy, and processes the emotion in the amygdala.

If you spot your chief rival, whom you suspect of stabbing you in the back, the emotion of fear is processed by the amygdala, which sends an urgent message to the prefrontal cortex, alerting it to possible danger. At the same time, signals are sent to your endocrine system to start pumping adrenaline and other hormones into the blood, thereby increasing your heartbeat and preparing you for a possible fight-or-flight response.

This is illustrated in Figure 8.

But beyond simply recognizing other people, the brain has the uncanny ability to guess what other people are thinking about. This is called the Theory of Mind, a theory first proposed by Dr. David Premack of the University of Pennsylvania, which is the ability to infer the thoughts of others. In any complex society, anyone with the ability to correctly guess the intentions, motives, and plans of other people has a tremendous survival advantage over those who can't. The Theory of Mind allows you to form alliances

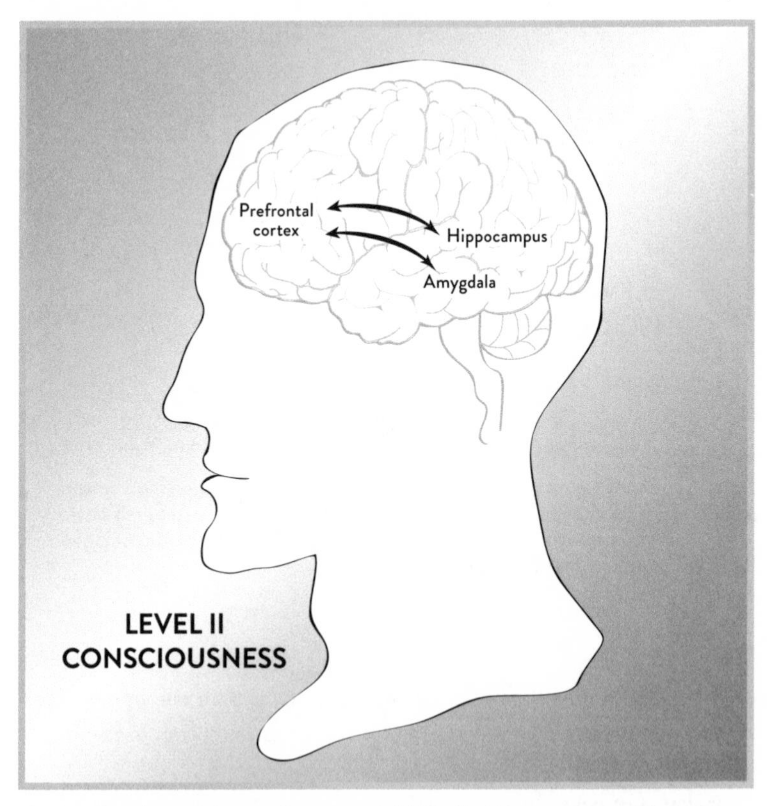

Figure 8. Emotions originate and are processed in the limbic system. In Level II consciousness, we are continually bombarded with sensory information, but emotions are rapid-fire responses to emergencies from the limbic system that do not need permission from the prefrontal cortex. The hippocampus is also important for processing memories. So Level II consciousness, at its core, involves the reaction of the amygdala, hippocampus, and prefrontal cortex.

with others, isolate your enemies, and solidify your friendships, which vastly increases your power and chances of survival and mating. Some anthropologists even believe that the mastery of the Theory of Mind was essential in the evolution of the brain.

But how is the Theory of Mind accomplished? One clue came in 1996, with the discovery of "mirror neurons" by Drs. Giacomo Rizzolatti, Leonardo Fogassi, and Vittorio Gallese. These neurons fire when you are performing a certain task and also when you see someone else performing that same task. (Mirror neurons also fire for emotions as well as physical acts. If you feel a certain emotion, and think another is feeling that same emotion, then the mirror neurons will fire.)

Mirror neurons are essential for mimicry and also for empathy, giving us the ability not only to copy the complex tasks performed by others but also to experience the emotions that person must be feeling. Mirror neurons were thus probably essential for our evolution as human beings, since cooperation is essential for holding the tribe together.

Mirror neurons were first found in the premotor areas of monkey brains. But since then, they have been found in humans in the prefrontal cortex. Dr. V. S. Ramachandran believes that mirror neurons were essential in giving us the power of self-awareness and concludes, "I predict that mirror neurons will do for psychology what DNA did for biology: they will provide a unifying framework and help explain a host of mental abilities that have hitherto remained mysterious and inaccessible to experiments." (We should point out, however, that all scientific results have to be tested and reconfirmed. There is no doubt that certain neurons are performing this crucial behavior involved with empathy, mimicry, etc., but there is some debate about the identity of these mirror neurons. For example, some critics claim that perhaps these behaviors are common to many neurons, and that there is not a single class of neurons dedicated to this behavior.)

LEVEL III: SIMULATING THE FUTURE

The highest level of consciousness, which is associated primarily with *Homo sapiens,* is Level III consciousness, in which we take our model of the world and then run simulations into the future. We do this by analyzing past memories of people and events, and then simulating the future by making many causal links to form a "causal" tree. As we look at the various faces at the cocktail party, we begin to ask ourselves simple questions: How can this individual help me? How will the gossip floating in the room play out in the future? Is anyone out to get me?

Let's say that you just lost your job and you are desperately looking for a new one. In this case, as you talk to various people at the cocktail party, your mind is feverishly simulating the future with each person you talk to. You ask yourself, How can I impress this person? What topics should I bring out to present my best case? Can he offer me a job?

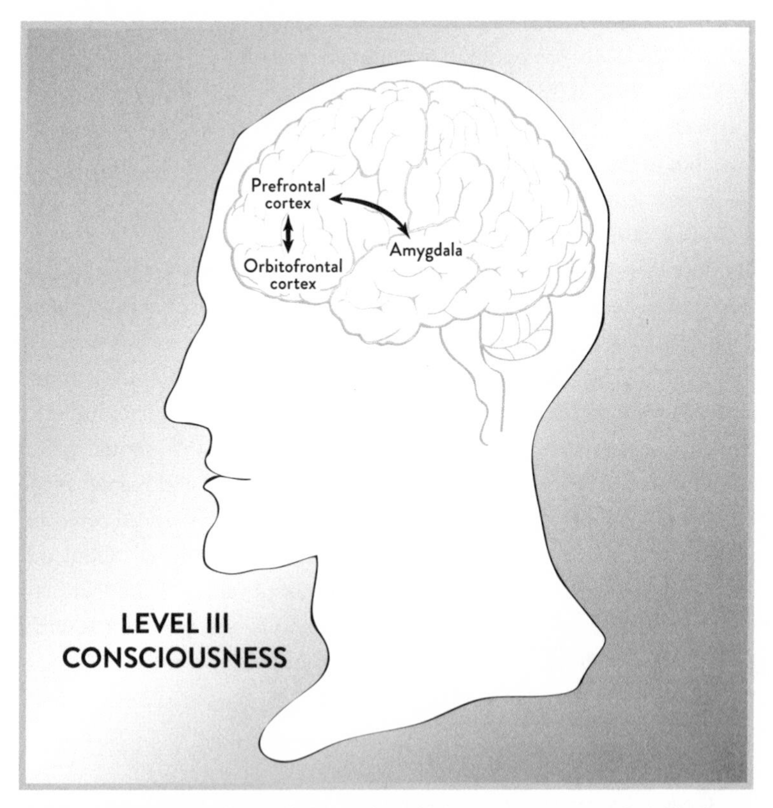

Figure 9. Simulating the future, the heart of Level III consciousness, is mediated by the dorsolateral prefrontal cortex, the CEO of the brain, with competition between the pleasure center and the orbitofrontal cortex (which acts to check our impulses). This roughly resembles the outline given by Freud of the struggle between our conscience and desires. The actual process of simulating the future takes place when the prefrontal cortex accesses the memories of the past in order to approximate future events.

Recent brain scans have shed partial light on how the brain simulates the future. These simulations are done mainly in the dorsolateral prefrontal cortex, the CEO of the brain, using memories of the past. On one hand, simulations of the future may produce outcomes that are desirable and pleasurable, in which case the pleasure centers of the brain light up (in the nucleus accumbens and hypothalamus). On the other hand, these outcomes may also have a downside to them, so the orbitofrontal cortex kicks in to warn us of possible dangers. There is a struggle, then, between different parts of the brain concerning the future, which may have desirable and undesirable outcomes. Ultimately it is the dorsolateral prefrontal cortex that mediates between these and makes the final decisions. (See Figure 9.) (Some neurologists have pointed out that this struggle resembles, in a crude way, the dynamics between Freud's ego, id, and superego.)

THE MYSTERY OF SELF-AWARENESS

If the space-time theory of consciousness is correct, then it also gives us a rigorous definition of self-awareness. Instead of vague, circular references, we should be able to give a definition that is testable and useful. We'll define self-awareness as follows:

> **Self-awareness is creating a model of the world and simulating the future in which you appear.**

Animals therefore have some self-awareness, since they have to know where they are located if they are going to survive and mate, but it is limited largely by instinct.

When most animals are placed in front of a mirror, they either ignore it or attack it, not realizing that it is an image of themselves. (This is called the "mirror test," which goes all the way back to Darwin.) However, animals like elephants, the great apes, bottlenose dolphins, orcas, and European magpies can figure out that the image they see in the mirror represents themselves.

Humans, however, take a giant step forward and constantly run future simulations in which we appear as a principal actor. We constantly imagine ourselves faced with different situations—going on a date, applying for a job, changing careers—none of which is determined by instinct. It is extremely difficult to stop your brain from simulating the future, though elaborate methods have been devised (for instance, meditation) to attempt to do so.

Daydreaming, as an example, consists largely of our acting out different possible futures to attain a goal. Since we pride ourselves in knowing our limitations and strengths, it is not hard to put ourselves inside the model and hit the "play" button so we begin to act out hypothetical scenarios, like being an actor in a virtual play.

WHERE AM "I"?

There is probably a specific part of the brain whose job it is to unify the signals from the two hemispheres to create a smooth, coherent sense of self. Dr. Todd Heatherton, a psychologist at Dartmouth College, believes that this region is located within the prefrontal cortex, in what is called the medial prefrontal cortex. Biologist Dr. Carl Zimmer writes, "The medial prefrontal cortex may play the same role for the self as the hippocampus plays in memory . . . [it] could be continually stitching together a sense of who we are." In other words, this may be the gateway to the concept of "I," the central region of the brain that fuses, integrates, and concocts a unified narrative of who we are. (This does not mean, however, that the medial prefrontal cortex is the homunculus sitting in our brain that controls everything.)

If this theory is true, then the resting brain, when we are idly daydreaming about our friends and ourselves, should be more active than normal, even when other parts of the brain's sensory regions are quiet. In fact, brain scans bear this out. Dr. Heatherton concludes, "Most of the time we daydream—we think about something that happened to us or what we think about other people. All this involves self-reflection."

The space-time theory says that consciousness is cobbled together from many subunits of the brain, each competing with the others to create a model of the world, and yet our consciousness feels smooth and continuous. How can this be, when we all have the feeling that our "self" is uninterrupted and always in charge?

In the previous chapter, we met the plight of split-brain patients, who sometimes struggle with alien hands that literally have a mind of their own. It does appear that there are two centers of consciousness living within the same brain. So how does all this create the sense that we have a unified, cohesive "self" existing within our brains?

I asked one person who may have the answer: Dr. Michael Gazzaniga, who has spent several decades studying the strange behavior of split-brain patients. He noticed that the left brain of split-brain patients, when confronted with the fact that there seem to be two separate centers of consciousness residing in the same skull, would simply make up strange explanations, no matter how silly. He told me that, when presented with an obvious paradox, the left brain will "confabulate" an answer to explain inconvenient facts. Dr. Gazzaniga believes that this gives us the false sense that we are unified and whole. He calls the left brain the "interpreter," which is constantly thinking up ideas to paper over inconsistencies and gaps in our consciousness.

For example, in one experiment, he flashed the word "red" to just the left brain of a patient, and the word "banana" to just the right brain. (Notice that the dominant left brain therefore does not know about the banana.) Then the subject was asked to pick up a pen with his left hand (which is governed by the right brain) and draw a picture. Naturally he drew a picture of a banana. Remember that the right brain could do this, because it had seen the banana, but the left brain had no clue that the banana had been flashed to the right brain.

Then he was asked why he had drawn the banana. Because only the left brain controls speech, and because the left brain did not know anything about a banana, the patient should have said, "I don't know." Instead he said, "It is easiest to draw with this hand because this hand can pull down easier." Dr. Gazzaniga noted that the left brain was trying to find some excuse for this inconvenient fact, even though the patient was clueless about why his right hand drew the banana.

Dr. Gazzaniga concludes, "It is the left hemisphere that engages in the human tendency to find order in chaos, that tries to fit everything into a story and put it into a context. It seems that it is driven to hypothesize about the structure of the world even in the face of evidence that no pattern exists."

This is where our sense of a unified "self" comes from. Although consciousness is a patchwork of competing and often contradictory tendencies, the left brain ignores inconsistencies and papers over obvious gaps in order to give us a smooth sense of a single "I." In other words, the left brain is constantly making excuses, some of them harebrained and preposterous, to make sense of the world. It is constantly asking "Why?" and dreaming up excuses even if the question has no answer.

(There is probably an evolutionary reason that we evolved our split brains. A seasoned CEO will often encourage his aides to take opposing sides of an issue, to encourage thorough and thoughtful debate. Oftentimes, the correct view emerges out of intense interaction with incorrect ideas. Similarly, the two halves of the brain complement each other, offering pessimistic/optimistic or analytical/holistic analysis of the same idea. The two halves of the brain therefore play off each other. Indeed, as we shall see, certain forms of mental illness may arise when this interplay between the two brains goes awry.)

Now that we have a working theory of consciousness, the time has come to utilize it to understand how neuroscience will evolve in the future. There is a vast and remarkable set of experiments now being done in neuroscience that are fundamentally altering the entire scientific landscape. Using the power of electromagnetism, scientists can now probe people's thoughts, send telepathic messages, telekinetically control objects around us, record memories, and perhaps enhance our intelligence.

Perhaps the most immediate and practical application of this new technology is something once considered to be hopelessly impossible: telepathy.

BOOK II MIND OVER MATTER

The brain, like it or not, is a machine. Scientists have come to that conclusion, not because they are mechanistic killjoys, but because they have amassed evidence that every aspect of consciousness can be tied to the brain.

—STEVEN PINKER

3 TELEPATHY A PENNY FOR YOUR THOUGHTS

Harry Houdini, some historians believe, was the greatest magician who ever lived. His breathtaking escapes from locked, sealed chambers and death-defying stunts left audiences gasping. He could make people disappear and then reemerge in the most unexpected places. And he could read people's minds.

Or at least it seemed that way.

Houdini took pains to explain that everything he did was an illusion, a series of clever sleight-of-hand tricks. Mind reading, he would remind people, was impossible. He was so outraged that unscrupulous magicians would cheat wealthy patrons by performing cheap parlor tricks and séances that he even went around the country exposing fakes by pledging he could duplicate any feat of mind reading performed by these charlatans. He was even on a committee organized by *Scientific American* that offered a generous reward to anyone who could positively prove they had psychic power. (No one ever picked up the reward.)

Houdini believed that telepathy was impossible. But science is proving Houdini wrong.

Telepathy is now the subject of intense research at universities around the world, where scientists have already been able to use advanced sensors to read individual words, images, and thoughts in a person's brain. This could alter the way we communicate with stroke and accident victims who are "locked in" their bodies, unable to articulate their thoughts except through blinks. But that's just the start. Telepathy might also radically change the way we interact with computers and the outside world.

Indeed, in a recent "Next 5 in 5 Forecast," which predicts five revolutionary developments in the next five years, IBM scientists claimed that we will be able to mentally communicate with computers, perhaps replacing the mouse and voice commands. This means using the power of the mind to call people on the phone, pay credit card bills, drive cars, make appointments, create beautiful symphonies and works of art, etc. The possibilities are endless, and it seems that everyone—from computer giants, educators, video game companies, and music studios to the Pentagon—is converging on this technology.

True telepathy, found in science-fiction and fantasy novels, is not possible without outside assistance. As we know, the brain is electrical. In general, anytime an electron is accelerated, it gives off electromagnetic radiation. The same holds true for electrons oscillating inside the brain, which broadcasts radio waves. But these signals are too faint to be detected by others, and even if we could perceive these radio waves, it would be difficult to make sense of them. Evolution has not given us the ability to decipher this collection of random radio signals, but computers can. Scientists have been able to get crude approximations of a person's thoughts using EEG scans. Subjects would put on a helmet with EEG sensors and concentrate on certain pictures—say, the image of a car. The EEG signals were then recorded for each image and eventually a rudimentary dictionary of thought was created, with a one-to-one correspondence between a person's thoughts and the EEG image. Then, when a person was shown a picture of another car, the computer would recognize the EEG pattern as being from a car.

The advantage of EEG sensors is that they are noninvasive and quick. You simply put a helmet containing many electrodes onto the surface of the brain and the EEG can rapidly identify signals that change every millisecond. But the problem with EEG sensors, as we have seen, is that electromagnetic waves deteriorate as they pass through the skull, and it is difficult to locate their precise source. This method can tell if you are thinking of a car or a house, but it cannot re-create an image of the car. That is where Dr. Jack Gallant's work comes in.

VIDEOS OF THE MIND

The epicenter for much of this research is the University of California at Berkeley, where I received my own Ph.D. in theoretical physics years ago. I had the pleasure of touring the laboratory of Dr. Gallant, whose group has accomplished a feat once considered to be impossible: videotaping people's thoughts. "This is a major leap forward reconstructing internal imagery. We are opening a window into the movies in our mind," says Gallant.

When I visited his laboratory, the first thing I noticed was the team of young, eager postdoctoral and graduate students huddled in front of their computer screens, looking intently at video images that were reconstructed from someone's brain scan. Talking to Gallant's team, you feel as though you are witnessing scientific history in the making.

Gallant explained to me that first the subject lies flat on a stretcher, which is slowly inserted headfirst into a huge, state-of-the-art MRI machine, costing upward of $3 million. The subject is then shown several movie clips (such as movie trailers readily available on YouTube). To accumulate enough data, the subject has to sit motionless for hours watching these clips, a truly arduous task. I asked one of the postdocs, Dr. Shinji Nishimoto, how they found volunteers who were willing to lie still for hours on end with only fragments of video footage to occupy the time. He said the people in the room, the grad students and postdocs, volunteered to be guinea pigs for their own research.

As the subject watches the movies, the MRI machine creates a 3-D image of the blood flow within the brain. The MRI image looks like a vast collection of thirty thousand dots, or voxels. Each voxel represents a pinpoint of neural energy, and the color of the dot corresponds to the intensity of the signal and blood flow. Red dots represent points of large neural activity, while blue dots represent points of less activity. (The final image looks very much like thousands of Christmas lights in the shape of the brain. Immediately you can see that the brain is concentrating most of its mental energy in the visual cortex, which is located at the back of the brain, while watching these videos.)

Gallant's MRI machine is so powerful it can identify two to three hundred distinct regions of the brain and, on average, can take snapshots that have one hundred dots per region of the brain. (One goal for future generations of MRI technology is to provide an even sharper resolution by increasing the number of dots per region of the brain.)

At first, this 3-D collection of colored dots looks like gibberish. But after years of research, Dr. Gallant and his colleagues have developed a mathematical formula that begins to find relationships between certain features of a picture (edges, textures, intensity, etc.) and the MRI voxels. For example, if you look at a boundary, you'll notice it's a region separating lighter and darker areas, and hence the edge generates a certain pattern of voxels. By having subject after subject view such a large library of movie clips, this mathematical formula is refined, allowing the computer to analyze how all sorts of images are converted into MRI voxels. Eventually the scientists were able to ascertain a direct correlation between certain MRI patterns of voxels and features within each picture.

At this point, the subject is then shown another movie trailer. The computer analyzes the voxels generated during this viewing and re-creates a rough approximation of the original image. (The computer selects images from one hundred movie clips that most closely resemble the one that the subject just saw and then merges images to create a close approximation.) In this way, the computer is able to create a fuzzy video of the visual imagery going through your mind. Dr. Gallant's mathematical formula is so versatile that it can take a collection of MRI voxels and convert it into a picture, or it can do the reverse, taking a picture and then converting it to MRI voxels.

I had a chance to view the video created by Dr. Gallant's group, and it was very impressive. Watching it was like viewing a movie with faces, animals, street scenes, and buildings through dark glasses. Although you could not see the details within each face or animal, you could clearly identify the kind of object you were seeing.

Not only can this program decode what you are looking at, it can also decode imaginary images circulating in your head. Let's say you are asked to think of the *Mona Lisa.* We know from MRI scans that even though you're not viewing the painting with your eyes, the visual cortex of your brain will light up. Dr. Gallant's program then scans your brain while you are thinking of the *Mona Lisa* and flips through its data files of pictures, trying to find the closest match. In one experiment I saw, the computer selected a picture of the actress Salma Hayek as the closest approximation to the *Mona Lisa.* Of course, the average person can easily recognize hundreds of faces, but the fact that the computer analyzed an image within a person's brain and then picked out this picture from millions of random pictures at its disposal is still impressive.

The goal of this whole process is to create an accurate dictionary that allows you to rapidly match an object in the real world with the MRI pattern in your brain. In general, a detailed match is very difficult and will take years, but some categories are actually easy to read just by flipping through some photographs. Dr. Stanislas Dehaene of the Collège de France in Paris was examining MRI scans of the parietal lobe, where numbers are recognized, when one of his postdocs casually mentioned that just by quickly scanning the MRI pattern, he could tell what number the subject was looking at. In fact, certain numbers created distinctive patterns on the MRI scan. He notes, "If you take 200 voxels in this area, and look at which of them are active and which are inactive, you can construct a machine-learning device that decodes which number is being held in memory."

This leaves open the question of when we might be able to have picture-quality videos of our thoughts. Unfortunately, information is lost when a person is visualizing an image. Brain scans corroborate this. When you compare the MRI scan of the brain as it is looking at a flower to an MRI scan as the brain is thinking about a flower, you immediately see that the second image has far fewer dots than the first. So although this technology will vastly improve in the coming years, it will never be perfect. (I once read a short story in which a man meets a genie who offers to create anything that the person can imagine. The man immediately asks for a luxury car, a jet plane, and a million dollars. At first, the man is ecstatic. But when he looks at these items in detail, he sees that the car and the plane have no engines, and the image on the cash is all blurred. Everything is useless. This is because our memories are only approximations of the real thing.)

But given the rapidity with which scientists are beginning to decode the MRI patterns in the brain, will we soon be able to actually read words and thoughts circulating in the mind?

READING THE MIND

In fact, in a building next to Gallant's laboratory, Dr. Brian Pasley and his colleagues are literally reading thoughts—at least in principle. One of the postdocs there, Dr. Sara Szczepanski, explained to me how they are able to identify words inside the mind.

The scientists used what is called ECOG (electrocorticogram) technology, which is a vast improvement over the jumble of signals that EEG scans produce. ECOG scans are unprecedented in accuracy and resolution, since signals are directly recorded from the brain and do not pass through the skull. The flipside is that one has to remove a portion of the skull to place a mesh, containing sixty-four electrodes in an eight-by-eight grid, directly on top of the exposed brain.

Luckily they were able to get permission to conduct experiments with ECOG scans on epileptic patients, who were suffering from debilitating seizures. The ECOG mesh was placed on the patients' brains while open-brain surgery was being performed by doctors at the nearby University of California at San Francisco.

As the patients hear various words, signals from their brains pass through the electrodes and are then recorded. Eventually a dictionary is formed, matching the word with the signals emanating from the electrodes in the brain. Later, when a word is uttered, one can see the same electrical pattern. This correspondence also means that if one is thinking of a certain word, the computer can pick up the characteristic signals and identify it.

With this technology, it might be possible to have a conversation that takes place entirely telepathically. Also, stroke victims who are totally paralyzed may be able to "talk" through a voice synthesizer that recognizes the brain patterns of individual words.

Not surprisingly, BMI (brain-machine interface) has become a hot field, with groups around the country making significant breakthroughs. Similar results were obtained by scientists at the University of Utah in 2011. They placed grids, each containing sixteen electrodes, over the facial motor cortex (which controls movements of the mouth, lips, tongue, and face) and Wernicke's area, which processes information about language.

The person was then asked to say ten common words, such as "yes" and "no," "hot" and "cold," "hungry" and "thirsty," "hello" and "good-bye," and "more" and "less." Using a computer to record the brain signals when these words were uttered, the scientists were able to create a rough one-to-one correspondence between spoken words and computer signals from the brain. Later, when the patient voiced certain words, they were able to correctly identify each one with an accuracy ranging from 76 percent to 90 percent. The next step is to use grids with 121 electrodes to get better resolution.

In the future, this procedure may prove useful for individuals suffering from strokes or paralyzing illnesses such as Lou Gehrig's disease, who would be able to speak using the brain-to-computer technique.

TYPING WITH THE MIND

At the Mayo Clinic in Minnesota, Dr. Jerry Shih has hooked up epileptic patients via ECOG sensors so they can learn how to type with the mind. The calibration of this device is simple. The patient is first shown a series of letters and is told to focus mentally on each symbol. A computer records the signals emanating from the brain as it scans each letter. As with the other experiments, once this one-to-one dictionary is created, it is then a simple matter for the person to merely think of the letter and for the letter to be typed on a screen, using only the power of the mind.

Dr. Shih, the leader of this project, says that the accuracy of his machine is nearly 100 percent. Dr. Shih believes that he can next create a machine to record images, not just words, that patients conceive in their minds. This could have applications for artists and architects, but the big drawback of ECOG technology, as we have mentioned, is that it requires opening up patients' brains.

Meanwhile, EEG typewriters, because they are noninvasive, are entering the marketplace. They are not as accurate or precise as ECOG typewriters, but they have the advantage that they can be sold over the counter. Guger Technologies, based in Austria, recently demonstrated an EEG typewriter at a trade show. According to their officials, it takes only ten minutes or so for people to learn how to use this machine, and they can then type at the rate of five to ten words per minute.

TELEPATHIC DICTATION AND MUSIC

The next step might be to transmit entire conversations, which could rapidly speed up telepathic transmission. The problem, however, is that it would require making a one-to-one map between thousands of words and their EEG, MRI, or ECOG signals. But if one can, for example, identify the brain signals of several hundred select words, then one might be able to rapidly transmit words found in a common conversation. This means that one would think of the words in entire sentences and paragraphs of a conversation and a computer would print them out.

This could be extremely useful for journalists, writers, novelists, and poets, who could simply think and have a computer take dictation. The computer would also become a mental secretary. You would mentally give instructions to the robo-secretary about a dinner, plane trip, or vacation, and it would fill in all the details about the reservations.

Not only dictation but also music may one day be transcribed in this way. Musicians would simply hum a few melodies in their head and a computer would print them out, in musical notation. To do this, you would ask someone to mentally hum a series of notes, which would generate certain electrical signals for each one. A dictionary would again be created in this way, so that when you think of a musical note, the computer would print it out in musical notation.

In science fiction, telepaths often communicate across language barriers, since thoughts are considered to be universal. However, this might not be true. Emotions and feelings may well be nonverbal and universal, so that one could telepathically send them to anyone, but rational thinking is so closely tied to language that it is very unlikely that complex thoughts could be sent across language barriers. Words will still be sent telepathically in their original language.

TELEPATHY HELMETS

In science fiction, we also often encounter telepathy helmets. Put them on, and—presto!—you can read other people's minds. The U.S. Army, in fact, has expressed interest in this technology. In a firefight, with explosions going off and bullets whizzing overhead, a telepathy helmet could be a lifesaver, since it can be difficult to communicate orders amid the sound and fury of the battlefield. (I can personally testify to this. Years ago, during the Vietnam War, I served in the U.S. Infantry at Fort Benning, outside Atlanta, Georgia. During machine-gun training, the sound of hand grenades and rounds of bullets going off on the battlefield next to my ear was deafening; it was so intense I could not hear anything else. Later, there was a loud ringing in my ear that lasted for three full days.) With a telepathy helmet, a soldier could mentally communicate with his platoon amid all the thunder and noise.

Recently, the army gave a $6.3 million grant to Dr. Gerwin Schalk at Albany Medical College, but it knows that a fully functional telepathy helmet is still years away. Dr. Schalk experiments with ECOG technology, which, as we have seen, requires placing a mesh of electrodes directly on top of the exposed brain. With this method, his computers have been able to recognize vowels and thirty-six individual words inside the thinking brain. In some of his experiments, he approached 100 percent accuracy. But at present, this is still impractical for the U.S. Army, since it requires removing part of the skull in the clean, sterile environment of a hospital. And even then, recognizing vowels and a handful of words is a far cry from sending urgent messages to headquarters in a firefight. But his ECOG experiments have demonstrated that it is possible to communicate mentally on the battlefield.

Another method is being explored by Dr. David Poeppel of New York University. Instead of opening up the skulls of his subjects, he employs MEG technology, using tiny bursts of magnetic energy rather than electrodes to create electrical charges in the brain. Besides being noninvasive, the advantage of MEG technology is that it can precisely measure fleeting neural activity, in contrast to the slower MRI scans. In his experiments, Poeppel has been able to successfully record electrical activity in the auditory cortex when people think silently of a certain word. But the drawback is that this recording still requires the use of large, table-size machines to generate a magnetic pulse.

Obviously, one wants a method that is noninvasive, portable, and accurate. Dr. Poeppel hopes his work with MEG technology will complement the work being done using EEG sensors. But true telepathy helmets are still many years away, because MEG and EEG scans lack accuracy.

MRI IN A CELL PHONE

At present, we are hindered by the relatively crude nature of the existing instruments. But, as time goes by, more and more sophisticated instruments will probe deeper into the mind. The next big breakthrough may be MRI machines that are handheld.

The reason why MRI machines have to be so huge right now is that one needs a uniform magnetic field to get good resolution. The larger the magnet, the more uniform one can make the field, and the better accuracy one finds in the final pictures. However, physicists know the exact mathematical properties of magnetic fields (they were worked out by physicist James Clerk Maxwell back in the 1860s). In 1993 in Germany, Dr. Bernhard Blümich and his colleagues created the world's smallest MRI machine, which is the size of a briefcase. It uses a weak and distorted magnetic field, but supercomputers can analyze the magnetic field and correct for this so that the device produces realistic 3-D pictures. Since computer power doubles roughly every two years, they are now powerful enough to analyze the magnetic field created by the briefcase-sized device and compensate for its distortion.

As a demonstration of their machine, in 2006 Dr. Blümich and his colleagues were able to take MRI scans of Ötzi, the "Iceman," who was frozen in ice about 5,300 years ago toward the end of the last ice age. Because Ötzi was frozen in an awkward position, with his arms spread apart, it was difficult to cram him inside the small cylinder of a conventional MRI machine, but Dr. Blümich's portable machine easily took MRI photographs.

These physicists estimate that, with increasing computer power, an MRI machine of the future might be the size of a cell phone. The raw data from this cell phone would be sent wirelessly to a supercomputer, which would process the data from the weak magnetic field and then create a 3-D image. (The weakness of the magnetic field is compensated for by the increase in computer power.) This then could vastly accelerate research. "Perhaps something like the *Star Trek* tricorder is not so far off after all," Dr. Blümich has said. (The tricorder is a small, handheld scanning device that gives an instant diagnosis of any illness.) In the future, you may have more computer power in your medicine cabinet than there is in a modern university hospital today. Instead of waiting to get permission from a hospital or university to use an expensive MRI machine, you could gather data in your own living room by simply waving the portable MRI over yourself and then e-mailing the results to a lab for analysis.

It could also mean that, at some point in the future, an MRI telepathy helmet might be possible, with vastly better resolution than an EEG scan. Here is how it may work in the coming decades. Inside the helmet, there would be electromagnetic coils to produce a weak magnetic field and radio pulses that probe the brain. The raw MRI signals would then be sent to a pocket-size computer placed in your belt. The information would then be radioed to a server located far from the battlefield. The final processing of the data would be done by a supercomputer in a distant city. Then the message would be radioed back to your troops on the battlefield. The troops would hear the message either through speakers or through electrodes placed in the auditory cortex of their brains.

DARPA AND HUMAN ENHANCEMENT

Given the costs of all this research, it is legitimate to ask: Who is paying for it? Private companies have only recently shown interest in this cutting-edge technology, but it's still a big gamble for many of them to fund research that may never pay off. Instead, one of the main backers is DARPA, the Pentagon's Defense Advanced Research Projects Agency, which has spearheaded some of the most important technologies of the twentieth century.

DARPA was originally set up by President Dwight Eisenhower after the Russians sent Sputnik into orbit in 1957 and shocked the world. Realizing that the United States might quickly be outpaced by the Soviets in high technology, Eisenhower hastily established this agency to keep the country competitive with the Russians. Over the years, the numerous projects it started grew so large that they became independent entities by themselves. One of its first spinoffs was NASA.

DARPA's strategic plan reads like something from science fiction: its "*only* charter is radical innovation." The only justification for its existence is "to accelerate the future into being." DARPA scientists are constantly pushing the boundaries of what is physically possible. As former DARPA official Michael Goldblatt says, they try not to violate the laws of physics, "or at least not knowingly. Or at least not more than one per program."

But what separates DARPA from science fiction is its track record, which is truly astounding. One of its early projects in the 1960s was Arpanet, which was a war-fighting telecommunications network that would electronically connect scientists and officials during and after World War III. In 1989, the National Science Foundation decided that, in light of the breakup of the Soviet bloc, it was unnecessary to keep it a secret, so it declassified this hush-hush military technology and essentially gave codes and blueprints away for free. Arpanet would eventually become the Internet.

When the U.S. Air Force needed a way to guide its ballistic missiles in space, DARPA helped create Project 57, a top-secret project that was designed to place H-bombs on hardened Soviet missile silos in a thermonuclear exchange. It would later become the foundation for the Global Positioning System (GPS). Instead of guiding missiles, today it guides lost motorists.

DARPA has been a key player in a series of inventions that have altered the twentieth and twenty-first centuries, including cell phones, night-vision goggles, telecommunications advances, and weather satellites. I have had a chance to interact with DARPA scientists and officials on several occasions. I once had lunch with one of the agency's former directors at a reception filled with many scientists and futurists. I asked him a question that had always bothered me: Why do we have to rely on dogs to sniff our luggage for the presence of high explosives? Surely our sensors are sensitive enough to pick up the telltale signature of explosive chemicals. He replied that DARPA had actively looked into this same question but had come up against some severe technical problems. The olfactory sensors of dogs, he said, had evolved over millions of years to be able to detect a handful of molecules, and that kind of sensitivity is extremely difficult to match, even with our most finely tuned sensors. It's likely that we will continue to rely on dogs at airports for the foreseeable future.

On another occasion, a group of DARPA physicists and engineers came to a talk I gave about the future of technology. Later I asked them if they had any concerns of their own. One concern, they said, was their public image. Most people have never heard of DARPA, but some link it to dark, nefarious government conspiracies, everything from UFO cover-ups, Area 51, and Roswell to weather control, etc. They sighed. If only these rumors were true, they could certainly use help from alien technology to jump-start their research!

With a budget of $3 billion, DARPA has now set its sights on the brain-machine interface. When discussing the potential applications, former DARPA official Michael Goldblatt pushes the boundary of the imagination. He says, "Imagine if soldiers could communicate by thought alone. . . . Imagine the threat of biological attack being inconsequential. And contemplate, for a moment, a world in which learning is as easy as eating, and the replacement of damaged body parts as convenient as a fast-food drive-through. As impossible as these visions sound or as difficult as you might think the task would be, these visions are the everyday work of the Defense Sciences Office [a branch of DARPA]."

Goldblatt believes that historians will conclude that the long-term legacy of DARPA will be human enhancement, "our future historical strength." He notes that the famous army slogan "Be All You Can Be" takes on a new meaning when contemplating the implications of human enhancement. Perhaps it is no accident that Michael Goldblatt is pushing human enhancement so vigorously at DARPA. His own daughter suffers from cerebral palsy and has been confined to a wheelchair all her life. Since she requires outside help, her illness has slowed her down, but she has always risen above adversity. She is going to college and dreaming of starting her own company. Goldblatt acknowledges that his daughter is his inspiration. As *Washington Post* editor Joel Garreau has noted, "What he is doing is spending untold millions of dollars to create what might well be the next step in human evolution. And yet, it has occurred to him that the technology he is helping create might someday allow his daughter not just to walk, but to transcend."

PRIVACY ISSUES

When hearing of mind-reading machines for the first time, the average person might be concerned about privacy. The idea that a machine concealed somewhere may be reading our intimate thoughts without our permission is unnerving. Human consciousness, as we have stressed, involves constantly running simulations of the future. In order for these simulations to be accurate, we sometimes imagine scenarios that wade into immoral or illegal territory, but whether or not we act on these plans, we prefer to keep them private.

For scientists, life would be easier if they could simply read people's thoughts from a distance using portable devices (rather than by using clumsy helmets or surgically opening up the skull), but the laws of physics make this exceedingly difficult.

When I asked Dr. Nishimoto, who works in Dr. Gallant's Berkeley lab, about the question of privacy, he smiled and replied that radio signals degrade quite rapidly outside the brain, so these signals would be too diffuse and weak to make any sense to anyone standing more than a few feet away. (In school, we learned about Newton's laws and that gravity diminishes as the square of the distance, so that if you doubled your distance from a star, the gravity field diminishes by a factor of four. But magnetic fields diminish much faster than the square of the distance. Most signals decrease by the cube or quartic of the distance, so if you double the distance from an MRI machine, the magnetic field goes down by a factor of eight or more.)

Furthermore, there would be interference from the outside world, which would mask the faint signals coming from the brain. This is one reason why scientists require strict laboratory conditions to do their work, and even then they are able to extract only a few letters, words, or images from the thinking brain at any given time. The technology is not adequate to record the avalanche of thoughts that often circulate in our brain as we simultaneously consider several letters, words, phrases, or sensory information, so using these devices for mind reading as seen in the movies is not possible today, and won't be for decades to come.

For the foreseeable future, brain scans will continue to require direct access to the human brain in laboratory conditions. But in the highly unlikely event that someone in the future finds a way to read thoughts from a distance, there are still countermeasures you can take. To keep your most important thoughts private, you might use a shield to block brain waves from entering the wrong hands. This can be done with something called a Faraday cage, invented by the great British physicist Michael Faraday in 1836, although the effect was first observed by Benjamin Franklin. Basically, electricity will rapidly disperse around a metal cage, such that the electric field inside the cage is zero. To demonstrate this, physicists (like myself) have entered a metallic cage on which huge electrical bolts are fired. Miraculously, we are unscratched. This is why airplanes can be hit by lightning bolts and not suffer damage, and why cable wires are covered with metallic threads. Similarly, a telepathy shield would consist of thin metal foil placed around the brain.

TELEPATHY VIA NANOPROBES IN THE BRAIN

There is another way to partially solve the privacy issue, as well as the difficulty of placing ECOG sensors into the brain. In the future, it may be possible to exploit nanotechnology, the ability to manipulate individual atoms, to insert a web of nanoprobes into the brain that can tap into your thoughts. These nanoprobes might be made of carbon nanotubes, which conduct electricity and are as thin as the laws of atomic physics allow. These nanotubes are made of individual carbon atoms arrayed in a tube a few molecules thick.

(They are the subject of intense scientific interest, and are expected in the coming decades to revolutionize the way scientists probe the brain.)

The nanoprobes would be placed precisely in those areas of the brain devoted to certain activities. In order to convey speech and language, they would be placed in the left temporal lobes. In order to process visual images, they would be placed in the thalamus and visual cortex. Emotions would be sent via nanoprobes in the amygdala and limbic system. The signals from these nanoprobes would be sent to a small computer, which would process the signals and wirelessly send information to a server and then the Internet.

Privacy issues would be partially solved, since you would completely control when your thoughts are being sent over cables or the Internet. Radio signals can be detected by any bystander with a receiver, but electrical signals sent along a cable cannot. The problem of opening up the skull to use messy ECOG meshes is also solved, because the nanoprobes can be inserted via microsurgery.

Some science-fiction writers have conjectured that when babies are born in the future, these nanoprobes might be painlessly implanted, so that telepathy becomes a way of life for them. In *Star Trek*, for example, implants are routinely placed into the children of the Borg at birth so that they can telepathically communicate with others. These children cannot imagine a world where telepathy does not exist. They take it for granted that telepathy is the norm.

Because these nanoprobes are tiny, they would be invisible to the outside world, so there would be no social ostracism. Although society might be repulsed at the idea of inserting probes permanently into the brain, these science-fiction writers assume that people will get used to the idea because the nanoprobes would be so useful, just like test-tube babies have been accepted by society today after the initial controversy surrounding them.

LEGAL ISSUES

For the foreseeable future, the question is not whether someone will be able to read our thoughts secretly from a remote, concealed device, but whether we will willingly allow our thoughts to be recorded. What happens, then, if some unscrupulous person gets unauthorized access to those files? This raises the issue of ethics, since we would not want our thoughts to be read against our will. Dr. Brian Pasley says, "There are ethical concerns, not with the current research, but with the possible extensions of it. There has to be a balance. If we are somehow able to decode someone's thoughts instantaneously that might have great benefits for the thousands of severely disabled people who are unable to communicate right now. On the other hand, there are great concerns if this were applied to people who didn't want that."

Once it becomes possible to read people's minds and make recordings, a host of other ethical and legal questions will arise. This happens whenever any new technology is introduced. Historically it often takes years before the law is fully able to address their implications.

For instance, copyright laws may have to be rewritten. What happens if someone steals your invention by reading your thoughts? Can you patent your thoughts? Who actually owns the idea?

Another problem occurs if the government is involved. As John Perry Barlow, poet and lyricist for the Grateful Dead, once said, "Relying on the government to protect your privacy is like asking a peeping tom to install your window blinds." Would the police be allowed to read your thoughts when you are being interrogated? Already courts have been ruling on cases where an alleged criminal refused to submit his DNA as evidence. In the future, will the government be allowed to read your thoughts without your consent, and if so, will they be admissible in court? How reliable would they be? In the same way that MRI lie detectors measure only increased brain activity, it's important to note that thinking about a crime and actually committing one are two different things. During cross-examination, a defense lawyer might argue that these thoughts were just random musings and nothing more.

Another gray area concerns the rights of people who are paralyzed. If they are drafting a will or legal document, can a brain scan be sufficient to create a legal document? Assume that a totally paralyzed person has a sharp, active mind and wants to sign a contract or manage his funds. Are these documents legal, given that the technology may not be perfect?

There is no law of physics that can resolve these ethical questions. Ultimately, as this technology matures, these issues will have to be settled in court by judges and juries.

Meanwhile, governments and corporations might have to invent new ways to prevent mental espionage. Industrial espionage is already a multimillion-dollar industry, with governments and corporations building expensive "safe rooms" that have been scanned for bugs and listening devices. In the future (assuming that a method can be devised to listen to brain waves from a distance), safe rooms may have to be designed so that brain signals are not accidentally leaked to the outside world. These safe rooms would be surrounded by metallic walls, which would form a Faraday cage shielding the interior of the room from the outside world.

Every time a new form of radiation has been exploited, spies have tried to use it for espionage, and brain waves are probably no exception. The most famous case involved a tiny microwave device hidden in the Great Seal of the United States in the U.S. embassy in Moscow. From 1945 until 1952, it was transmitting top-secret messages from U.S. diplomats directly to the Soviets. Even during the Berlin Crisis of 1948 and the Korean War, the Soviets used this bug to decipher what the United States was planning. It might have continued to leak secrets even today, changing the course of the Cold War and world history, but it was accidentally discovered when a British engineer heard secret conversations on an open radio band. U.S. engineers were shocked when they picked apart the bug; they failed to detect it for years because it was passive, requiring no energy source. (The Soviets cleverly evaded detection because the bug was energized by microwave beams from a remote source.) It is possible that future espionage devices will be made to intercept brain waves as well.

Although much of this technology is still primitive, telepathy is slowly becoming a fact of life. In the future, we may interact with the world via the mind. But scientists want to go beyond just reading the mind, which is passive. They want to take an active role—to move objects with the mind. Telekinesis is a power usually ascribed to the gods. It is the divine power to shape reality to your wishes. It is the ultimate expression of our thoughts and desires.

We will soon have it.

It is the business of the future to be dangerous. . . . The major advances in civilization are processes that all but wreck the societies in which they occur.

—ALFRED NORTH WHITEHEAD

4 TELEKINESIS MIND CONTROLLING MATTER

Cathy Hutchinson is trapped inside her body.

She was paralyzed fourteen years ago by a massive stroke. A quadriplegic, she is like thousands of "locked-in" patients who have lost control over most of their muscles and bodily functions. Most of the day, she lies helpless, requiring continual nursing care, yet her mind is clear. She is a prisoner in her own body.

But in May 2012, her fortunes changed radically. Scientists at Brown University placed a tiny chip on top of her brain, called Braingate, which is connected by wires to a computer. Signals from her brain are relayed through the computer to a mechanical robotic arm. By simply thinking, she gradually learns to control the motion of the arm so that it can, for instance, grab a bottled drink and bring it to her mouth. For the first time, she is able to have some control of the world around her.

Because she is paralyzed and cannot talk, she had to communicate her excitement by making eye movements. A device tracks her eyes and then translates her movements into a typed message. When she was asked how she felt, after years of being imprisoned inside a shell called her body, she replied, "Ecstatic!" Looking forward to the day when her other limbs are connected to her brain via computer, she added, "I would love to have a robotic leg support." Before her stroke, she loved to cook and tend her garden. "I know that someday this will happen again," she added. At the rate at which the field of cyber prosthetics is moving, she might have her wish soon.

Professor John Donoghue and his colleagues at Brown University and also at the University of Utah have created a tiny sensor that acts like a bridge to the outside world for those who can no longer communicate. When I interviewed him, he told me, "We have taken a tiny sensor, the size of a baby aspirin, or four millimeters, and implanted it onto the surface of the brain. Because of ninety-six little 'hairs' or electrodes that pick up brain impulses, it can pick up signals of your intention to move your arm. We target the arm because of its importance." Because the motor cortex has been carefully mapped over the decades, it is possible to place the chip directly on top of the neurons that control specific limbs.

The key to Braingate lies in translating neural signals from the chip into meaningful commands that can move objects in the real world, starting with the cursor of a computer screen. Donoghue told me that he does this by asking the patient to imagine moving the cursor of a computer screen in a certain way, e.g., moving it to the right. It takes only a few minutes to record the brain signals corresponding to this task. In this way, the computer recognizes that whenever it detects a brain signal like that, it should move the cursor to the right.

Then, whenever that person thinks of moving the cursor to the right, the computer actually moves the cursor in that direction. In this way, there is a one-to-one map between certain actions that the patient imagines and the actual action itself. A patient can immediately start to control the movement of the cursor, practically on the first try.

Braingate opens the door to a new world of neuroprosthetics, allowing a paralyzed person to move artificial limbs with the mind. In addition, it lets the patient communicate directly with their loved ones. The first version of this chip, tested in 2004, was designed so that paralyzed patients could communicate with a laptop computer. Soon afterward, these patients were surfing the web, reading and writing e-mails, and controlling their wheelchairs.

More recently, the cosmologist Stephen Hawking had a neuroprosthetic device attached to his glasses. Like an EEG sensor, it can connect his thoughts to a computer so that he can maintain some contact with the outside world.

It is rather primitive, but eventually devices similar to it will become much more sophisticated, with more channels and greater sensitivity.

All this, Dr. Donoghue told me, could have a profound impact on the lives of these patients: "Another useful thing is that you can connect this computer to any device—a toaster, a coffee maker, an air conditioner, a light switch, a typewriter. It's really quite easy to do these things these days, and it's very inexpensive. For a quadriplegic who can't get around, they will be able to change the TV channel, turn the lights on, and do all those things without anybody coming into the room and doing it for them." Eventually, they will be able to do anything a normal person can do, via computers.

FIXING SPINAL CORD INJURIES

A number of other groups are entering the fray. Another breakthrough was made by scientists at Northwestern University who have connected a monkey's brain directly to his own arm, bypassing an injured spinal cord. In 1995, there was the sad story of Christopher Reeve, who soared into outer space in the *Superman* movies but was completely paralyzed due to an injury to his spinal cord. Unfortunately, he was thrown off a horse and landed on his neck, so the spinal cord was damaged just beneath his head. If he had lived longer, he might have seen the work of scientists who want to use computers to replace broken spinal cords. In the United States alone, more than two hundred thousand people have some form of spinal cord injury. In an earlier age, these individuals might have died soon after the accident, but because of advances in acute trauma care, the number of people who survive these sorts of injuries has actually grown in recent years. We are also haunted by the images of thousands of wounded warriors who were victims of roadside bombs in Iraq and Afghanistan. And if you include the number of patients paralyzed by strokes and other illnesses, like amyotropic lateral sclerosis (ALS), the number of patients swells to two million.

The scientists at Northwestern used a one-hundred-electrode chip, which was placed directly on the brain of a monkey. The signals from the brain were carefully recorded as the monkey grasped a ball, lifted it, and released it into a tube. Since each task corresponds to a specific firing of neurons, the scientists could gradually decode these signals.

When the monkey wanted to move his arm, the signals were processed by a computer using this code, and, instead of sending the messages to a mechanical arm, they sent the signals directly to the nerves of the monkey's real arm. "We are eavesdropping on the natural electrical signals from the brain that tell the arm and hand how to move, and sending those signals directly to the muscles," says Dr. Lee Miller.

By trial and error, the monkey learned to coordinate the muscles in his arm. "There is a process of motor learning that is very similar to the process you go through when you learn to use a new computer, mouse, or a different tennis racquet," adds Dr. Miller.

(It is remarkable that the monkey was able to master so many motions of his arm, given the fact that there are only one hundred electrodes on this brain chip. Dr. Miller points out that millions of neurons are involved in controlling the arm. The reason that one hundred electrodes can give a reasonable approximation to the output of millions of neurons is that the chip connects to the output neurons, after all the complex processing has already been done by the brain. With the sophisticated analysis out of the way, the one hundred electrodes are responsible simply for feeding that information to the arm.)

This device is one of several being devised at Northwestern that will allow patients to bypass their injured spinal cords. Another neural prosthesis uses the motion of the shoulders to control the arm. An upward shrug causes the hand to close. A downward shrug causes the hand to open. The patient also has the ability to curl his fingers around an object like a cup, or manipulate a key that is grasped between the thumb and index finger.

Dr. Miller concludes, "This connection from brain to muscles might someday be used to help patients paralyzed due to spinal cord injury perform activities of daily living and achieve greater independence."

REVOLUTIONIZING PROSTHETICS

Much of the funding driving these remarkable developments comes from a DARPA project called Revolutionizing Prosthetics, a $150 million effort that has been bankrolling these efforts since 2006. One of the driving forces behind Revolutionizing Prosthetics is retired U.S. Army colonel Geoffrey Ling, who is a neurologist with several tours of duty in Iraq and Afghanistan. He was appalled at the human carnage he witnessed on the battlefield caused by roadside bombs. In previous wars, many of these brave service members would have died on the spot. But today, with helicopters and an extensive medical evacuation infrastructure, many of them survive but still suffer from serious bodily injuries. More than 1,300 service members have lost limbs after coming back from the Middle East.

Dr. Ling asked himself whether there was a scientific way to replace these lost limbs. Backed by funding from the Pentagon, he asked his staff to come up with concrete solutions within five years. When he made that request, he was met with incredulity. He recalled, "They thought we were crazy. But it's in insanity that things happen."

Spurred into action by Dr. Ling's boundless enthusiasm, his crew has created miracles in the laboratory. For example, Revolutionary Prosthetics funded scientists at the Johns Hopkins Applied Physics Laboratory who have created the most advanced mechanical arm on Earth, which can duplicate nearly all the delicate motions of the fingers, hand, and arm in three dimensions. It is the same size and has the same strength and agility as a real arm. Although it is made of steel, if you covered it up with flesh-colored plastic, it would be nearly indistinguishable from a real arm.

This arm was attached to Jan Sherman, a quadriplegic who had suffered from a genetic disease that damaged the connection between her brain and her body, leaving her completely paralyzed from the neck down. At the University of Pittsburgh, electrodes were placed directly on top of her brain, which were then connected to a computer and then to a mechanical arm. Five months after surgery to attach the arm, she appeared on *60 Minutes*. Before a national audience, she cheerfully used her new arm to wave, greet the host, and shake his hand. She even gave him a fist bump to show how sophisticated the arm was.

Dr. Ling says, "In my dream, we will be able to take this into all sorts of patients, patients with strokes, cerebral palsy, and the elderly."

TELEKINESIS IN YOUR LIFE

Not only scientists but also entrepreneurs are looking at brain-machine interface (BMI). They wish to incorporate many of these dazzling inventions as a permanent part of their business plans. BMI has already penetrated the youth market, in the form of video games and toys that use EEG sensors so that you can control objects with the mind in both virtual reality and the real world. In 2009, NeuroSky marketed the first toy, Mindflex, specifically designed to use EEG sensors to move a ball through a maze. Concentrating while wearing the Mindflex EEG device increases the speed of a fan within the maze and propels a tiny ball down a pathway.

Mind-controlled video games are also blossoming. Seventeen hundred software developers are working with NeuroSky, many of them on the company's $129 million Mindwave Mobile headset. These video games use a small, portable EEG sensor wrapped around your forehead that allows you to navigate in virtual reality, where the movements of your avatar are controlled mentally. As you maneuver your avatar on the video screen, you can fire weapons, evade enemies, rise to new levels, score points, etc., as in an ordinary video game, except that everything is hands-free.

"There's going to be a whole ecosystem of new players, and NeuroSky is very well positioned to be like the Intel of this new industry," claims Alvaro Fernandez of SharpBrains, a market research firm.

Besides firing virtual weapons, the EEG helmet can also detect when your attention begins to flatten out. NeuroSky has been getting inquiries from companies concerned about injuries to workers who lose concentration while operating a dangerous machine or who fall asleep at the wheel. This technology could be a lifesaver, alerting the worker or driver that he is losing his focus. The EEG helmet would set off an alarm when the wearer dozes off. (In Japan, this headset is already creating a fad among partygoers. The EEG sensors look like cat ears when you put them on your head. The ears suddenly rise when your attention is focused and then flatten out when it fades. At parties, people can express romantic interest just by thinking, so you know if you are impressing someone.)

But perhaps the most novel applications of this technology are being pursued by Dr. Miguel Nicolelis of Duke University. When I interviewed him, he told me that he thinks he can duplicate many of the devices found only in science fiction.

SMART HANDS AND MIND MELDS

Dr. Nicolelis has shown that this brain-machine interface can be done across continents. He places a monkey on a treadmill. A chip is positioned on the monkey's brain, which is connected to the Internet. On the other side of the planet, in Kyoto, Japan, signals from the monkey are used to control a robot that can walk. By walking on the treadmill in North Carolina, the monkey controls a robot in Japan, which executes the same walking motion. Using only his brain sensors and the reward of a food pellet, Dr. Nicolelis has trained these monkeys to control a humanoid robot called CB-1 halfway around the world.

He is also tackling one of the main problems with brain-machine interface: the lack of feeling. Today's prosthetic hands don't have a sense of touch, and hence they feel foreign; because there's no feedback, they might accidentally crush someone's fingers while engaging in a handshake. Picking up an eggshell with a mechanical arm would be nearly impossible.

Nicolelis hopes to circumvent this problem by having a direct brain-to-brain interface. Messages would be sent from the brain to a mechanical arm that has sensors, which would then send messages directly back to the brain, thereby bypassing the stem altogether. This brain-machine-brain interface (BMBI) could enable a clean, direct feedback mechanism to allow for the sensation of touch.

Dr. Nicolelis started by connecting the motor cortex of rhesus monkeys to mechanical arms. These mechanical arms have sensors on them, which then send signals back to the brain by electrodes connected to the somatosensory cortex (which registers the sensation of touch). The monkeys were given a reward after every successful trial; they learned how to use this apparatus within four to nine trials.

To do this, Dr. Nicolelis had to invent a new code that would represent different surfaces (which were rough or smooth). "After a month of practice," he told me, "this part of the brain learns this new code, and starts to associate this new artificial code that we created with different textures. So this is the first demonstration that we can create a sensory channel" that can simulate sensations of the skin.

I mentioned to him that this idea sounds like the "holodeck" of *Star Trek*, where you wander in a virtual world but feel sensations when you bump into virtual objects, just as if they were real. This is called "haptic technology," which uses digital technology to simulate the sense of touch. Nicolelis replied, "Yes, I think this is the first demonstration that something like the holodeck will be possible in the near future."

The holodeck of the future might use a combination of two technologies. First, people in the holodeck would wear Internet contact lenses, so that they would see an entirely new virtual world everywhere they looked. The scenery in your contact lens would change instantly with the push of a button. And if you touched any object in this world, signals sent into the brain would simulate the sensation of touch, using BMBI technology. In this way, objects in the virtual world you see inside your contact lens would feel solid.

Brain-to-brain interface would make possible not only haptic technology, but also an "Internet of the mind," or brain-net, with direct brain-to-brain contact. In 2013, Dr. Nicolelis was able to accomplish something straight out of *Star Trek,* a "mind meld" between two brains. He started with two groups of rats, one at Duke University, the other in Natal, Brazil. The first group learned to press a lever when seeing a red light. The second group learned to press a lever when their brains were stimulated by a signal sent via an implant. Their reward for pressing the lever was a sip of water. Then Dr. Nicolelis connected the motor cortices of the brains of both groups via a fine wire through the Internet.

When the first group of rats saw the red light, a signal was sent over the Internet to Brazil to the second group, which then pressed the lever. In seven out of ten trials, the second group of rats correctly responded to the signals sent by the first group. This was the first demonstration that signals could be transferred and also interpreted correctly between two brains. It's still a far cry from the mind meld of science fiction, where two minds merge into one, because this is still primitive and the sample size is small, but it is a proof of principle that a brain-net might be possible.

In 2013, the next important step was taken when scientists went beyond animal studies and demonstrated the first direct human brain-to-brain communication, with one human brain sending a message to another via the Internet.

This milestone was achieved at the University of Washington, with one scientist sending a brain signal (move your right arm) to another scientist. The first scientist wore an EEG helmet and played a video game. He fired a cannon by imagining moving his right arm, but was careful not to move it physically.

The signal from the EEG helmet was sent over the Internet to another scientist, who was wearing a transcranial magnetic helmet carefully placed over the part of his brain that controlled his right arm. When the signal reached the second scientist, the helmet would send a magnetic pulse into his brain, which made his right arm move involuntarily, all by itself. Thus, by remote control, one human brain could control the movement of another.

This breakthrough opens up a number of possibilities, such as exchanging nonverbal messages via the Internet. You might one day be able to send the experience of dancing the tango, bungee jumping, or skydiving to the people on your e-mail list. Not just physical activity, but emotions and feelings as well might be sent via brain-to-brain communication.

Nicolelis envisions a day when people all over the world could participate in social networks not via keyboards, but directly through their minds. Instead of just sending e-mails, people on the brain-net would be able to telepathically exchange thoughts, emotions, and ideas in real time. Today a phone call conveys only the information of the conversation and the tone of voice, nothing more. Video conferencing is a bit better, since you can read the body language of the person on the other end. But a brain-net would be the ultimate in communications, making it possible to share the totality of mental information in a conversation, including emotions, nuances, and reservations. Minds would be able to share their most intimate thoughts and feelings.

TOTAL IMMERSION ENTERTAINMENT

Developing a brain-net may also have an impact on the multibillion-dollar entertainment industry. Back in the 1920s, the technology of tape-recording sound as well as light was perfected. This set off a transformation in the entertainment industry as it made the transition from silent movies to the "talkies." This basic formula of combining sound and sight hasn't changed much for the past century. But in the future, the entertainment industry may make the next transition, recording all five senses, including smell, taste, and touch, as well as the full range of emotions. Telepathic probes would be able to handle the full range of senses and emotions that circulate in the brain, producing a complete immersion of the audience in the story. Watching a romantic movie or an action thriller, we would be swimming in an ocean of sensations, as if we were really there, experiencing all the rush of feelings and the emotions of the actors. We would smell the perfume of the heroine, feel the terror of the victims in a horror movie, and relish the vanquishing of the bad guys.

This immersion would involve a radical shift in how movies are made. First, actors would have to be trained to act out their roles with EEG/MRI sensors and nanoprobes recording their sensations and emotions. (This would place an added burden on the actors, who would have to act out each scene by simulating all five senses. In the same way that some actors could not make the transition from silent movies to the talkies, perhaps a new generation of actors will emerge who can act out scenes with all five senses.) Editing would require not just cutting and splicing film, but also combining tapes of the various sensations within each scene. And finally the audience, as they sit in their seats, would have all these electrical signals fed into their brains. Instead of 3-D glasses, the audience would wear brain sensors of some sort. Movie theaters would also have to be retrofitted to process this data and then send it to the people in the audience.

CREATING A BRAIN-NET

Creating a brain-net that can transmit such information would have to be done in stages. The first step would be inserting nanoprobes into important parts of the brain, such as the left temporal lobe, which governs speech, and the occipital lobe, which governs vision. Then computers would analyze these signals and decode them. This information in turn could be sent over the Internet by fiber-optic cables.

More difficult would be to insert these signals back into another person's brain, where they could be processed by the receiver. So far, progress in this area has focused only on the hippocampus, but in the future it should be possible to insert messages directly into other parts of the brain corresponding to our sense of hearing, light, touch, etc. So there is plenty of work to be done as scientists try to map the cortices of the brain involved in these senses. Once these cortices have been mapped—such as the hippocampus, which we'll discuss in the next chapter—it should be possible to insert words, thoughts, memories, and experiences into another brain.

Dr. Nicolelis writes, "It is not inconceivable that our human progeny may indeed muster the skills, technology, and ethics needed to establish a functional brain-net, a medium through which billions of human beings consensually establish temporary direct contacts with fellow human beings through thought alone. What such a colossus of collective consciousness may look like, feel like, or do, neither I nor anyone in our present time can possibly conceive or utter."

THE BRAIN-NET AND CIVILIZATION

A brain-net may even change the course of civilization itself. Each time a new communication system has been introduced, it has irrevocably accelerated changes in society, lifting us from one era to the next. In prehistoric times, for thousands of years our ancestors were nomads wandering in small tribes, communicating with one another through body language and grunts. The coming of language allowed us for the first time to communicate symbols and complex ideas, which facilitated the rise of villages and eventually cities. Within the last few thousand years, written language has enabled us to accumulate knowledge and culture across generations, allowing for the rise of science, the arts, architecture, and huge empires. The coming of the telephone, radio, and TV extended the reach of communication across continents. The Internet now makes possible the rise of a planetary civilization that will link all the continents and peoples of the world. The next giant step might be a planetary brain-net, in which the full spectrum of senses, emotions, memories, and thoughts are exchanged on a global scale.

"WE WILL BE PART OF THEIR OPERATING SYSTEM"

When I interviewed Dr. Nicolelis, he told me that he became interested in science at an early age while growing up in his native Brazil. He remembers watching the Apollo moon shot, which captured the world's attention. To him, it was an amazing feat. And now, he told me, his own "moon shot" is making it possible to move any object with the mind.

He became interested in the brain while still in high school, where he came across a 1964 book by Isaac Asimov titled *The Human Brain*. But he was disappointed by the end of the book. There was no discussion about how all these structures interacted with one another to create the mind (because no one knew the answer back then). It was a life-changing moment and he realized that his own destiny might lie in trying to understand the secrets of the brain.

About ten years ago, he told me, he began to look seriously into doing research on his childhood dream. He started by taking a mouse and letting it control a mechanical device. "We placed sensors into the mouse which read the electrical signals from the brain. Then we transmitted these signals to a little robotic lever that could bring water from a fountain back to the mouse's mouth. So the animal had to learn how to mentally move the robotic device to bring the water back. That was the first-ever demonstration that you could connect an animal to a machine so that it could operate a machine without moving its own body," he explained to me.

Today he can analyze not just fifty but one thousand neurons in the brain of a monkey, which can reproduce various movements in different parts of the monkey's body. Then the monkey can control various devices, such as mechanical arms, or even virtual images in cyberspace. "We even have a monkey avatar that can be controlled by the monkey's thoughts without the monkey making any movement," he told me. This is done by having the monkey watch a video in which he sees an avatar that represents his body. Then, by mentally commanding his body to move, the monkey makes the avatar move in the corresponding way.

Nicolelis envisions a day in the very near future when we will play video games and control computers and appliances with our minds. "We will be part of their operating system. We will be immersed in them with mechanisms that are very similar to the experiments that I am describing."

EXOSKELETONS

The next undertaking for Dr. Nicolelis is the Walk Again Project. Its goal is nothing less than a complete exoskeleton for the body controlled by the mind. At first, an exoskeleton conjures up an image of something from the *Iron Man* movies. Actually, it is a special suit that encases the entire body so that the arms and legs can move via motors. He calls it a "wearable robot." (See Figure 10.)

His goal, he said, is to help the paralyzed "walk by thinking." He plans to use wireless technology, "so there's nothing sticking out of the head. . . . We are going to record twenty to thirty thousand neurons, to command a whole body robotic vest, so he can think and walk again and move and grab objects."

Nicolelis realizes that a series of hurdles must be overcome before the

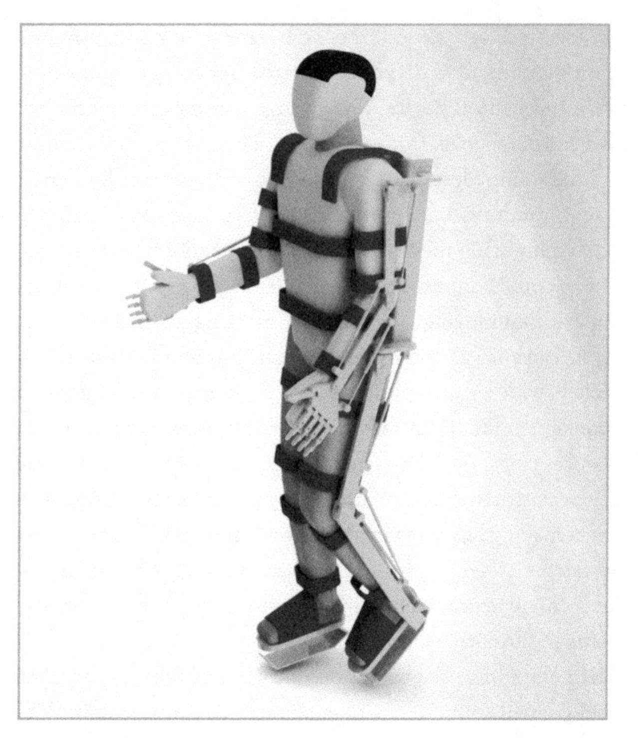

Figure 10. This is the exoskeleton that Dr. Nicolelis hopes will be controlled by the mind of a totally paralyzed person.

exoskeleton becomes a reality. First, a new generation of microchips must be created that can be placed in the brain safely and reliably for years at a time. Second, wireless sensors must be created so the exoskeleton can roam freely. The signals from the brain would be received wirelessly by a computer the size of a cell phone that would probably be attached to your belt. Third, new advances must be made in deciphering and interpreting signals from the brain via computers. For the monkeys, a few hundred neurons were necessary to control the mechanical arms. For a human, you need, at minimum, several thousand neurons to control an arm or leg. And fourth, a power supply must be found that is portable and powerful enough to energize the entire exoskeleton.

Nicolelis's goal is a lofty one: to have a working exoskeleton suit ready for the 2014 World Cup in Brazil, where a quadriplegic Brazilian will deliver the opening kick. He told me proudly, "This is our Brazilian moon shot."

AVATARS AND SURROGATES

In the movie *Surrogates,* Bruce Willis plays an FBI agent who is investigating mysterious murders. Scientists have created exoskeletons so perfect that they exceed human capabilities. These mechanical creatures are super strong, with perfect bodies. In fact, they are so perfect that humanity has become dependent on them. People live their entire life in pods, mentally controlling their handsome, beautiful surrogate with wireless technology. Everywhere you go, you see busy "people" at work, except they are all perfectly shaped surrogates. Their aging masters are conveniently hidden from view. The plot takes a sharp twist, however, when Bruce Willis discovers that the person behind these murders might be linked to the same scientist who invented these surrogates in the first place. That forces him to wonder whether the surrogates are a blessing or a curse.

And in the blockbuster movie *Avatar,* in the year 2154 Earth has depleted most of its minerals, so a mining company has journeyed to a distant moon called Pandora in the Alpha Centauri star system in search of a rare metal, unobtanium. There are native people who inhabit this distant moon, called the Na'vi, who live in harmony with their lush environment. In order to communicate with the native people, specially trained workers are placed in pods, where they learn to mentally control the body of a genetically engineered native. Although the atmosphere is poisonous and the environment differs radically from Earth's, avatars have no difficulty living in this alien world. This uneasy relationship, however, soon collapses when the mining company finds a rich deposit of unobtainium underneath the Na'vi's sacred ceremonial tree. Inevitably a conflict arises between the mining company, which wants to destroy the sacred tree and strip-mine the land for its rare metal, and the natives, who worship it. It looks like a lost cause for the natives until one of the specially trained workers switches sides and leads the Na'vi to victory.

Avatars and surrogates are the stuff of science fiction today, but one day they may become an essential tool for science. The human body is frail, perhaps too delicate for the rigors of many dangerous missions, including space travel. Although science fiction is filled with the heroic exploits of brave astronauts traveling to the farthest reaches of our galaxy, the reality is much different. Radiation in deep space is so intense that our astronauts will have to be shielded or else face premature aging, radiation sickness, and even cancer. Solar flares shot from the sun can bathe a spacecraft in lethal radiation. A simple transatlantic flight from the United States to Europe exposes you to a millirem of radiation per hour, or roughly the same as a dental X-ray. But in outer space, the radiation could be many times more intense, especially in the presence of cosmic rays and solar bursts. (During intense solar storms, NASA has actually warned astronauts in the space station to move to sections where there is more shielding against radiation.)

In addition, there are many other dangers awaiting us in outer space, such as micrometeorites, the effects of prolonged weightlessness, and the problems of adjusting to different gravity fields. After just a few months in weightlessness, the body loses a large fraction of its calcium and minerals, leaving the astronauts incredibly weak, even if they exercise every day. After a year in outer space, Russian astronauts had to crawl out of their space capsules like worms. Furthermore, it is believed that some of the effects of muscle and bone loss are permanent, so that astronauts will feel the consequences of prolonged weightlessness for the rest of their lives.

The dangers of micrometeorites and intense radiation fields on the moon are so great that many scientists have proposed using a gigantic underground cave as a permanent lunar space station to protect our astronauts. These caves form naturally as lava tubes near extinct volcanoes. But the safest way of building a moon base is to have our astronauts sit in the comfort of their living rooms. This way they would be shielded from all the hazards found on the moon, yet through surrogates they would be able to perform the same tasks. This could vastly reduce the cost of manned space travel, since providing life support for human astronauts is very expensive.

Perhaps when the first interplanetary ship reaches a distant planet, and an astronaut's surrogate sets foot on this alien terrain, he or she might start with "One small step for the mind . . ."

One possible problem with this approach is that it takes time for messages to go to the moon and beyond. In a little over a second, a radio message can travel from Earth to the moon, so surrogates on the moon could be easily controlled by astronauts on Earth. More difficult would be communicating with surrogates on Mars, since it can take twenty minutes or more for radio signals to reach the Red Planet.

But surrogates have practical implications closer to home. In Japan, the Fukushima reactor accident in 2011 caused billions of dollars in damages.

Because workers can't enter areas with lethal levels of radiation for more than a few minutes, the final cleanup may take up to forty years. Unfortunately, robots are not sufficiently advanced to go into these blistering radiation fields and make needed repairs. In fact, the only robots used at Fukushima are quite primitive, basically simple cameras placed on top of a computer sitting on wheels. A full-blown automaton that can think for itself (or be controlled by a remote operator) and make repairs in high-radiation fields is many decades away.

The lack of industrial robots caused an acute problem for the Soviets as well during the 1986 Chernobyl accident in the Ukraine. Workers sent directly to the accident site to put out the flames died horrible deaths due to lethal exposure to radiation. Eventually Mikhail Gorbachev ordered the air force to "sand bag" the reactor, dropping five thousand tons of borated sand and cement by helicopter. Radiation levels were so high that 250,000 workers were recruited to finally contain the accident. Each worker could spend only a few minutes inside the reactor building doing repairs. Many received the maximum lifetime allowed dose of radiation. Each one got a medal. This massive project was the largest civil engineering feat ever undertaken. It could not have been done by today's robots.

The Honda Corporation has, in fact, built a robot that may eventually go into deadly radioactive environments, but it is not ready yet. Honda's scientists have placed an EEG sensor on the head of a worker, which is connected to a computer that analyzes his brain waves. The computer is then connected to a radio that sends messages to the robot, called ASIMO (Advanced Step in Innovative Mobility). Hence, by altering his own brain waves, a worker can control ASIMO by pure thought.

Unfortunately, this robot is incapable of making repairs at Fukushima right now, since it can execute only four basic motions (all of which involve moving its head and shoulders) while hundreds of motions are required to make repairs at a shattered nuclear power plant. This system is not developed enough to handle simple tasks such as turning a screwdriver or swinging a hammer.

Other groups have also explored the possibility of mentally controlled robots. At the University of Washington, Dr. Rajesh Rao has created a similar robot that is controlled by a person wearing an EEG helmet. This shiny humanoid robot is two feet tall and is called Morpheus (after a character in the movie *The Matrix*, as well as the Greek god of dreams). A student puts on the EEG helmet and then makes certain gestures, such as moving a hand, which creates an EEG signal that is recorded by a computer. Eventually the computer has a library of such EEG signals, each one corresponding to a specific motion of a limb. Then the robot is programmed to move its hand whenever that EEG signal is sent to it. In this way, if you think about moving your hand, the robot Morpheus moves its hand as well. When you put on the EEG helmet for the first time, it takes about ten minutes for the computer to calibrate to your brain signals. Eventually you get the hang of making gestures with your mind that control the robot. For example, you can have it walk toward you, pick up a block from a table, walk six feet to another table, and then place the block there.

Research is also progressing rapidly in Europe. In 2012, scientists in Switzerland at the École Polytechnique Fédérale de Lausanne unveiled their latest achievement, a robot controlled telepathically by EEG sensors whose controller is located sixty miles away. The robot itself looks like the Roomba robotic vacuum cleaner now found in many living rooms. But it is actually a highly sophisticated robot equipped with a camera that can navigate its way through a crowded office. A paralyzed patient can, for example, look at a computer screen, which is connected to a video camera on the robot many miles away, and see through the eyes of the robot. Then, by thinking, the patient is able to control the motion of the robot as it moves past obstacles.

In the future, one can imagine the most dangerous jobs being done by robots controlled by humans in this fashion. Dr. Nicolelis says, "We will likely be able to operate remotely controlled envoys and ambassadors, robots and airships of many shapes and sizes, sent on our behalf to explore other planets and stars in distant corners of the universe."

For example, in 2010 the world looked on in horror as 5 million barrels of crude oil spilled unabated into the Gulf of Mexico. The Deepwater Horizon spill was one of the largest oil disasters in history, yet engineers were largely helpless for three months. Robotic subs, which are controlled remotely, floundered for weeks trying to cap the well because they lacked the dexterity and versatility necessary for this underwater mission. If surrogate subs, which are much more sensitive in manipulating tools, had been available, they might have capped the well in the first few days of the spill, preventing billions in property damage and lawsuits.

Another possibility is that surrogate submarines might one day enter the human body and perform delicate surgery from the inside. This idea was explored in the movie *Fantastic Voyage,* starring Raquel Welch, in which a submarine was shrunk down to the size of a blood cell and then injected into the bloodstream of someone who had a blood clot in his brain. Shrinking atoms violates the laws of quantum physics, but one day MEMS (micro-electrical-mechanical systems) the size of cells might be able to enter a person's bloodstream. MEMS are incredibly small machines that can easily fit on a pinpoint. MEMS employ the same etching technology used in Silicon Valley, which can put hundreds of millions of transistors on a wafer the size of your fingernail. An elaborate machine with gears, levers, pulleys, and even motors can be made smaller than the period at the end of this sentence. One day a person may be able to put on a telepathy helmet and then command a MEMS submarine using wireless technology to perform surgery inside a patient.

So MEMS technology may open up an entirely new field of medicine, based on microscopic machines entering the body. These MEMS submarines might even guide nanoprobes as they enter the brain so that they connect precisely to the neurons that are of interest. In this way, nanoprobes might be able to receive and transmit signals from the handful of neurons that are involved in specific behaviors. The hit-or-miss approach of inserting electrodes into the brain will be eliminated.

THE FUTURE

In the short term, all these remarkable advances taking place in laboratories around the world may alleviate the suffering of those afflicted by paralysis and other disabilities. Using the power of their minds, they will be able to communicate with loved ones, control their wheelchairs and beds, walk by mentally guiding mechanical limbs, manipulate household appliances, and lead seminormal lives.

But in the long term, these advances could have profound economic and practical implications for the world. By mid-century, it could become commonplace to interact with computers directly with the mind. Since the computer business is a multitrillion-dollar industry that can create young billionaires and corporations almost overnight, advances in the mind-computer interface will reverberate on Wall Street—and also in your living room.

All the devices we use to communicate with computers (the mouse, keyboards, etc.) may eventually disappear. In the future, we may simply give mental commands and our wishes will be silently carried out by tiny chips hidden in the environment. While sitting in our offices, taking a stroll in the park, doing window-shopping, or just relaxing, our minds could be interacting with scores of hidden chips, allowing us to mentally balance our finances, arrange for theater tickets, or make a reservation.

Artists may also make good use of this technology. If they can visualize their artwork in their minds, then the image can be displayed via EEG sensors on a holographic screen in 3-D. Since the image in the mind is not as precise as the original object, the artist could then make improvements on the 3-D image and dream up the next iteration. After several cycles, the artist could print out the final image on a 3-D printer.

Similarly, engineers would be able to create scale models of bridges, tunnels, and airports by simply using their imagination. They could also rapidly make changes in their blueprints through thought alone. Machine parts could fly off the computer screen and into a 3-D printer.

Some critics, however, have claimed that these telekinetic powers have one great limitation: the lack of energy. In the movies, super beings have the power to move mountains using their thoughts. In the movie *X-Men: The Last Stand,* the super villain Magneto had the ability to move the Golden Gate Bridge simply by pointing his fingers, but the human body can muster only about one-fifth of a horsepower on average, which is much too little power to perform the feats we see in the comic books. Therefore, all the herculean feats of telekinetic super beings appear to be pure fantasy.

There is one solution to this energy problem, however. You may be able to connect your thoughts to a power source, which would then magnify your power millions of times. In this way, you could approximate the power of a god. In one episode of *Star Trek,* the crew journeys to a distant planet and meets a godlike creature who claims to be Apollo, the Greek god of the sun. He can perform feats of magic that dazzle the crew. He even claims to have visited Earth eons ago, where the earthlings worshipped him. But the crew, not believing in gods, suspect a fraud. Later they figure out that this "god" just mentally controls a hidden power source, which then performs all the magic tricks. When this power source is destroyed, he becomes a mere mortal.

Similarly, in the future our minds may mentally control a power source that will then give us superpowers. For example, a construction worker might telepathically exploit a power source that energizes heavy machinery. Then a single worker might be able to build complex buildings and houses just by using the power of his mind. All the heavy lifting would be done by the power source, and the construction worker would resemble a conductor, able to orchestrate the motion of colossal cranes and powerful bulldozers through thought alone.

Science is beginning to catch up to science fiction in yet another way. The *Star Wars* saga was supposed to take place in a time when civilizations span the entire galaxy. The peace of the galaxy, in turn, is maintained by the Jedi Knights, a highly trained cadre of warriors who use the power of the "Force" to read minds and guide their lightsabers.

However, one need not wait until we have colonized the entire galaxy to begin contemplating the Force. As we've seen, some aspects of the Force are possible today, such as being able to tap into the thoughts of others using ECOG electrodes or EEG helmets. But the telekinetic powers of the Jedi Knights will also become a possibility as we learn to harness a power source with our minds. The Jedi Knights, for example, can summon a lightsaber simply by waving their hands, but we can already accomplish the same feat by exploiting the power of magnetism (much as the magnet in an MRI machine can hurl a hammer across a room). By mentally activating the power source, you can grab lightsabers from across the room with today's technology.

THE POWER OF A GOD

Telekinesis is a power usually reserved for a deity or a superhero. In the universe of superheros appearing in blockbuster Hollywood movies, perhaps the most powerful character is Phoenix, a telekinetic woman who can move any object at will. As a member of the X-Men, she can lift heavy machinery, hold back floods, or raise jet airplanes via the power of her mind. (However, when she is finally consumed by the dark side of her power, she goes on a cosmic rampage, capable of incinerating entire solar systems and destroying stars. Her power is so great and uncontrollable that it leads to her eventual self-destruction.)

But how far can science go in harnessing telekinetic powers?

In the future, even with an external power source to magnify our thoughts, it is unlikely that people with telekinetic powers will be able to move basic objects like a pencil or mug of coffee on command. As we mentioned, there are only four known forces that rule the universe, and none of them can move objects unless there is an external power source. (Magnetism comes close, but magnetism can move only magnetic objects. Objects made of plastic, water, or wood can easily pass through magnetic fields.) Simple levitation, a trick found in most magicians' shows, is beyond our scientific capability.

So even with an external power supply, is it unlikely that a telekinetic person would be able to move the objects around them at will. However, there is a technology that may come close, and that involves the ability to change one object into another.

The technology is called "programmable matter," and it has become a subject of intense research for the Intel Corporation. The idea behind programmable matter is to create objects made of tiny "catoms," which are microscopic computer chips. Each catom can be controlled wirelessly; it can be programmed to change the electrical charge on its surface so it can bind with other catoms in different ways. By programming the electric charges one way, the catoms bind together to form, say, a cell phone. Push a button to change their programming, and the catoms rearrange themselves to re-form into another object, like a laptop.

I saw a demonstration of this technology at Carnegie Mellon University in Pittsburgh, where scientists have been able to create a chip the size of a pinpoint. To exam these catoms, I had to enter a "clean room" wearing a special white uniform, plastic boots, and a cap to prevent even the smallest dust particle from entering. Then, under a microscope, I could see the intricate circuitry inside each catom, which makes it possible to program it wirelessly to change the electrical charge on its surface. In the same way we can program software today, in the future it may be possible to program hardware.

The next step is to determine if these catoms can combine to form useful objects, and to see if they can be changed or morphed into another object at will. It may take until mid-century before we have working prototypes of programmable matter. Because of the complexity of programming billions of catoms, a special computer would have to be created to orchestrate the charge on each catom. Perhaps by the end of this century, it will be possible to mentally control this computer so that we can change one object into another. We would not have to memorize the charges and configuration within an object. We would just give the mental command to the computer to change one object into another.

Eventually we might have catalogs listing all the various objects that are programmable, such as furniture, appliances, and electronics. Then by telepathically communicating with the computer, it should be possible to change one object into another. Redecorating your living room, remodeling your kitchen, and buying Christmas presents could all be done mentally.

A MORALITY TALE

Having every wish come true is something that only a divinity can accomplish. However, there is also a downside to this celestial power. All technologies can be used for good or for evil. Ultimately, science is a double-edged sword. One side of the sword can cut against poverty, disease, and ignorance. But the other side can cut against people, in several ways.

These technologies could conceivably make wars even more vicious. Perhaps one day, all hand-to-hand combat will be between two surrogates, armed with a battery of high-tech weapons. The actual warriors, sitting safely thousands of miles away, would unleash a barrage of the latest high-tech weaponry with little regard for the collateral damage they are inflicting on civilians. Although wars fought with surrogates may preserve the lives of the soldiers themselves, they might also cause horrendous civilian and property damage.

The bigger problem is that this power may also be too great for any common mortal to control. In the novel *Carrie,* Stephen King explored the world of a young girl who was constantly taunted by her peers. She was ostracized by the in-crowd and her life became a never-ending series of insults and humiliations. However, her tormentors did not know one thing about her: she was telekinetic.

After enduring the taunts and having blood splashed all over her dress at the prom, she finally cracks. She summons all her telekinetic power to trap her classmates and then annihilate them one by one. In a final gesture, she decides to burn the entire school down. But her telekinetic power was too great to control. She ultimately perishes in the fire that she started.

Not only can the awesome power of telekinesis backfire, but there is another problem as well. Even if you have taken all the precautions to understand and harness this power, it could still destroy you if, ironically enough, it is too obedient to your thoughts and commands. Then the very thoughts you conceive may spell your doom.

The movie *Forbidden Planet* (1956) is based on a play by William Shakespeare, *The Tempest,* which begins with a sorcerer and his daughter stranded on a deserted island. But in *Forbidden Planet,* a professor and his daughter are stranded on a distant planet that was once the home of the Krell, a civilization millions of years more advanced than ours. Their greatest achievement was to create a device that gave them the ultimate power of telekinesis, the power to control matter in all its forms by the mind. Anything they desired suddenly materialized before them. This was the power to reshape reality itself to their whims.

Yet on the eve of their greatest triumph, as they were turning on this device the Krell disappeared without a trace. What could have possibly destroyed this most advanced civilization?

When a crew of earthmen land on the planet to rescue the man and his daughter, they find that there is a hideous monster haunting the planet, slaughtering crew members at will. Finally, one crew member discovers the secret behind both the Krell and the monster. Before he dies, he gasps, "Monsters from the id."

Then the shocking truth suddenly dawns on the professor. The very night that the Krell turned on their telekinesis machine, they fell asleep. All the repressed desires from their ids then suddenly materialized. Buried in the subconscious of these highly developed creatures were the long-suppressed animal urges and desires of their ancient past. Every fantasy, every dream of revenge suddenly came true, so this great civilization destroyed itself overnight. They had conquered many worlds, but there was one thing they could not control: their own subconscious minds.

That is a lesson for anyone who desires to unleash the power of the mind. Within the mind, you find the noblest achievements and thoughts of humanity. But you will also find monsters from the id.

CHANGING WHO WE ARE: OUR MEMORIES AND INTELLIGENCE

So far, we have discussed the power of science to extend our mental abilities via telepathy and telekinesis. We basically remain the same; these developments do nothing to change the essence of who we are. However, there is an entirely new frontier opening up that alters the very nature of what it means to be human. Using the very latest in genetics, electromagnetics, and drug therapy, it may become possible in the near future to alter our memories and even enhance our intelligence. The idea of downloading a memory, learning complex skills overnight, and becoming super intelligent is slowly leaving the realm of science fiction.

Without our memories, we are lost, cast adrift in an aimless sea of pointless stimuli, unable to understand the past or ourselves. So what happens if one day we can input artificial memories into our brains? What happens when we can become a master of any discipline simply by downloading the file into our memory? And what happens if we cannot tell the difference between real and fake memories? Then who are we?

Scientists are moving past being passive observers of nature to actively shaping and molding nature. This means that we might be able to manipulate memories, thoughts, intelligence, and consciousness. Instead of simply witnessing the intricate mechanics of the mind, in the future it will be possible to orchestrate them.

So let us now answer this question: Can we download memories?

If our brains were simple enough to be understood, we wouldn't be smart enough to understand them.

—ANONYMOUS

5 MEMORIES AND THOUGHTS MADE TO ORDER

Neo is The One. Only he can lead a defeated humanity to victory against the Machines. Only Neo can destroy the Matrix, which has implanted false memories into our brains as a means to control us.

In a now-classic scene from the film *The Matrix*, the evil Sentinels, who guard the Matrix, have finally cornered Neo. It looks like humanity's last hope is about to be terminated. But previously Neo had had an electrode jacked into the back of his neck that could instantly download martial-arts skills into his brain. In seconds, he becomes a karate master able to take down the Sentinels with breathtaking aerial kicks and well-placed strikes.

In *The Matrix*, learning the amazing skills of a black-belt karate master is no harder than slipping an electrode into your brain and pushing the "download" button. Perhaps one day we, too, may be able to download memories, which will vastly increase our abilities.

But what happens when the memories downloaded into your brain are false? In the movie *Total Recall*, Arnold Schwarzenegger has fake memories placed into his brain, so that the distinction between reality and fiction becomes totally blurred. He valiantly fights off the bad guys on Mars until the end of the movie, when he suddenly realizes that he himself is their leader.

He is shocked to find that his memories of being a normal, law-abiding citizen are totally manufactured.

Hollywood is fond of movies that explore the fascinating but fictional world of artificial memories. All this is impossible, of course, with today's technology, but one can envision a day, a few decades from now, when artificial memories may indeed be inserted into the brain.

HOW WE REMEMBER

Like Phineas Gage's, the strange case of Henry Gustav Molaison, known in the scientific literature as simply HM, created a sensation in the field of neurology that led to many fundamental breakthroughs in understanding the importance of the hippocampus in formulating memories.

At the age of nine, HM suffered head injuries in an accident that caused debilitating convulsions. In 1953, when he was twenty-five years old, he underwent an operation that successfully relieved his symptoms. But another problem surfaced because surgeons mistakenly cut out part of his hippocampus. At first, HM appeared normal, but it soon became apparent that something was terribly wrong; he could not retain new memories. Instead, he constantly lived in the present, greeting the same people several times a day with the same expressions, as if he were seeing them for the first time. Everything that went into his memory lasted only a few minutes before it disappeared. Like Bill Murray in the movie *Groundhog Day,* HM was doomed to relive the same day, over and over, for the rest of his life. But unlike Bill Murray's character, he was unable to recall the previous iterations. His long-term memory, however, was relatively intact and could remember his life before the surgery. But without a functioning hippocampus, HM was unable to record new experiences. For example, he would be horrified when looking in a mirror, since he saw the face of an old man but thought he was still twenty-five. But mercifully, the memory of being horrified would also soon disappear into the fog. In some sense, HM was like an animal with Level II consciousness, unable to recall the immediate past or simulate the future. Without a functioning hippocampus, he regressed from Level III down to Level II consciousness.

Today, further advances in neuroscience have given us the clearest picture yet of how memories are formed, stored, and then recalled. "It has all come together just in the past few years, due to two technical developments—

computers and modern brain scanning," says Dr. Stephen Kosslyn, a neuroscientist at Harvard.

As we know, sensory information (e.g., vision, touch, taste) must first pass through the brain stem and onto the thalamus, which acts like a relay station, directing the signals to the various sensory lobes of the brain, where they are evaluated. The processed information reaches the prefrontal cortex, where it enters our consciousness and forms what we consider our short-term memory, which can range from several seconds to minutes. (See Figure 11.)

To store these memories for a longer duration, the information must then run through the hippocampus, where memories are broken down into different categories. Rather than storing all memories in one area of the

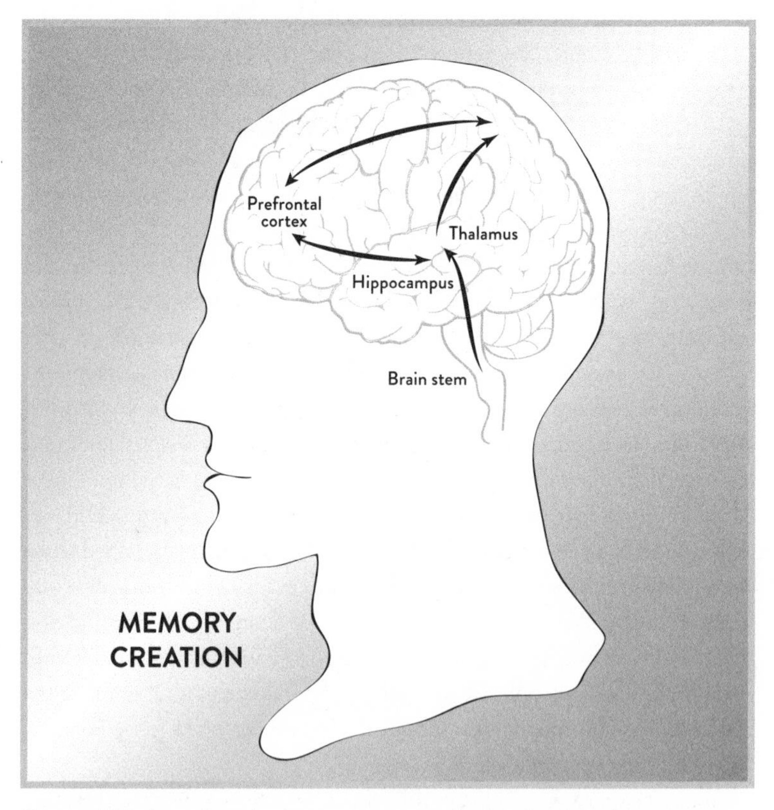

Figure 11. This shows the path taken to create memories. Impulses from the senses pass through the brain stem, to the thalamus, out to the various cortices, and then to the prefrontal cortex. They then pass to the hippocampus to form long-term memories.

brain like a tape recorder or hard drive, the hippocampus redirects the fragments to various cortices. (Storing memories in this way is actually more efficient than storing them sequentially. If human memories were stored sequentially, like on computer tape, a vast amount of memory storage would br required. In fact, in the future, even digital storage systems may adopt this trick from the living brain, rather than storing whole memories sequentially.) For instance, emotional memories are stored in the amygdala, but words are recorded in the temporal lobe. Meanwhile, colors and other visual information are collected in the occipital lobe, and the sense of touch and movement reside in the parietal lobe. So far, scientists have identified more than twenty categories of memories that are stored in different parts of the brain, including fruits and vegetables, plants, animals, body parts, colors, numbers, letters, nouns, verbs, proper names, faces, facial expressions, and various emotions and sounds.

A single memory—for instance, a walk in the park—involves information that is broken down and stored in various regions of the brain, but reliving just one aspect of the memory (e.g., the smell of freshly cut grass) can suddenly send the brain racing to pull the fragments together to form a cohesive recollection. The ultimate goal of memory research is, then, to figure out how these scattered fragments are somehow reassembled when we recall an experience. This is called the "binding problem," and a solution could potentially explain many puzzling aspects of memory. For instance, Dr. Antonio Damasio has analyzed stroke patients who are incapable of identifying a single category, even though they are able to recall everything else. This is because the stroke has affected just one particular area of the brain, where that certain category was stored.

The binding problem is further complicated because all our memories and experiences are highly personal. Memories might be customized for the individual, so that the categories of memories for one person may not correlate with the categories of memories for another. Wine tasters, for example, may have many categories for labeling subtle variations in taste, while physicists may have other categories for certain equations. Categories, after all, are by-products of experience, and different people may therefore have different categories.

One novel solution to the binding problem uses the fact that there are electromagnetic vibrations oscillating across the entire brain at roughly forty cycles per second, which can be picked up by EEG scans. One fragment of memory might vibrate at a very precise frequency and stimulate another fragment of memory stored in a distant part of the brain. Previously it was thought that memories might be stored physically close to one another, but this new theory says that memories are not linked spatially but rather temporally, by vibrating in unison. If this theory holds up, it means that there are electromagnetic vibrations constantly flowing through the entire brain, linking up different regions and thereby re-creating entire memories. Hence the constant flow of information between the hippocampus, the prefrontal cortex, the thalamus, and the different cortices might not be entirely neural after all. Some of this flow may be in the form of resonance across different brain structures.

RECORDING A MEMORY

Sadly, HM died in 2008 at the age of eighty-two, before he could take advantage of some sensational results achieved by science: the ability to create an artificial hippocampus and then insert memories into the brain. This is something straight out of science fiction, but scientists at Wake Forest University and the University of Southern California made history in 2011 when they were able to record a memory made by mice and store it digitally in a computer. This was a proof-of-principle experiment, in which they showed that the dream of downloading memories into the brain might one day become reality.

At first, the very idea of downloading memories into the brain seems like an impossible dream, because, as we have seen, memories are created by processing a variety of sensory experiences, which are then stored in multiple places in the neocortex and limbic system. But as we know from HM, there is one place through which all memories flow and are converted into long-term memories: the hippocampus. Team leader Dr. Theodore Berger of USC says, "If you can't do it with the hippocampus, you can't do it anywhere."

The scientists at Wake Forest and USC first started with the observation, garnered from brain scans, that there are at least two sets of neurons in a mouse's hippocampus, called CA1 and CA3, which communicate with each other as a new task is learned. After training mice to press two bars, one after the other, in order to get water, the scientists reviewed the findings and attempted to decode these messages, which proved frustrating at first since the signals between these two sets of neurons didn't appear to follow a pattern. But by monitoring the signals millions of times, they were eventually able to determine which electrical input created which output. With the use of probes in the mice's hippocampi, the scientists were able to record the signals between CA1 and CA3 when the mice learned to press the two bars in sequence.

Then the scientists injected the mice with a special chemical, making them forget the task. Finally they played back the memory into the same mouse's brain. Remarkably, the memory of the task returned, and the mice could successfully reproduce the original task. Essentially, they had created an artificial hippocampus with the ability to duplicate digital memory. "Turn the switch on, the animal has the memory; turn it off and they don't," says Dr. Berger. "It's a very important step because it's the first time we have put all the pieces together."

As Joel Davis of the Office of the Chief of Naval Operations, which sponsored this work, said, "Using implantables to enhance competency is down the road. It's only a matter of time."

Not surprisingly, with so much at stake, this area of research is moving very rapidly. In 2013, yet another breakthrough was made, this time at MIT, by scientists who were able to implant not just ordinary memories into a mouse, but false ones as well. This means that, one day, memories of events that never took place may be implanted into the brain, which would have a profound impact on fields like education and entertainment.

The MIT scientists used a technique called optogenetics (which we will discuss more in Chapter 8), which allows you to shine a light on specific neurons to activate them. Using this powerful method, scientists can identify the specific neurons responsible for certain memories.

Let's say that a mouse enters a room and is given a shock. The neurons responsible for the memory of that painful event can actually be isolated and recorded by analyzing the hippocampus. Then the mouse is placed in an entirely different room that is totally harmless. By turning on a light on an optical fiber, one can use optogenetics to activate the memory of the shock, and the mouse exhibits a fear response, although the second room is totally safe.

In this way, the MIT scientists were able not only to implant ordinary memories, but also memories of events that never took place. One day, this technique may give educators the ability to implant memories of new skills to retrain workers, or give Hollywood an entirely new form of entertainment.

AN ARTIFICIAL HIPPOCAMPUS

At present, the artificial hippocampus is primitive, able to record only a single memory at a time. But these scientists plan to increase the complexity of their artificial hippocampus so that it can store a variety of memories and record them for different animals, eventually working up to monkeys. They also plan to make this technology wireless by replacing the wires with tiny radios so that memories can be downloaded remotely without the need for clumsy electrodes implanted into the brain.

Because the hippocampus is involved with memory processing in humans, scientists see a vast potential application in treating strokes, dementia, Alzheimer's, and a host of other problems that occur when there is damage or deterioration in this region of the brain.

Many hurdles have to be negotiated, of course. Despite all we have learned about the hippocampus since HM, it still remains something of a black box whose inner workings are largely unknown. As a result, it is not possible to construct a memory from scratch, but once a task has been performed and the memory processed, it is possible to record it and play it back.

FUTURE DIRECTIONS

Working with the hippocampus of primates and even humans will be more difficult, since their hippocampi are much larger and more complex. The first step is to create a detailed neural map of the hippocampus. This means placing electrodes at different parts of the hippocampus to record the signals that are constantly being exchanged between different regions. This will establish the flow of information that constantly moves across the hippocampus. The hippocampus has four basic divisions, CA1 to CA4, and hence scientists will record the signals that are exchanged between them.

The second step involves the subject performing certain tasks, after which scientists will record the impulses that flow across the various regions of the hippocampus, thereby recording the memory. For example, the memory of learning a certain task, such as jumping through a hoop, will create electrical activity in the hippocampus that can be recorded and carefully analyzed. Then a dictionary can be created matching the memory with the flow of information across the hippocampus.

Finally, step three involves making a recording of this memory and feeding the electrical signal into the hippocampus of another subject via electrodes, to see if that memory can been uploaded. In this fashion, the subject may learn to jump through a hoop although it has never done so before. If successful, scientists would gradually create a library containing recordings of certain memories.

It may take decades to work all the way up to human memories, but one can envision how it might work. In the future, people may be hired to create certain memories, like a luxury vacation or a fictitious battle. Nanoelectrodes will be placed at various places in their brain to record the memory. These electrodes must be extremely small so that they do not interfere with the formation of the memory.

The information from these electrodes will then be sent wirelessly to a computer and then recorded. Later a subject who wants to experience these memories will have similar electrodes placed in his hippocampus, and the memory will be inserted into the brain.

(There are complications to this idea, of course. If we try to insert the memory of physical activity, such as a martial art, we have the problem of "muscle memory." For example, when walking, we do not consciously think about putting one leg in front of the other. Walking has become second nature to us because we do it so often, and from an early age. This means that the signals controlling our legs no longer originate entirely in the hippocampus, but also in the motor cortex, the cerebellum, and the basal ganglia. In the future, if we wish to insert memories involving sports, scientists may have to decipher the way in which memories are partially stored in other areas of the brain as well.)

VISION AND HUMAN MEMORIES

The formation of memories is quite complex, but the approach we have been discussing takes a shortcut by eavesdropping on the signals moving through the hippocampus, where the sensory impulses have already been processed. In *The Matrix*, however, an electrode is placed in the back of the head to upload memories directly into the brain. This assumes that one can decode the raw, unprocessed impulses coming in from the eyes, ears, skin, etc., that are moving up the spinal cord and brain stem and into the thalamus. This is much more elaborate and difficult than analyzing the processed messages circulating in the hippocampus.

To give you a sense of the sheer volume of unprocessed information that comes up the spinal cord into the thalamus, let's consider just one aspect: vision, since many of our memories are encoded this way. There are roughly 130 million cells in the eye's retina, called cones and rods; they process and record 100 million bits of information from the landscape at any time.

This vast amount of data is then collected and sent down the optic nerve, which transports 9 million bits of information per second, and on to the thalamus. From there, the information reaches the occipital lobe, at the very back of the brain. This visual cortex, in turn, begins the arduous process of analyzing this mountain of data. The visual cortex consists of several patches at the back of the brain, each of which is designed for a specific task. They are labeled V1 to V8.

Remarkably, the area called V1 is like a screen; it actually creates a pattern on the back of your brain very similar in shape and form to the original image. This image bears a striking resemblance to the original, except that the very center of your eye, the fovea, occupies a much larger area in V1 (since the fovea has the highest concentration of neurons). The image cast on V1 is therefore not a perfect replica of the landscape but is distorted, with the central region of the image taking up most of the space.

Besides V1, other areas of the occipital lobe process different aspects of the image, including:

- Stereo vision. These neurons compare the images coming in from each eye. This is done in area V2.
- Distance. These neurons calculate the distance to an object, using shadows and other information from both eyes. This is done in area V3.
- Colors are processed in area V4.
- Motion. Different circuits can pick out different classes of motion, including straight-line, spiral, and expanding motion. This is done in area V5.

More than thirty different neural circuits involved with vision have been identified, but there are probably many more.

From the occipital lobe, the information is sent to the prefrontal cortex, where you finally "see" the image and form your short-term memory. The information is then sent to the hippocampus, which processes it and stores it for up to twenty-four hours. The memory is then chopped up and scattered among the various cortices.

The point here is that vision, which we think happens effortlessly, requires billions of neurons firing in sequence, transmitting millions of bits of information per second. And remember that we have signals from five sense organs, plus emotions associated with each image. All this information is processed by the hippocampus to create a simple memory of an image. At present, no machine can match the sophistication of this process, so replicating it presents an enormous challenge for scientists who want to create an artificial hippocampus for the human brain.

REMEMBERING THE FUTURE

If encoding the memory of just one of the senses is such a complex process, then how did we evolve the ability to store such vast amounts of information in our long-term memory? Instinct, for the most part, guides the behavior of animals, which do not appear to have much of a long-term memory. But as neurobiologist Dr. James McGaugh of the University of California at Irvine says, "The purpose of memory is to predict the future," which raises an interesting possibility. Perhaps long-term memory evolved *because* it was useful for simulating the future. In other words, the fact that we can remember back into the distant past is due to the demands and advantages of simulating the future.

Indeed, brain scans done by scientists at Washington University in St. Louis indicate that areas used to recall memories are the same as those involved in simulating the future. In particular, the link between the dorsolateral prefrontal cortex and the hippocampus lights up when a person is engaged in planning for the future *and* remembering the past. In some sense, the brain is trying to "recall the future," drawing upon memories of the past in order to determine how something will evolve into the future. This may also explain the curious fact that people who suffer from amnesia—such as HM—are often unable to visualize what they will be doing in the future or even the very next day.

"You might look at it as mental time travel—the ability to take thoughts about ourselves and project them either into the past or into the future," says Dr. Kathleen McDermott of Washington University. She also notes that their study proves a "tentative answer to a longstanding question regarding the evolutionary usefulness of memory. It may just be that the reason we can recollect the past in vivid detail is that this set of processes is important for being able to envision ourselves in future scenarios. This ability to envision the future has clear and compelling adaptive significance." For an animal, the past is largely a waste of precious resources, since it gives them little evolutionary advantage. But simulating the future, given the lessons of the past, is an essential reason why humans became intelligent.

AN ARTIFICIAL CORTEX

In 2012 the same scientists from Wake Forest Baptist Medical Center and the University of Southern California who created an artificial hippocampus in mice announced an even more far-reaching experiment. Instead of recording a memory in the mouse hippocampus, they duplicated the much more sophisticated thinking process of the cortex of a primate.

They took five rhesus monkeys and inserted tiny electrodes into two layers of their cortex, called the L2/3 and L5 layers. They then recorded neural signals that went between these two layers as the monkeys learned a task. (This task involved the monkeys seeing a set of pictures, and then being rewarded if they could pick out these same pictures from a much larger set.) With practice, the monkeys could perform the task with 75 percent accuracy. But if the scientists fed the signal back into the cortex as the monkey was performing the test, its performance increased by 10 percent. When certain chemicals were given to the monkey, its performance dropped by 20 percent. But if the recording was fed back into the cortex, its performance exceeded its normal level. Although this was a small sample size and there was only a modest improvement in performance, the study still suggests that the scientists' recording accurately captured the decision-making process of the cortex.

Because this study was done on primates rather than mice and involved the cortex and not the hippocampus, it could have vast implications when human trials begin. Dr. Sam A. Deadwyler of Wake Forest says, "The whole idea is that the device would generate an output pattern that bypasses the damaged area, proving an alternative connection" in the brain. This experiment has a possible application for patients whose neocortex has been damaged. Like a crutch, this device would perform the thinking operation of the damaged area.

AN ARTIFICIAL CEREBELLUM

It should also be pointed out that the artificial hippocampus and neocortex are but the first steps. Eventually, other parts of the brain will have artificial counterparts. For example, scientists at Tel Aviv University in Israel have already created an artificial cerebellum for a rat. The cerebellum is an essential part of the reptilian brain that controls our balance and other basic bodily functions.

Usually when a puff of air is directed at a rat's face, it blinks. If a sound is made at the same time, the rat can be conditioned to blink just by hearing the sound. The goal of the Israeli scientists was to create an artificial cerebellum that could duplicate this feat.

First the scientists recorded the signals entering the brain stem when the puff of air hit the rat's face and the sound was heard. Then the signal was processed and sent back to the brain stem at another location. As expected, the rats blinked upon receiving the signal. Not only is this the first time that an artificial cerebellum functioned correctly, it is the first time that messages were received from one part of the brain, processed, and then uploaded into a different part of the brain.

Commenting on this work, Francesco Sepulveda of the University of Essex says, "This demonstrates how far we have come towards creating circuitry that could one day replace damaged brain areas and even enhance the power of the healthy brain."

He also sees great potential for artificial brains in the future, adding, "It will likely take us several decades to get there, but my bet is that specific, well-organized brain parts such as the hippocampus or the visual cortex will have synthetic correlates before the end of the century."

Although progress in creating artificial replacements for the brain is moving remarkably fast given the complexity of the process, it is a race against time when one considers the greatest threat facing our public health system, the declining mental abilities of people with Alzheimer's.

ALZHEIMER'S—DESTROYER OF MEMORY

Alzheimer's disease, some people claim, might be the disease of the century. There are 5.3 million Americans who currently have Alzheimer's, and the number is expected to quadruple by 2050. Five percent of people from age sixty-five to seventy-four have Alzheimer's, but more than 50 percent of those over eighty-five have it, even if they have no obvious risk factors. (Back in 1900, life expectancy in the United States was forty-nine, so Alzheimer's was not a significant problem. But now, people over eighty are one of the fastest-growing demographic groups in the country.)

In the early stages of Alzheimer's, the hippocampus, the part of the brain through which memories are processed, begins to deteriorate. Indeed, brain scans clearly show that the hippocampus shrinks in Alzheimer's patients, but the wiring linking the prefrontal cortex to the hippocampus also thins, leaving the brain unable to properly process short-term memories. Long-term memories already stored throughout the cortices of the brain remain relatively intact, at least at first. This creates a situation where you may not remember what you just did a few minutes ago but can clearly recall events that took place decades ago.

Eventually, the disease progresses to the point where even basic long-term memories are destroyed. The person is unable to recognize their children or spouse and to remember who they are, and can even fall into a comalike vegetative state.

Sadly, the basic mechanisms for Alzheimer's have only recently begun to be understood. One major breakthrough came in 2012, when it was revealed that Alzheimer's begins with the formation of tau amyloid proteins, which in turn accelerates the formation of beta amyloid, a gummy, gluelike substance that clogs up the brain. (Before, it was not clear if Alzheimer's was caused by these plaques or whether perhaps these plaques were by-products of a more fundamental disorder.)

What makes these amyloid plaques so difficult to target with drugs is that they are most likely made of "prions," which are misshapen protein molecules. They are not bacteria or viruses, but nevertheless they can reproduce. When viewed atomically, a protein molecule resembles a jungle of ribbons of atoms tied together. This tangle of atoms must fold onto itself correctly for the protein to assume the proper shape and function. But prions are misshapen proteins that have folded incorrectly. Worse, when they bump into healthy proteins, they cause them to fold incorrectly as well. Hence one prion can cause a cascade of misshapen proteins, creating a chain reaction that contaminates billions more.

At present, there is no known way to stop the inexorable progression of Alzheimer's. Now that the basic mechanics behind Alzheimer's are being unraveled, however, one promising method is to create antibodies or a vaccine that might specifically target these misshapen protein molecules. Another way might be to create an artificial hippocampus for these individuals so that their short-term memory can be restored.

Yet another approach is to see if we can directly increase the brain's ability to create memories using genetics. Perhaps there are genes that can improve our memory. The future of memory research may lie in the "smart mouse."

THE SMART MOUSE

In 1999, Dr. Joseph Tsien and colleagues at Princeton, MIT, and Washington University found that adding a single extra gene dramatically boosted a mouse's memory and ability. These "smart mice" could navigate mazes faster, remember events better, and outperform other mice in a wide variety of tests. They were dubbed "Doogie mice," after the precocious character on the TV show *Doogie Howser, M.D.*

Dr. Tsien began by analyzing the gene NR2B, which acts like a switch controlling the brain's ability to associate one event with another. (Scientists know this because when the gene is silenced or rendered inactive, mice lose this ability.) All learning depends on NR2B, because it controls the communication between memory cells of the hippocampus. First Dr. Tsien created a strain of mice that lacked NR2B, and they showed impaired memory and learning disabilities. Then he created a strain of mice that had more copies of NR2B than normal, and found that the new mice had superior mental capabilities. Placed in a shallow pan of water and forced to swim, normal mice would swim randomly about. They had forgotten from just a few days before that there was a hidden underwater platform. The smart mice, however, went straight to the hidden platform on the first try.

Since then, researchers have been able to confirm these results in other labs and create even smarter strains of mice. In 2009, Dr. Tsien published a paper announcing yet another strain of smart mice, dubbed "Hobbie-J" (named after a character in Chinese cartoons). Hobbie-J was able to remember novel facts (such as the location of toys) three times longer than the genetically modified strain of mouse previously thought to be the smartest. "This adds to the notion that NR2B is a universal switch for memory formation," remarked Dr. Tsien. "It's like taking Michael Jordon and making him a super Michael Jordan," said graduate student Deheng Wang.

There are limits, however, even to this new mice strain. When these mice were given a choice to take a left or right turn to get a chocolate reward, Hobbie-J was able to remember the correct path for much longer than the normal mice, but after five minutes he, too, forgot. "We can never turn it into a mathematician. They are rats, after all," says Dr. Tsien.

It should also be pointed out that some of the strains of smart mice were exceptionally timid compared to normal mice. Some suspect that, if your memory becomes too great, you also remember all the failures and hurts as well, perhaps making you hesitant. So there is also a potential downside to remembering too much.

Next, scientists hope to generalize their results to dogs, since we share so many genes, and perhaps also to humans.

SMART FLIES AND DUMB MICE

The NR2B gene is not the only gene being studied by scientists for its impact on memory. In yet another groundbreaking series of experiments, scientists have been able to breed a strain of fruit flies with "photographic memory," and also a strain of mice that are amnesiac. These experiments may eventually explain many mysteries of our long-term memory, such as why cramming for an exam is not the best way to study, and why we remember events if they are emotionally charged. Scientists have found that there are two important genes, the CREB activator (which stimulates the formation of new connections between neurons) and the CREB repressor (which suppresses the formation of new memories).

Dr. Jerry Yin and Timothy Tully of Cold Spring Harbor have been doing interesting experiments with fruit flies. Normally it takes ten trials for them to learn a certain task (e.g., detecting an odor, avoiding a shock). Fruit flies with an extra CREB repressor gene could not form lasting memories at all, but the real surprise came when they tested fruit flies with an extra CREB activator gene. They learned the task in just one session. "This implies these flies have a photographic memory," says Dr. Tully. He said they are just like students "who could read a chapter of a book once, see it in their mind, and tell you that the answer is in paragraph three of page two seventy-four."

This effect is not just restricted to fruit flies. Dr. Alcino Silva, also at Cold Spring Harbor, has been experimenting with mice. He found that mice with a defect in their CREB activator gene were virtually incapable of forming long-term memories. They were amnesiac mice. But even these forgetful mice could learn a bit if they had short lessons with rest in between. Scientists theorize that we have a fixed amount of CREB activator in the brain that can limit the amount we can learn in any specific time. If we try to cram before a test, it means that we quickly exhaust the amount of CREB activators, and hence we cannot learn any more—at least until we take a break to replenish the CREB activators.

"We can now give you a biological reason why cramming doesn't work," says Dr. Tully. The best way to prepare for a final exam is to mentally review the material periodically during the day, until the material becomes part of your long-term memory.

This may also explain why emotionally charged memories are so vivid and can last for decades. The CREB repressor gene is like a filter, cleaning out useless information. But if a memory is associated with a strong emotion, it can either remove the CREB repressor gene or increase levels of the CREB activator gene.

In the future, we can expect more breakthroughs in understanding the genetic basis of memory. Not just one but a sophisticated combination of genes is probably required to shape the enormous capabilities of the brain. These genes, in turn, have counterparts in the human genome, so it is a distinct possibility that we can also enhance our memory and mental skills genetically.

However, don't think that you will be able to get a brain boost anytime soon. Many hurdles still remain. First, it is not clear if these results apply to humans. Often therapies that show great promise in mice do not translate well to our species. Second, even if these results can be applied to humans, we do not know what their impact will be. For example, these genes may help improve our memory but not affect our general intelligence. Third, gene therapy (i.e., fixing broken genes) is more difficult than previously thought. Only a small handful of genetic diseases can be cured with this method. Even though scientists use harmless viruses to infect cells with the "good" gene, the body still sends antibodies to attack the intruder, often rendering the therapy useless. It's possible that the insertion of a gene to enhance memory would face a similar fate. (In addition, the field of gene therapy suffered a major setback a few years ago when a patient died at the University of Pennsylvania during a gene therapy procedure. The work of modifying human genes therefore faces many ethical and even legal questions.)

Human trials, then, will progress much more slowly than animal trials. However, one can foresee the day when this procedure might be perfected and become a reality. Altering our genes in this way would require no more than a simple shot in the arm. A harmless virus would then enter our blood, which would then infect normal cells by injecting its genes. Once the "smart gene" is successfully incorporated into our cells, the gene becomes active and releases proteins that would increase our memory and cognitive skills by affecting the hippocampus and memory formation.

If the insertion of genes becomes too difficult, another possibility is to insert the proper proteins directly into the body, bypassing the use of gene therapy. Instead of getting a shot, we would swallow a pill.

A SMART PILL

Ultimately, one goal of this research is to create a "smart pill" that could boost concentration, improve memory, and maybe increase our intelligence. Pharmaceutical companies have experimented with several drugs, such as MEM 1003 and MEM 1414, that do seem to enhance mental function.

Scientists have found that in animal studies, long-term memories are made possible by the interaction of enzymes and genes. Learning takes place when certain neural pathways are reinforced as specific genes are activated, such as the CREB gene, which in turn emits a corresponding protein. Basically, the more CREB proteins circulating in the brain, the faster long-term memories are formed. This has been verified in studies on sea mollusks, fruit flies, and mice. The key property of MEM 1414 is that it accelerates the production of the CREB proteins. In lab tests, aged animals given MEM 1414 were able to form long-term memories significantly faster than a control group.

Scientists are also beginning to isolate the precise biochemistry required in the formation of long-term memories, at both the genetic and the molecular level. Once the process of memory formation is completely understood, therapies will be devised to accelerate and strengthen this key process. Not only the aged and Alzheimer's patients but eventually the average person may well benefit from this "brain boost."

CAN MEMORIES BE ERASED?

Alzheimer's may destroy memories indiscriminately, but what about selectively erasing them? Amnesia is one of Hollywood's favorite plot devices. In *The Bourne Identity*, Jason Bourne (played by Matt Damon), a skilled CIA agent, is found floating in the water, left for dead. When he is revived, he has severe memory loss. He is being relentlessly chased by assassins who want to kill him, but he does not know who he is, what happened, or why they want him dead. The only clue to his memory is his uncanny ability to instinctively engage in combat like a secret agent.

It is well documented that amnesia can occur accidentally through trauma, such as a blow on the head. But can memories be selectively erased? In the film *Eternal Sunshine of the Spotless Mind*, starring Jim Carrey, two people meet accidentally on a train and are immediately attracted to each other. However, they are shocked to find that they were actually lovers years ago but have no memory of it. They learn that they paid a company to wipe memories of each other after a particularly bad fight. Apparently, fate has given them a second chance at love.

Selective amnesia was taken to an entirely new level in *Men in Black*, in which Will Smith plays an agent from a shadowy, secret organization that uses the "neuralizer" to selectively erase inconvenient memories of UFOs and alien encounters. There is even a dial to determine how far back the memories should be erased.

All these make for thrilling plot lines and box-office hits, but are any of them really possible, even in the future?

We know that amnesia is, indeed, possible, and that there are two basic types, depending on whether short- or long-term memory has been affected. "Retrograde amnesia" occurs when there is some trauma or damage to the brain and preexisting memories are lost, usually dating from the event that caused the amnesia. This would be similar to the amnesia faced by Jason Bourne, who lost all memories from before he was left for dead in the water. Here the hippocampus is still intact, so new memories can be formed even though long-term memory has been damaged. "Anterograde amnesia" occurs when short-term memory is damaged, so the person has difficulty forming new memories after the event that caused the amnesia. Usually, amnesia may last for minutes to hours due to damage to the hippocampus. (Anterograde amnesia was featured prominently in the movie *Memento,* where a man is bent on revenge for the death of his wife. The problem, however, is that his memory lasts only about fifteen minutes, so he has to continually write messages on scraps of papers, photos, and even tattoos in order to remember the clues he has uncovered about the murderer. By painfully reading this trail of messages he has written to himself, he can accumulate crucial evidence that he would have soon forgotten.)

The point here is that memory loss dates back to the time of the trauma or disease, which would make the selective amnesia of Hollywood highly improbable. Movies like *Men in Black* assume that memories are stored sequentially, as in a hard disk, so you just hit the "erase" button after a designated point in time. However, we know that memories are actually broken up, with separate pieces stored in different places in the brain.

A FORGETFUL DRUG

Meanwhile, scientists are studying certain drugs that may erase traumatic memories that continue to haunt and disturb us. In 2009, Dutch scientists, led by Dr. Merel Kindt, announced that they had found new uses for an old drug called propranolol, which could act like a "miracle" drug to ease the pain associated with traumatic memories. The drug did not induce amnesia that begins at a specific point in time, but it did make the pain more manageable—and in just three days, the study claimed.

The discovery caused a flurry of headlines, in light of the thousands of victims who suffer from PTSD (post-traumatic stress disorder). Everyone from war veterans to victims of sexual abuse and horrific accidents could apparently find relief from their symptoms. But it also seemed to fly in the face of brain research, which shows that long-term memories are encoded not electrically, but at the level of protein molecules. Recent experiments, however, suggest that recalling memories requires both the retrieval and then the reassembly of the memory, so that the protein structure might actually be rearranged in the process. In other words, recalling a memory actually changes it. This may be the reason why the drug works: propranolol is known to interfere with adrenaline absorption, a key in creating the long-lasting, vivid memories that often result from traumatic events. "Propranolol sits on that nerve cell and blocks it. So adrenaline can be present, but it can't do its job," says Dr. James McGaugh of the University of California at Irvine. In other words, without adrenaline, the memory fades.

Controlled tests done on individuals with traumatic memories showed very promising results. But the drug hit a brick wall when it came to the ethics of erasing memory. Some ethicists did not dispute its effectiveness, but they frowned on the very idea of a forgetfulness drug, since memories are there for a purpose: to teach us the lessons of life. Even unpleasant memories, they said, serve some larger purpose. The drug got a thumbs-down from the President's Council on Bioethics. Its report concluded that "dulling our memory of terrible things [would] make us too comfortable with the world, unmoved by suffering, wrongdoing, or cruelty. . . . Can we become numb to life's sharpest sorrows without also becoming numb to its greatest joys?"

Dr. David Magus of Stanford University's Center for Biomedical Ethics says, "Our breakups, our relationships, as painful as they are, we learn from some of those painful experiences. They make us better people."

Others disagree. Dr. Roger Pitman of Harvard University says that if a doctor encounters an accident victim who is in intense pain, "should we deprive them of morphine because we might be taking away the full emotional experience? Who would ever argue with that? Why should psychiatry be different? I think that somehow behind this argument lurks the notion that mental disorders are not the same as physical disorders."

How this debate is ultimately resolved could have direct bearing on the next generation of drugs, since propranolol is not the only one involved.

In 2008, two independent groups, both working with animals, announced other drugs that could actually erase memories, not just manage the pain they cause. Dr. Joe Tsien of the Medical College of Georgia and his colleagues in Shanghai stated that they had actually eliminated a memory in mice using a protein called CaMKII, while scientists at SUNY Downstate Medical Center in Brooklyn found that the molecule PKMzeta could also erase memories. Dr. Andre Fenson, one of the authors of this second study, said, "If further work confirms this view, we can expect to one day see therapies based on PKMzeta memory erasure." Not only may the drug erase painful memories, it also "might be useful in treating depression, general anxiety, phobias, post-traumatic stress, and addictions," he added.

So far, research has been limited to animals, but human trials will begin soon. If the results transfer from animals to humans, then a forgetful pill may be a real possibility. It will not be the kind of pill seen in Hollywood movies (which conveniently creates amnesia at a precise, opportune time) but could have vast medical applications in the real world for people haunted by traumatic memories. It remains to be seen, though, how selective this memory erasure might be in humans.

WHAT CAN GO WRONG?

There may come a day, however, when we can carefully register *all* the signals passing through the hippocampus, thalamus, and the rest of the limbic system and make a faithful record. Then, by feeding this information into our brains, we might be able to reexperience the totality of what another person went through. Then the question is: What can go wrong?

In fact, the implications of this idea were explored in a movie, *Brainstorm* (1983), starring Natalie Wood, which was far ahead of its time. In the movie, scientists create the Hat, a helmet full of electrodes that can faithfully record all the sensations a person is experiencing. Later, a person can have precisely the same sensory experience by playing that tape back into his brain. For fun, one person puts on the Hat when he is making love and tape-records the experience. Then the tape is put into a loop so the experience is greatly magnified. But when another person unknowingly inserts the experience into his brain, he nearly dies because of a sensory overload. Later, one of the scientists experiences a fatal heart attack. But before she dies, she records her final moments on tape. When another person plays the death tape into his brain, he, too, has a sudden heart attack and dies.

When news of this powerful machine finally leaks out, the military wants to seize control. This sets off a power struggle between the military, which views it as a powerful weapon, and the original scientists, who want to use it to unlock the secrets of the mind.

Brainstorm prophetically highlighted not only the promise of this technology but also its potential pitfalls. It was meant to be science fiction, but some scientists believe that sometime in the future, these very issues may play out in our headlines and in our courts.

Earlier, we saw that there have been promising developments in recording a single memory created by a mouse. It may take until mid-century before we can reliably record a variety of memories in primates and humans. But creating the Hat, which can record the totality of stimulation entering into the brain, requires tapping into the raw, sensory data surging up the spinal cord and into the thalamus. It may be late in this century before this can be done.

SOCIAL AND LEGAL ISSUES

Some aspects of this dilemma may play out in our lifetimes. On one hand, we may reach a point where we can learn calculus by simply uploading the skill. The educational system would be turned upside down; perhaps it would free teachers to spend more time mentoring students and giving them one-on-one attention in areas of cognition that are less skill-based and cannot be mastered by hitting a button. The rote memorization necessary to become a professional doctor, lawyer, or scientist could also be drastically reduced through this method.

In principle, it might even give us memories of vacations that never happened, prizes that we never won, lovers whom we never loved, or families that we never had. It could make up for deficiencies, creating perfect memories of a life never lived. Parents would love this, since they could teach their children lessons taken from real memories. The demand for such a device could be enormous. Some ethicists fear that these fake memories would be so vivid that we would prefer to relive imaginary lives rather than experiencing our real ones.

The unemployed may also benefit from being able to learn new marketable skills by having memories implanted. Historically, millions of workers were left behind every time a new technology was introduced, often without any safety net. That's why we don't have many blacksmiths or wagon makers anymore. They turned into autoworkers and other industrial workers. But retraining requires a large amount of time and commitment. If skills can be implanted into the brain, there would be an immediate impact on the world economic system, since we wouldn't have to waste so much human capital. (To some degree, the value of a certain skill may be devalued if memories can be uploaded into anyone, but this is compensated for by the fact that the number and quality of skilled workers would vastly increase.)

The tourism industry will also experience a tremendous boost. One barrier to foreign travel is the pain of learning new customs and conversing with new phrases. Tourists would be able to share in the experience of living in a foreign land, rather than getting bogged down trying to master the local currency and the details of the transportation system. (Although uploading an entire language, with tens of thousands of words and expressions, would be difficult, it might be possible to upload enough information to carry on a decent conversation.)

Inevitably, these memory tapes will find their way onto social media. In the future, you might be able to record a memory and upload it to the Internet for millions to feel and experience. Previously, we discussed a brain-net through which you can send thoughts. But if memories can be recorded and created, you might also be able to send entire experiences. If you just won a gold medal at the Olympic Games, why not share the agony and the ecstasy of victory by putting your memories on the web? Maybe the experience will go viral and billions can share in your moment's glory. (Children, who are often at the forefront of video games and social media, may make a habit of recording memorable experiences and uploading them onto the Internet. Like taking a picture with a cell phone, it would be second nature to them to record entire memories. This would require both the sender and the receiver to have nearly invisible nanowires connecting to their hippocampus. The information would then be sent wirelessly to a server, which would convert the message to a digital signal that can be carried by the Internet. In this way, you could have blogs, message boards, social media, and chat rooms where, instead of uploading pictures and videos, you would upload memories and emotions.)

A LIBRARY OF SOULS

People may also want to have a geneology of memories. When searching records of our ancestors, we see only a one-dimensional portrait of their lives. Throughout human history, people have lived, loved, and died without leaving a substantial record of their existence. Mostly we just find the birth and death dates of our relatives, with little in between. Today we leave a long trail of electronic documents (credit card receipts, bills, e-mails, bank statements, etc.). By default, the web is becoming a giant repository of all the documents that describe our lives, but this still doesn't tell anyone much about what we were thinking or feeling. Perhaps in the far future, the web could become a giant library chronicling not just the details of our lives but also our consciousness.

In the future, people might routinely record their memories so their descendants can share the same experiences. Visiting the library of memories for your clan, you would be able to see *and* feel how they lived, and also how you fit into the larger scheme of things.

This means that anyone could replay our lives, long after we have died, by hitting the "play" button. If this vision is correct, it means that we might be able to "bring back" our ancestors for an afternoon chat, simply by inserting a disk into the library and pushing a button.

Meanwhile, if you want to share in the experiences of your favorite historical figures, you might be able to have an intimate look into how they felt as they confronted major crises in their lives. If you have a role model and wish to know how they negotiated and survived the great defeats of their life, you could experience their memory tapes and gain valuable insight. Imagine being able to share the memories of a Nobel Prize–winning scientist. You might get clues about how great discoveries are made. Or you might be able to share the memories of great politicians and statesmen as they made crucial decisions that affected world history.

Dr. Miguel Nicolelis believes all this will one day become reality. He says, "Each of these perennial records would be revered as a uniquely precious jewel, one among billions of equally exclusive minds that once lived, loved, suffered, and prospered, until they, too, become immortalized, not clad in cold and silent gravestones, but released through vivid thoughts, intensely lived loves, and mutually endured sorrows."

THE DARK SIDE OF TECHNOLOGY

Some scientists have pondered the ethical implications of this technology. Almost every new medical discovery caused ethical concerns when it was introduced. Some of them had to be restricted or banned when proven harmful (like the sleeping drug thalidomide, which caused birth defects). Others have been so successful they changed our conception of who we are, such as test-tube babies. When Louise Brown, the first test-tube baby, was born in 1978, it created such a media storm that even the pope issued a document critical of this technology. But today, perhaps your sibling, child, spouse, or even you may be a product of in vitro fertilization. Like many technologies, eventually the public will simply get used to the idea that memories can be recorded and shared.

Other bioethicists have different worries. What happens if memories are given to us without our permission? What happens if these memories are painful or destructive? Or what about Alzheimer's patients, who are eligible for memory uploads but are too sick to give permission?

The late Bernard Williams, a philosopher at Oxford University, worried that this device might disturb the natural order of things, which is to forget. "Forgetting is the most beneficial process we possess," he says.

If memories can be implanted like uploading computer files, it could also shake the foundation of our legal system. One of the pillars of justice is the eyewitness account, but what would happen if fake memories were implanted? Also, if the memory of a crime can be created, then it might secretly be implanted into the brain of an innocent person. Or, if a criminal needs an alibi, he could secretly implant a memory into another person's brain, convincing him that they were together when the crime was being committed. Furthermore, not just verbal testimony but also legal documents would be suspect, since when we sign affidavits and legal documents, we depend on our memory to clarify what is true and false.

Safeguards would have to be introduced. Laws will have to be passed that clearly define the limits of granting or denying access to memories. Just as there are laws limiting the ability of the police or third parties to enter your home, there would be laws to prevent people from accessing your memories without your permission. There would also have to be a way to mark these memories so that the person realizes that they are fake. Thus, he would still be able to enjoy the memory of a nice vacation, but he would also know that it never happened.

Taping, storing, and uploading our memories may allow us to record the past and master new skills. But doing so will not alter our innate ability to digest and process this large body of information. To do that, we need to enhance our intelligence. Progress in this direction is hindered by the fact that there is no universally accepted definition of intelligence. However, there is one example of genius and intelligence that no one can dispute, and that is Albert Einstein. Remarkably, sixty years after his death, his brain is still yielding invaluable clues to the nature of intelligence.

Some scientists believe that, using a combination of electromagnetics, genetics, and drug therapy, it may be possible to boost our intelligence to the genius level. They cite the fact that random injuries to the brain have been documented that can suddenly change a person of normal ability into a "savant," one whose spectacular mental and artistic ability is off the scale. This can be achieved now by random accidents, but what happens when science intervenes and illuminates the secret of this process?

The brain is wider than the sky
For, put them side by side
The one the other will contain
With ease, and you beside.

—EMILY DICKINSON

Talent hits a target no one else can hit. Genius hits a target no one else can see.

—ARTHUR SCHOPENHAUER

6 EINSTEIN'S BRAIN AND ENHANCING OUR INTELLIGENCE

Albert Einstein's brain is missing.

Or, at least it was for fifty years, until the heirs of the doctor who spirited it away shortly after his death in 1955 finally returned it to the National Museum of Health and Medicine in 2010. Analysis of his brain may help clarify these questions: What is genius? How do you measure intelligence and its relationship to success in life? There are also philosophical questions: Is genius a function of our genes, or is it more a question of personal struggle and achievement?

And, finally, Einstein's brain may help answer the key question: Can we boost our own intelligence?

The word "Einstein" is no longer a proper noun that refers to a specific person. It now simply means "genius." The picture that the name conjures up (baggy pants, flaming white hair, disheveled looks) is equally iconic and instantly recognizable.

The legacy of Einstein has been enormous. When some physicists in 2011 raised the possibility that he was wrong, that particles could break the light barrier, it created a firestorm of controversy in the physics world that spilled over into the popular press. The very idea that relativity, which forms the cornerstone of modern physics, could be wrong had physicists around the world shaking their heads. As expected, once the result was recalibrated, Einstein was shown to be right once again. It is always dangerous to go up against Einstein.

One way to gain insight into the question "What is genius?" is to analyze Einstein's brain. Apparently on the spur of the moment, Dr. Thomas Harvey, the doctor at the Princeton hospital who was performing the autopsy on Einstein, decided to secretly preserve his brain, against the knowledge and wishes of Einstein's family.

Perhaps he preserved Einstein's brain with the vague notion that one day it might unlock the secret of genius. Perhaps he thought, like many others, that there was a peculiar part of Einstein's brain that was the seat of his vast intelligence. Brian Burrell, in his book *Postcards from the Brain Museum,* speculates that perhaps Dr. Harvey "got caught up in the moment and was transfixed in the presence of greatness. What he quickly discovered was that he had bitten off more than he could chew."

What happened to Einstein's brain after that sounds more like a comedy than a science story. Over the years, Dr. Harvey promised to publish his results of analyzing Einstein's brain. But he was no brain specialist, and kept making excuses. For decades, the brain sat in two large mason jars filled with formaldehyde and placed in a cider box, under a beer cooler. He had a technician slice the brain into 240 pieces, and on rare occasions he would mail a few to scientists who wanted to study them. Once, pieces were mailed to a scientist at Berkeley in a mayonnaise container.

Forty years later, Dr. Harvey drove across the country in a Buick Skylark carrying Einstein's brain in a Tupperware container, hoping to return it to Einstein's granddaughter Evelyn. She refused to accept it. After Dr. Harvey's death in 2007, it was left to his heirs to properly donate his collection of slides and portions of Einstein's brain to science. The history of Einstein's brain is so unusual that a TV documentary was filmed about it.

(It should be pointed out that Einstein's brain was not the only one to be preserved for posterity. The brain of one of the greatest geniuses of mathematics, Carl Friedrich Gauss, often called the Prince of Mathematicians, was also preserved by a doctor a century earlier. Back then, the anatomy of the brain was largely unexplored, and no conclusions could be drawn other than the fact that it had unusually large convolutions or folds.)

One might expect that Einstein's brain was far beyond an ordinary human's, that it must have been huge, perhaps with areas that were abnormally large. In fact, the opposite has been found (it is slightly smaller, not larger, than normal). Overall, Einstein's brain is quite ordinary. If a neurologist did not know that this was Einstein's brain, he probably would not give it a second thought.

The only differences found in Einstein's brain were rather minor. A certain part of his brain, called the angular gyri, was larger than normal, with the inferior parietal regions of both hemispheres 15 percent wider than average. Notably, these parts of the brain are involved in abstract thought, in the manipulation of symbols such as writing and mathematics, and in visual-spatial processing. But his brain was still within the norm, so it is not clear whether the genius of Einstein lay in the organic structure of his brain or in the force of his personality, his outlook, and the times. In a biography that I once wrote of Einstein, titled *Einstein's Cosmos,* it was clear to me that certain features of his life were just as important as any anomaly in his brain. Perhaps Einstein himself said it best when he said, "I have no special talents. . . . I am only passionately curious." In fact, Einstein would confess that he had to struggle with mathematics in his youth. To one group of schoolchildren, he once confided, "No matter what difficulties you may have with mathematics, mine were greater." So why was Einstein Einstein?

First, Einstein spent most of his time thinking via "thought experiments." He was a theoretical physicist, not an experimental one, so he was continually running sophisticated simulations of the future in his head. In other words, his laboratory was his mind.

Second, he was known to spend up to ten years or more on a single thought experiment. From the age of sixteen to twenty-six, he focused on the problem of light and whether it was possible to outrace a light beam. This led to the birth of special relativity, which eventually revealed the secret of the stars and gave us the atomic bomb. From the age of twenty-six to thirty-six, he focused on a theory of gravity, which eventually gave us black holes and the big-bang theory of the universe. And then from the age of thirty-six to the end of his life, he tried to find a theory of everything to unify all of physics. Clearly, the ability to spend ten or more years on a single problem showed the tenacity with which he would simulate experiments in his head.

Third, his personality was important. He was a bohemian, so it was natural for him to rebel against the establishment in physics. Not every physicist had the nerve or the imagination to challenge the prevailing theory of Isaac Newton, which had held sway for two hundred years before Einstein.

Fourth, the time was right for the emergence of an Einstein. In 1905, the old physical world of Newton was crumbling in light of experiments that clearly suggested a new physics was about to be born, waiting for a genius to show the way. For example, the mysterious substance called radium glowed in the dark all by itself indefinitely, as if energy was being created out of thin air, violating the theory of conservation of energy. In other words, Einstein was the right man for the times. If somehow it becomes possible to clone Einstein from the cells in his preserved brain, I suspect that the clone would not be the next Einstein. The historic circumstances must also be right to create a genius.

The point here is that genius is perhaps a combination of being born with certain mental abilities and also the determination and drive to achieve great things. The essence of Einstein's genius was probably his extraordinary ability to simulate the future through thought experiments, creating new physical principles via pictures. As Einstein himself once said, "The true sign of intelligence is not knowledge, but imagination." And to Einstein, imagination meant shattering the boundaries of the known and entering the domain of the unknown.

All of us are born with certain abilities that are programmed into our genes and the structure of our brains. That is the luck of the draw. But how we arrange our thoughts and experiences and simulate the future is something that is totally within our control. Charles Darwin himself once wrote, "I have always maintained that, excepting fools, men did not differ much in intellect, only in zeal and hard work."

CAN GENIUS BE LEARNED?

This rekindles the question, Are geniuses made or born? How does the nature/nurture debate solve the mystery of intelligence? Can an ordinary person become a genius?

Since brain cells are notoriously hard to grow, it was once thought that intelligence was fixed by the time we became young adults. But one thing is becoming increasingly clear with new brain research: the brain itself can change when it learns. Although brain cells are not being added in the cortex, the connections between neurons are changing every time a new task is learned.

For example, scientists in 2011 analyzed the brains of London's famous taxicab drivers, who have to laboriously memorize twenty-five thousand streets in the dizzying maze that makes up modern London. It takes three to four years to prepare for this arduous test, and only half the trainees pass.

Scientists at University College London studied the brains of these drivers before they took the test, and then tested them again three to four years afterward. Those trainees who passed the test had a larger volume of gray matter than before, in an area called the posterior and the anterior hippocampus. The hippocampus, as we've seen, is where memories are processed. (Curiously, tests also showed that these taxicab drivers scored less than normal on processing visual information, so perhaps there is a trade-off, a price to pay for learning this volume of information.)

"The human brain remains 'plastic,' even in adult life, allowing it to adapt when we learn new tasks," says Eleanor Maguire of the Wellcome Trust, which funded the study. "This offers encouragement for adults who want to learn new skills later in life."

Similarly, the brains of mice that have learned many tasks are slightly different from the brains of other mice that have not learned these tasks. It is not so much that the number of neurons has changed, but rather that the nature of the neural connections has been altered by the learning process. In other words, learning actually changes the structure of the brain.

This raises the old adage "practice makes perfect." Canadian psychologist Dr. Donald Hebb discovered an important fact about the wiring of the brain: the more we exercise certain skills, the more certain pathways in our brains become reinforced, so the task becomes easier. Unlike a digital computer, which is just as dumb today as it was yesterday, the brain is a learning machine with the ability to rewire its neural pathways every time it learns something. This is a fundamental difference between a digital computer and the brain.

This lesson applies not only to London taxicab drivers, but also to accomplished concert musicians as well. According to psychologist Dr. K. Anders Ericsson and colleagues, who studied master violinists at Berlin's elite Academy of Music, top concert violinists could easily rack up ten thousand hours of grueling practice by the time they were twenty years old, practicing more than thirty hours per week. By contrast, he found that students who were merely exceptional studied only eight thousand hours or fewer, and future music teachers practiced only a total of four thousand hours. Neurologist Daniel Levitin says, "The emerging picture from such studies is that ten thousand hours of practice is required to achieve the level of mastery associated with being a world-class expert—in anything. . . . In study after study, of composers, basketball players, fiction writers, ice skaters, concert pianists, chess players, master criminals, and what have you, this number comes up again and again." Malcolm Gladwell, writing in the book *Outliers,* calls this the "10,000-hour rule."

HOW DO YOU MEASURE INTELLIGENCE?

But how do you measure intelligence? For centuries, any discussion of intelligence relied on hearsay and anecdote. But now MRI studies have shown that the principal activity of the brain while performing these mathematical puzzles involves the pathway connecting the prefrontal cortex (which engages in rational thought) with the parietal lobes (which processes numbers). This correlates with the anatomical studies of Einstein's brain, which showed that his inferior parietal lobes were larger than normal. So it is conceivable that mathematical ability correlates with increased information flows between the prefrontal cortex and the parietal lobes. But did the brain increase in size in this area because of hard work and study, or was Einstein born that way? The answer is still not clear.

The key problem is that there is no uniformly accepted definition of intelligence, let alone a consensus among scientists as to its origin. But the answer may prove critical if we wish to enhance it.

IQ EXAMS AND DR. TERMAN

By default, the most widely used measure of intelligence is the IQ exam, pioneered by Dr. Lewis Terman of Stanford University, who in 1916 revised an earlier test devised by Alfred Binet for the French government. For the next several decades, it became the gold standard by which to measure intelligence. Terman, in fact, dedicated his life to the proposition that intelligence could be measured and inherited, and was the strongest predictor of success in life.

Five years later, Terman started a landmark study on schoolchildren, *The Genetic Studies of Genius.* It was an ambitious project, whose scope and duration were unprecedented back in the 1920s. It set the tone for research in this field for an entire generation. He methodically chronicled the successes and failures of these individuals throughout their lives, compiling thick files of their achievements. These high-IQ students were dubbed the "Termites."

At first, Dr. Terman's idea seemed to be a resounding success. It became the standard by which both children and other tests were measured. During World War I, 1.7 million soldiers were given this test. But over the years, a different profile began to slowly emerge. Decades later, children who scored high on the IQ exam were only moderately more successful than those who did not. Terman could proudly point to some of his students who went on to win awards and secure well-paying jobs. But he became increasingly disturbed by the large number of his brightest students whom society would consider to be failures, taking menial, dead-end jobs, engaging in crime, or leading lives on the margins of society. These results were quite upsetting to Dr. Terman, who had dedicated his life to proving that high IQ meant success in life.

SUCCESS IN LIFE AND DELAYED GRATIFICATION

A different approach was taken in 1972 by Dr. Walter Mischel, also of Stanford, who analyzed yet another characteristic among children: the ability to delay gratification. He pioneered the use of the "marshmallow test," that is, would children prefer one marshmallow now, or the prospect of two marshmallows twenty minutes later? Six hundred children, aged four to six, participated in this experiment. When Mischel revisited the participants in 1988, he found that those who could delay gratification were more competent than those who could not.

In 1990, another study showed a direct correlation between those who could delay gratification and SAT scores. And a study done in 2011 indicated that this characteristic continued throughout a person's life. The results of these and other studies were eye-opening. The children who exhibited delayed gratification scored higher on almost every measure of success in life: higher-paying jobs, lower rates of drug addiction, higher test scores, higher educational attainment, better social integration, etc.

But what was most intriguing was that brain scans of these individuals revealed a definite pattern. They showed a distinct difference in the way the prefrontal cortex interacted with the ventral striatum, a region involved in addiction. (This is not surprising, since the ventral striatum contains the nucleus accumbens, known as the "pleasure center." So there seems to be a struggle here between the pleasure-seeking part of the brain and the rational part to control temptation, as we saw in Chapter 2.)

This difference was no fluke. The result has been tested by many independent groups over the years, with nearly identical results. Other studies have also verified the difference in the frontal-striatal circuitry of the brain, which appears to govern delayed gratification. It seems that the one characteristic most closely correlated with success in life, which has persisted over the decades, is the ability to delay gratification.

Although this is a gross simplification, what these brain scans show is that the connection between the prefrontal and parietal lobes seems to be important for mathematical and abstract thought, while the connection between the prefrontal and limbic system (involving the conscious control of our emotions and pleasure center) seems to be essential for success in life.

Dr. Richard Davidson, a neuroscientist at the University of Wisconsin–Madison, concludes, "Your grades in school, your scores on the SAT, mean less for life success than your capacity to co-operate, your ability to regulate your emotions, your capacity to delay your gratification, and your capacity to focus your attention. Those skills are far more important—all the data indicate—for life success than your IQ or your grades."

NEW MEASURES OF INTELLIGENCE

Clearly there have to be new ways to measure intelligence and success in life. IQ exams are not useless, but they measure only one limited form of intelligence. Dr. Michael Sweeney, author of *Brain: The Complete Mind,* notes, "Tests don't measure motivation, persistence, social skills, and a host of other attributes of a life that's well lived."

The problem with many of these standardized tests is that there may also be a subconscious bias due to cultural influences. In addition, these tests are evaluating only one particular form of intelligence, which some psychologists call "convergent" intelligence. Convergent intelligence focuses on one line of thought, ignoring the more complex "divergent" form of intelligence, which involves measuring differing factors. For example, during World War II, the U.S. Army Air Forces asked scientists to devise a psychological exam that would measure a pilot's intelligence and ability to handle difficult, unexpected situations. One question was: If you are shot down deep in enemy territory and must somehow make it back to friendly lines, what do you do? The results contradicted conventional thinking.

Most psychologists expected that the air force study would show that pilots with high IQs would score highly on this test as well. Actually, the reverse was true. The pilots who scored highest were the ones with higher levels of divergent thinking, who could see through many different lines of thought. Pilots who excelled at this, for example, were able to think up a variety of unorthodox and imaginative methods to escape after they were captured behind enemy lines.

The difference between convergent and divergent thinking is also reflected in studies on split-brain patients, which clearly show that each hemisphere of the brain is principally hardwired for one or the other. Dr. Ulrich Kraft of Fulda, Germany, writes, "The left hemisphere is responsible for convergent thinking and the right hemisphere for divergent thinking. The left side examines details and processes them logically and analytically but lacks a sense of overriding, abstract connections. The right side is more imaginative and intuitive and tends to work holistically, integrating pieces of an informational puzzle into a whole."

In this book, I take the position that human consciousness involves the ability to create a model of the world and then simulate the model into the future, in order to attain a goal. Pilots who demonstrated divergent thinking were able to simulate many possible future events accurately and with more complexity. Similarly, the children who mastered delayed gratification in the famous marshmallow test appear to be the ones who had the most ability to simulate the future, to see the long-term rewards and not just the short-term, get-rich-quick schemes.

A more sophisticated intelligence exam that directly quantifies a person's ability to simulate the future would be difficult but not impossible to create. A person could be asked to create as many realistic scenarios for the future as possible to win a game, with a score assigned depending on the number of simulations the person can imagine and the number of causal links involved with each one. Instead of measuring a person's ability to simply assimilate information, this new method would measure a person's ability to manipulate and mold this information to achieve a higher goal. For example, a person might be asked to figure out how to escape from a deserted island full of hungry wild animals and poisonous snakes. He would have to list all the various ways to survive, fend off the dangerous animals, and leave the island, creating an elaborate causal tree of possible outcomes and futures.

So we see that there is a common thread running through all this discussion, and that is that intelligence seems to be correlated with the complexity with which we can simulate future events, which correlates with our earlier discussion of consciousness.

But given the rapid advances taking place in the world's laboratories concerning electromagnetic fields, genetics, and drug therapies, is it possible not just to measure our intelligence, but to enhance it as well—to become another Einstein?

BOOSTING OUR INTELLIGENCE

This possibility was explored in the novel *Flowers for Algernon* (1958), later made into the Academy Award–winning movie *Charly* (1968). In it, we follow the sad life of Charly Gordon, who has an IQ of 68 and a menial job in a bakery. He lives a simple life, fails to understand that his fellow workers are constantly making fun of him, and does not even know how to spell his own name.

His only friend is Alice, a teacher who takes pity on him and tries to teach him to read. But one day, scientists discover a new procedure that can suddenly make ordinary mice intelligent. Alice hears about this and decides to introduce Charly to these scientists, who agree to perform the procedure on their first human subject. Within weeks, Charly has noticeably changed. His vocabulary increases, he devours books from the library, he becomes something of a ladies' man, and his room explodes with modern art. Soon he begins to read about relativity and the quantum theory, pushing the boundaries of advanced physics. He and Alice even become lovers.

But then the doctors notice that the mice have slowly lost their ability and died. Realizing that he, too, might lose everything, Charly furiously tries to use his superior intellect to find a cure, but instead he's forced to witness his own inexorable decline. His vocabulary shrinks, he forgets mathematics and physics, and he slowly reverts back to his old self. In the final scene, a heartbroken Alice watches as Charly plays with children.

The novel and movie, although poignant and critically acclaimed, were dismissed as sheer science fiction. The plot was moving and original, but the idea of boosting one's intelligence was considered preposterous. Brain cells cannot regenerate, scientists said, so this movie's plot was obviously impossible.

But not anymore.

Although it is still impossible to boost your intelligence, rapid advances are being made in electromagnetic sensors, genetics, and stem cells that may one day make this a real possibility. In particular, scientific interest has focused on "autistic savants," who possess phenomenal, superhuman abilities that stagger the imagination. More important, due to specific injuries to the brain, normal people can rapidly acquire such near-miraculous powers. Some scientists even believe that these uncanny abilities might be induced using electromagnetic fields.

SAVANTS: SUPER GENIUSES?

A bullet went crashing through the skull of Mr. Z when he was nine years old. It did not kill him, as his doctors feared, but wreaked extensive damage to the left side of his brain, causing paralysis of the right side of his body and leaving him permanently deaf and mute.

However, the bullet also had a bizarre side effect. Mr. Z developed supernormal mechanical abilities and a prodigious memory, typical of "savants."

Mr. Z is not alone. In 1979, a ten-year-old boy named Orlando Serrell was knocked unconscious by a baseball that hit the left side of his head. At first, he complained of severe headaches. But after the pain subsided, he was able to do remarkable mathematical calculations and had a near-photographic memory of certain events happening in his life. He could calculate dates thousands of years into the future.

In the entire world of roughly seven billion people, there are only about one hundred documented cases of these astounding savants. (The number is much larger if we include those whose mental skills are still extraordinary but not superhuman. It is believed that about 10 percent of autistic individuals show some savant capabilities.) These extraordinary savants possess abilities far beyond our current scientific understanding.

There are several types of savants that have recently elicited the curiosity of scientists. About half of savants have some form of autism (the other half display other forms of mental illness or psychological disorder). They often have profound problems interacting socially, leading to deep isolation.

Then there is the "acquired savant syndrome," in which people who appear perfectly normal suffer from some extreme trauma later in life (e.g., hitting their head on the bottom of a swimming pool or being struck by a baseball or a bullet), almost always on the left side of their brain. Some scientists, however, suggest that this distinction is misleading, that perhaps *all* savant skills are acquired. Since autistic savants begin to show their abilities around age three or four, perhaps their autism (like a blow to their head) is the origin of their abilities.

There is scientific disagreement about the origin of these extraordinary abilities. Some believe that these individuals are simply born this way and hence are unique, one-of-a-kind anomalies. Their skills, even if awakened by a bullet, are hardwired into their brains from birth. If so, then perhaps this skill can never be learned or transferred.

Others claim that such hardwiring violates the theory of evolution, which takes place incrementally over long periods of time. If savant geniuses exist, then the rest of us must also possess similar abilities, although they are latent. Does this mean, then, that one day we might be able to turn on these miraculous powers at will? Some believe so, and there are even published papers claiming that some savant skills are latent in all of us and can be brought to light using the magnetic fields generated by an electromagnetic scanner (TES). Or perhaps there is a genetic basis to this skill, in which case gene therapy might re-create these astonishing abilities. It might also be possible to cultivate stem cells that would allow neurons to grow in the prefrontal cortex and other key centers of the brain. Then we might be able to increase our mental abilities.

All these avenues are the source of much speculation and research. Not only might they allow doctors to reverse the ravages of diseases like Alzheimer's, but they could also enable us to enhance our own intelligence. The possibilities are intriguing.

The first documented case of a savant was recorded in 1789 by Dr. Benjamin Rush, who studied an individual who seemed to be mentally handicapped. Yet when he was asked how many seconds a man had lived (who was seventy years, seventeen days, and twelve hours old), it took him only ninety seconds to give the correct answer of 2,210,500,800.

Dr. Darold Treffert, a Wisconsin physician, has studied these savants at length. He recites one story of a blind savant who was asked a simple question. If you put one corn kernel in the first square of a chess board, two kernels in the second, four in the next, and keep doubling after that, how many kernels would you have on the sixty-fourth square? It took him just forty-five seconds to correctly reply: 18,446,744,073,709,551,616.

Perhaps the best-known example of a savant was the late Kim Peek, who was the inspiration for the movie *Rain Man,* starring Dustin Hoffman and Tom Cruise. Although Kim Peek was severely mentally handicapped (he was incapable of living by himself and could barely tie his shoelaces or button his shirt), he memorized about twelve thousand books and could recite lines from them, word for word, on any particular page. It took him about eight seconds to read a page. (He could memorize a book in about half an hour, but he read them in an unusual way. He could read both pages simultaneously, using each eye to read a different page at the same time.) Although incredibly shy, he eventually began to enjoy performing dazzling feats of mathematics before curious onlookers, who would try to challenge him with tricky questions.

Scientists, of course, have to be careful in distinguishing true savant skills from simple memorization tricks. Their skills are not just mathematical—they also extend to incredible musical, artistic, and mechanical capabilities. Since autistic savants have great difficulty verbally expressing their mental processes, another avenue is to investigate individuals who have Asperger's syndrome, which is a milder form of autism. Only in 1994 was Asperger's syndrome recognized as a distinct psychological condition, so there is very little solid research in this area. Like autistic individuals, people with Asperger's have a difficult time interacting socially with others. However, with proper training, they can learn enough social skills to hold down a job and articulate their mental processes. And a fraction of them have remarkable savant skills. Some scientists believe that many great scientists had Asperger's syndrome. This might explain the strange, reclusive nature of physicists like Isaac Newton and Paul Dirac (one of the founders of the quantum theory). Newton, in particular, was pathologically incapable of small talk.

I had the pleasure of interviewing one such individual, Daniel Tammet, who has written a best seller, *Born on a Blue Day.* Almost alone among these remarkable savants, he is able to articulate his thoughts in books, on the radio, and in TV interviews. For someone who had such difficulty relating to others as a child, he now has a superb grasp of communication skills.

Daniel has the distinction of setting a world record for memorizing pi, a fundamental number in geometry. He was able to memorize it to 22,514 decimal places. I asked him how he prepared for such a herculean feat. Daniel told me that he associates a color or texture with every number. Then I asked him the key question: If every digit has a color or texture, then how does he remember tens of thousands of them? Sadly, at that point he said he doesn't know. It just comes to him. Numbers have been his life ever since he was a child, and hence they simply appear in his mind. His mind is a constant mixture of numbers and colors.

ASPERGER'S AND SILICON VALLEY

So far, this discussion may seem abstract, without any direct bearing on our daily lives. But the impact of people with mild autism and Asperger's may be more widespread than previously thought, especially in certain high-tech fields.

In the hit television series *The Big Bang Theory,* we follow the antics of several young scientists, mainly nerdy physicists, in their awkward quest for female companionship. In every episode, there is a hilarious incident that reveals how clueless and pathetic they are in this endeavor.

There is a tacit assumption running through the series that their intellectual brilliance is matched only by their geekiness. And anecdotally, people have noticed that among the high-tech gurus in Silicon Valley, a higher percentage than normal seem to lack some social skills. (There is a saying among women scientists who attend highly specialized engineering universities, where the girl-to-guy ratio is decidedly in their favor: "The odds are good, but the goods are odd.")

Scientists set out to investigate this suspicion. The hypothesis is that people with Asperger's and other mild forms of autism have mental skills ideally suited for certain fields, like the information technology industry. Scientists at University College London examined sixteen people who were diagnosed with a mild form of autism and compared them with sixteen normal individuals. Both groups were shown slides containing random numbers and letters arranged in increasingly complex patterns.

Their results showed that people with autism had a superior ability to focus on the task. In fact, as the tasks became harder, the gap between the intellectual skills of both groups began to widen, with the autistic individuals performing significantly better than the control group. (The test, however, also showed that these individuals were more easily distracted by outside noises and blinking lights than the control group.)

Dr. Nilli Lavie says, "Our study confirms our hypothesis that people with autism have higher perceptual capacity compared to the typical population. . . . People with autism are able to perceive significantly more information than the typical adult."

This certainly does not prove that all people who are intellectually brilliant have some form of Asperger's. But it does indicate that fields requiring the ability to focus intellectually might have a higher proportion of people with Asperger's.

BRAIN SCANS OF SAVANTS

The subject of savants has always been shrouded in hearsay and amazing anecdotal stories. But recently, the entire field has been turned upside down with the development of MRI and other brain scans.

Kim Peek's brain, for example, was unusual. MRI scans show that it lacked the corpus callosum connecting the left and right brain, which is probably why he could read two pages at the same time. His poor motor skills were reflected in a deformed cerebellum, the area that controls balance. Unfortunately, MRI scans could not reveal the exact origin of his extraordinary abilities and photographic memory. But in general, brain scans have shown that many suffering from acquired savant syndrome have experienced damage to their left brain.

In particular, interest has focused on the left anterior temporal and orbitofrontal cortices. Some believe that perhaps *all* savant skills (autistic, acquired, and Asperger's) arise from damage to this very specific spot in the left temporal lobe. This area can act like a "censor" that periodically flushes out irrelevant memories. But after damage occurs to the left hemisphere, the right hemisphere starts to take over. The right brain is much more precise than the left brain, which often distorts reality and confabulates. In fact, it is believed that the right brain must work extra hard because of damage to the left brain, and hence savant skills develop as a consequence. For example, the right brain is much more artistic than the left brain. Normally, the left brain restricts this talent and holds it in check. But if the left brain is injured in a certain way, it may unleash the artistic abilities latent in the right brain, causing an explosion of artistic talent. So the key to unleashing savant capabilities might be to dampen the left brain so that it can no longer restrain the natural talents of the right brain. This is sometimes referred to as "left brain injury, right brain compensation."

In 1998, Dr. Bruce Miller of the University of California at San Francisco performed a series of studies that seem to back this idea up. He and coworkers studied five normal individuals who began to show signs of frontotemporal dementia (FTD). As their dementia started to progress, savant abilities gradually began to emerge. As their dementia got worse, several of these individuals began to exhibit even more extraordinary artistic ability, although none had shown gifts in this area before. Moreover, the abilities they exhibited were typical of savant behavior. Their abilities were visual, not auditory, and their artworks, remarkable as they were, were just copies lacking any original, abstract, or symbolic qualities. (One patient actually got better during the study. But her emerging savant skills were also reduced as a consequence. This suggests a close relationship between emerging disorders of the left temporal lobe and emerging savant skills.)

Dr. Miller's analysis seemed to show that degeneration of the left anterior temporal and orbitofrontal cortices probably decreased inhibition of the visual systems in the right hemisphere, thereby increasing artistic abilities. Again, damaging the left hemisphere in a particular location forced the right hemisphere to take over and develop.

In addition to the savants, MRI scans have also been done on people with hyperthymestic syndrome, who also have photographic memories. These people do not suffer from autism and mental disorders, but they share some of their skills. In the entire United States, there are only four documented cases of true photographic memory. One of them is Jill Price, a school administrator in Los Angeles. She can recall precisely what she was doing on any particular day going back decades. But she complains that she finds it difficult to erase certain thoughts. Indeed, her brain seems to be "stuck on autopilot." She compares her memory to watching the world through a split screen, in which the past and present are constantly competing for her attention.

Since 2000, scientists at the University of California at Irvine have scanned her brain, and they've found it to be unusual. Several regions were larger than normal, such as the caudate nuclei (which is involved with forming habits) and the temporal lobe (which stores facts and figures). It is theorized that these two areas work in tandem to create her photographic memory. Her brain is therefore different from the brains of savants who suffer an injury or damage to their left temporal lobe. The reason is unknown, but it points to another path by which one may obtain these fantastic mental abilities.

CAN WE BECOME SAVANTS?

All this raises the intriguing possibility that one might be able to deliberately deactivate parts of the left brain and thereby increase the activity of the right brain, forcing it to acquire savant capabilities.

We recall that transcranial magnetic stimulation, or TMS, allows one to effectively silence parts of the brain. If so, then why can't we silence this part of the left anterior temporal and orbitofrontal cortices using the TMS and turn on a savantlike genius at will?

This idea has actually been tried. Dr. Allan Snyder of the University of Sydney, Australia, made headlines a few years ago when he claimed that, by applying the TMS to a certain part of the left brain, his subjects could suddenly perform savantlike feats. By directing low-frequency magnetic waves into the left hemisphere, one can in principle turn off this dominant region of the brain so that the right hemisphere takes over. Dr. Synder and his colleagues did an experiment with eleven male volunteers. They applied the TMS to the subjects' left frontotemporal region while the subjects were performing tests involving reading and drawing. This did not produce savant skills among the subjects, but two of them had significant improvements in their ability to proofread words and recognize duplicated words. In another experiment, Dr. R. L. Young and his colleagues gave a battery of psychological tests to seventeen individuals. The tests were specifically designed to test for savant skills. (Tests of this sort analyze a person's ability to memorize facts, manipulate numbers and dates, create artwork, or perform music.) Five of the subjects reported improvement in savantlike skills after treatment with TMS.

Dr. Michael Sweeney has observed, "When applied to the prefrontal lobes, TMS has been shown to enhance the speed and agility of cognitive processing. The TMS bursts are like a localized jolt of caffeine, but nobody knows for sure how the magnets actually do their work." These experiments hint, but by no means prove, that silencing a part of the left frontotemporal region could initiate some enhanced skills. These skills are a far cry from savant abilities, and we should also be careful to point out that other groups have looked into these experiments, and the results have been inconclusive. More experimental work must be done, so it is still too early to render a final judgment one way or the other.

TMS probes are the easiest and most convenient instrument to use for this purpose, since they can selectively silence various parts of the brain at will without relying on brain damage and traumatic accidents. But it should also be noted that TMS probes are still crude, silencing millions of neurons at a time. Magnetic fields, unlike electrical probes, are not precise but spread out over several centimeters. We know that the left anterior temporal and orbitofrontal cortices are damaged in savants and likely responsible, at least in some part, for their unique abilities, but perhaps the specific area that must be dampened is an even smaller subregion. So each jolt of TMS might inadvertently deactivate some of the areas that need to remain intact in order to produce savantlike skills.

In the future, with TMS probes we might be able to narrow down the region of the brain involved with eliciting savant skills. Once this region is identified, the next step would be to use highly accurate electrical probes, like those used in deep brain stimulation, to dampen these areas even more precisely. Then, with the push of a button, it might be possible to use these probes to silence this tiny portion of the brain in order to bring out savant-like skills.

FORGETTING TO FORGET AND PHOTOGRAPHIC MEMORY

Although savant skills may be initiated by some sort of injury to the left brain (leading to right brain compensation), this still does not explain precisely how the right brain can perform these miraculous feats of memory. By what neural mechanism does photographic memory emerge? The answer to this question may determine whether we can become savants.

Until recently, it was thought that photographic memory was due to the special ability of certain brains to remember. If so, then it might be difficult for the average person to learn these memory skills, since only exceptional brains are capable of them. But in 2012, a new study showed that precisely the opposite may be true.

The key to photographic memory may not be the ability of remarkable brains to learn; on the contrary, it may be their inability to forget. If this is true, then perhaps photographic memory is not such a mysterious thing after all.

The new study was done by scientists at the Scripps Research Institute in Florida who were working with fruit flies. They found an interesting way in which these fruit flies learn, which may overturn a cherished idea of how memories are formed and forgotten. The fruit flies were exposed to different smells and were given positive reinforcement (with food) or negative reinforcement (with electric shocks).

The scientists knew that the neurotransmitter dopamine was important to forming memories. To their surprise, they found that dopamine actively regulates *both* the formation and the forgetting of new memories. In the process of creating new memories, the dCA1 receptor was activated. By contrast, forgetting was initiated by the activation of the DAMB receptor.

Previously, it was thought that forgetting might be simply the degradation of memories with time, which happens passively by itself. This new study shows that forgetting is an active process, requiring intervention by dopamine.

To prove their point, they showed that by interfering with the action of the dCA1 and DAMB receptors, they could, at will, increase or decrease the ability of fruit flies to remember and forget. A mutation in the dCA1 receptor, for example, impaired the ability of the fruit flies to remember. A mutation in the DAMB receptor decreased their ability to forget.

The researchers speculate that this effect, in turn, may be partially responsible for savants' skills. Perhaps there is a deficiency in their ability to forget. One of the graduate students involved in the study, Jacob Berry, says, "Savants have a high capacity for memory. But maybe it isn't memory that gives them this capacity; maybe they have a bad forgetting mechanism. This might also be the strategy for developing drugs to promote cognition and memory—what about drugs that inhibit forgetting as a cognitive enhancers?"

Assuming that this result holds up in human experiments as well, it could encourage scientists to develop new drugs and neurotransmitters that are able to dampen the forgetting process. One might thus be able to selectively turn on photographic memories when needed by neutralizing the forgetting process. In this way, we wouldn't have the continuous overflow of extraneous, useless information, which hinders the thinking of people with savant syndrome.

What is also exciting is the possibility that the BRAIN project, which is being championed by the Obama administration, might be able to identify the specific pathways involved with acquired savant syndrome. Transcranial magnetic fields are still too crude to pin down the handful of neurons that may be involved. But using nanoprobes and the latest in scanning technologies, the BRAIN project might be able to isolate the precise neural pathways that make possible photographic memory and incredible computational, artistic, and musical skills. Billions of research dollars will be channeled into identifying the specific neural pathways involved with mental disease and other afflictions of the brain, and the secret of savant skills may be revealed in the process. Then it might be possible to take normal individuals and make savants out of them. This has happened many times in the past because of random accidents. In the future, this may become a precise medical process. Time will tell.

So far, the methods analyzed here do not alter the nature of the brain or the body. The hope is that through the use of magnetic fields, we will be able to unleash the potential that already exists in our brains but is latent. The philosophy underlying this idea is that we are all savants waiting to happen, and it will just take some slight alteration of our neural circuits to unleash this hidden talent.

Yet another tactic is to directly alter the brain and the genes, using the latest in brain science and also genetics. One promising method is to use stem cells.

STEM CELLS FOR THE BRAIN

It was dogma for many decades that brain cells do not regenerate. It seemed impossible that you could repair old, dying brain cells, or grow new ones to boost your abilities, but all this changed in 1998. That year, it was discovered that adult stem cells could be found in the hippocampus, the olfactory bulb, and the caudate nucleus. In brief, stem cells are the "mother of all cells." Embryonic stem cells, for instance, can readily develop into any other cell.

Although each of our cells contains all the genetic material necessary to construct a human being, only embryonic stem cells have the ability to actually differentiate into any type of cell in the body.

Adult stem cells have lost that chameleon-like ability, but they can still reproduce and replace old, dying cells. As far as memory enhancement goes, interest has focused on adult stem cells in the hippocampus. It turns out that thousands of new hippocampus cells are born naturally each day, but most die soon afterward. However, it was shown that rats that learned new skills retained more of their new cells. A combination of exercise and mood-elevating chemicals can also boost the survival rate of new hippocampus cells. It turns out that stress, on the contrary, accelerates the death of new neurons.

In 2007, a breakthrough occurred when scientists in Wisconsin and Japan were able to take ordinary human skin cells, reprogram their genes, and turn them into stem cells. The hope is that these stem cells, either found naturally or converted using genetic engineering, can one day be injected into the brains of Alzheimer's patients to replace dying cells. (These new brain cells, because they do not yet have the proper connections, would not be integrated into the brain's neural architecture. This means that a person would have to relearn certain skills to incorporate these fresh new neurons.)

Stem cell research is naturally one of the most active areas in brain research. "Stem cell research and regenerative medicine are in an extremely exciting phase right now. We are gaining knowledge very fast and many companies are being formed and are starting clinical trials in different areas," says Sweden's Jonas Frisén of the Karolinska Institute.

GENETICS OF INTELLIGENCE

In addition to stem cells, another avenue of exploration involves isolating the genes responsible for human intelligence. Biologists note that we are about 98.5 percent genetically identical to a chimpanzee, yet we live twice as long and have exploded in intellectual skills in the past six million years. So among a handful of genes there must be the ones responsible for giving us the human brain. Within a few years, scientists will have a complete map of all these genetic differences, and the secret to human longevity and enhanced intelligence may be found within this tiny set. Scientists have focused on a few genes that possibly drove the evolution of the human brain.

So perhaps the clue to revealing the secret of intelligence lies in our understanding of our apelike ancestors. This raises another question: Can this research make possible the *Planet of the Apes*?

In this long-running series of movies, a nuclear war destroys modern civilization. Humanity is reduced to barbarism, but the radiation somehow accelerates the evolution of the other primates, so that they become the dominant species on the planet. They create an advanced civilization, while humans are reduced to scruffy, smelly savages roaming half naked in the forest. At best, humans become zoo animals. The tables have turned on the humans, so the apes gawk at us outside the bars of our cages.

In the latest installment, *The Rise of the Planet of the Apes*, scientists are looking for a cure for Alzheimer's disease. Along the way, they stumble on a virus that has the unintended consequence of increasing a chimpanzee's intelligence. Unfortunately, one of these enhanced apes is treated cruelly when placed in a shelter for primates. Using his increased intelligence, the ape breaks free, infects the other lab animals with the virus to raise their intelligence, and then frees all of them from their cages. Soon a caravan of shouting, intelligent apes runs amok on the Golden Gate Bridge, completely overwhelming local and state police. After a spectacular, harrowing confrontation with the authorities, the movie ends with the apes peacefully finding refuge in a redwood forest north of the bridge.

Is such a scenario realistic? In the short term, no, but it can't be ruled out in the future, since scientists in the coming years should be able to catalog all the genetic changes that created *Homo sapiens*. But many more mysteries have to be solved before we have intelligent apes.

One scientist who has been fascinated not by science fiction, but by the genetics of what makes us "human," is Dr. Katherine Pollard, an expert in a field called "bioinformatics," which barely existed a decade ago. In this field of biology, instead of cutting open animals to understand how they are put together, researchers use the vast power of computers to mathematically analyze the genes in animals' bodies. She has been at the forefront of finding the genes that define the essence of what separates us from the apes. Back in 2003, as a freshly minted Ph.D. from the University of California at Berkeley, she got her chance.

"I jumped at the opportunity to join the international team that was identifying the sequence of DNA bases, or 'letters,' in the genome of the common chimpanzee," she recalled. Her goal was clear. She knew that only fifteen million base pairs, or "letters," that make up our genome (out of three billion base pairs) separate us from the chimps, our closest genetic neighbor. (Each "letter" in our genetic code refers to a nucleic acid, of which there are four, labeled A,T,C, and G. So our genome consists of three billion letters, arranged like ATTCCAGGG. . . .)

"I was determined to find them," she wrote.

Isolating these genes could have enormous implications for our future. Once we know the genes that gave rise to *Homo sapiens,* it becomes possible to determine how humans evolved. The secret of intelligence might lie in these genes. It might even be possible to accelerate the path taken by evolution and even enhance our intelligence. But even fifteen million base pairs is a huge number to analyze. How can you find a handful of genetic needles out of this genetic haystack?

Dr. Pollard knew that most of our genome is made of "junk DNA" that does not contain any genes and was largely unaffected by evolution. This junk DNA slowly mutates at a known rate (roughly 1 percent of it changes over four million years). Since we differ from the chimps in our DNA by 1.5 percent, this means that we probably separated from the chimpanzees about six million years ago. Hence there is a "molecular clock" in each of our cells. And since evolution accelerates this mutation rate, analyzing where this acceleration took place allows you to tell which genes are driving evolution.

Dr. Pollard reasoned that if she could write a computer program that could find where most of these accelerated changes are located in our genome, she could isolate precisely the genes that gave birth to *Homo sapiens.* After months of hard work and debugging, she finally placed her program into the giant computers located at the University of California at Santa Cruz. Anxiously she awaited the results.

When the computer printout finally arrived, it showed what she was looking for: there are 201 regions of our genome showing accelerated change. But the first one on her list caught her attention.

"With my mentor David Haussler leaning over my shoulder, I looked at the top hit, a stretch of 118 bases that together became known as human accelerated region 1 (HAR1)," she recalled.

She was ecstatic. Bingo!

"We had hit the jackpot," she would write. It was a dream come true.

She was staring at an area of our genome containing only 118 base pairs, with the largest divergence of mutations separating us from the apes. Of these base pairs, only eighteen mutations were altered since we became human. Her remarkable discovery showed that a small handful of mutations could be responsible for raising us from the swamp of our genetic past.

Next she and her colleagues tried to decipher the precise nature of this mysterious cluster called HAR1. They found that HAR1 was remarkably stable across millions of years of evolution. Primates separated from chickens about three hundred million years ago, yet only two base pairs differ between chimps and chickens. So HAR1 was virtually unchanged for several hundred million years, with only two changes, in the letters G and C. Yet in just six million years, HAR1 mutated eighteen times, representing a huge acceleration in our evolution.

But what was more intriguing was the role HAR1 played in controlling the overall layout of the cerebral cortex, which is famous for its wrinkled appearance. A defect in the HAR1 region causes a disorder called "lissencephaly," or "smooth brain," causing the cortex to fold incorrectly. (Defects in this region are also linked to schizophrenia.) Besides the large size of our cerebral cortex, one of its main characteristics is that it is highly wrinkled and convoluted, vastly increasing its surface area and hence its computational power. Dr. Pollard's work showed that changing just eighteen letters in our genome was partially responsible for one of the major, defining genetic changes in human history, vastly increasing our intelligence. (Recall that the brain of Carl Friedrich Gauss, one of the greatest mathematicians in history, was preserved after his death and showed unusual wrinkling.)

Dr. Pollard's list went even further and identified a few hundred other areas that also showed accelerated change, some of which were already known. FOX2, for example, is crucial for the development of speech, another key characteristic of humans. (Individuals with a defective FOX2 gene have difficulty making the facial movements necessary for speech.) Another region called HAR2 gives our fingers the dexterity required to manipulate delicate tools.

Furthermore, since the genome of the Neanderthal has been sequenced, it is possible to compare our genetic makeup with a species even closer to us than the chimpanzees. (When analyzing the FOX2 gene in Neanderthals, scientists found that we shared the same gene with them. This means that there is a possibility that the Neanderthal could vocalize and create speech, as we do.)

Another crucial gene is called ASPM, which is thought to be responsible for the explosive growth of our brain capacity. Some scientists believe that this and other genes may reveal why humans became intelligent but the apes did not. (People with a defective version of the ASPM gene often suffer from microcephaly, a severe form of mental retardation, because they have a tiny skull, about the size of one of our ancestors, Australopithecus.)

Scientists have tracked the number of mutations within the ASPM gene and found that it has mutated about fifteen times in the last five to six million years, since we separated from the chimpanzee. More recent mutations in these genes seem to be correlated with milestones in our evolution. For example, one mutation occurred over one hundred thousand years ago, when modern humans emerged in Africa, indistinguishable in appearance from us. And the last mutation was 5,800 years ago, which coincides with the introduction of the written language and agriculture.

Because these mutations coincide with periods of rapid growth in intellect, it is tantalizing to speculate that ASPM is among the handful of genes responsible for our increased intelligence. If this is true, then perhaps we can determine whether these genes are still active today, and whether they will continue to shape human evolution into the future.

All this research raises a question: Can manipulating a handful of genes increase our intelligence?

Quite possibly.

Scientists are rapidly determining the precise mechanism by which these genes gave rise to intelligence. In particular, genetic regions and genes like HAR1 and ASPM could help solve a mystery concerning the brain. If there are roughly twenty-three thousand genes in your genome, then how can they possibly control the connections linking one hundred billion neurons, containing a quadrillion total connections (1 with fifteen zeros after it)? It seems mathematically impossible. The human genome is about a trillion times too small to code for all our neural connections. So our very existence seems to be a mathematical impossibility.

The answer may be that nature takes numerous shortcuts in creating the brain. First, many neurons are connected randomly, so that a detailed blueprint is not necessary, which means that these randomly connected regions organize themselves after a baby is born and starts to interact with the environment.

And second, nature also uses modules that repeat themselves over and over again. Once nature discovers something useful, she often repeats it. This may explain why only a handful of genetic changes are responsible for most of our explosive growth in intelligence in the last six million years.

Size does matter in this case, then. If we tweak the ASPM and a few other genes, the brain might become larger and more complex, thereby making it possible to increase our intelligence. (Increasing our brain size is not sufficient to do this, since how the brain is organized is also crucially important. But increasing the gray matter of our brain is a necessary precondition to increasing our intelligence.)

APES, GENES, AND GENIUS

Dr. Pollard's research focused on areas of our genome that we share with the chimpanzees but that are mutated. It is also possible that there are areas in our genome found only in humans, independent of the apes. One such gene was discovered recently, in November 2012. Scientists, led by a team at the University of Edinburgh, isolated the RIM-941 gene, which is the only gene ever discovered that is found strictly in *Homo sapiens* and not in other primates. Also, geneticists can show that the gene emerged between one and six million years ago (after the time when humans and chimpanzees split about six million years ago).

Unfortunately, this discovery also set off a huge firestorm in science newsletters and blogs as misleading headlines blared across the Internet. Breathless articles appeared claiming that scientists had found a single gene that could, in principle, make chimpanzees intelligent. The essence of "humanness" had finally been isolated at the genetic level, the headlines shouted.

Reputable scientists soon stepped in and tried to calm things down. In all likelihood, a series of genes, acting together in complex ways, is responsible for human intelligence. No single gene can make a chimp suddenly have human intelligence, they said.

Although these headlines were highly exaggerated, they did raise a serious question: How realistic is *Planet of the Apes*?

There are a series of complications. If the HAR1 and ASPM genes are tweaked so that the size and structure of the chimp brain suddenly expand, then a series of other genes would have to be modified as well. First, you would have to strengthen the chimp's neck muscles and increase its body size to support the larger head. But a large brain would be useless unless it could control fingers capable of exploiting tools. So the HAR2 gene would also have to be altered to increase their dexterity. But since chimps often walk on their hands, another gene would have to be altered so that the backbone would straighten out and an upright posture would free up the hands. Intelligence is also useless unless chimps can communicate with other members of the species. So the FOX2 gene would also have to be mutated so that humanlike speech would become possible. And lastly, if you want to create a species of intelligent apes, you would have to modify the birth canal, since it is not large enough to accommodate the larger skull. You could either perform caesarians to cut the fetus out or genetically alter the birth canal of the chimps to accommodate the larger brain.

After all these necessary genetic adjustments, we are left with a creature that would look very much like us. In other words, it may be anatomically impossible to create intelligent apes, as in the movies, without their also mutating into something closely resembling human beings.

Clearly, creating intelligent apes is no simple matter, then. The intelligent apes we see in Hollywood movies are actually monkey suits with humans inside, or are computer-generated graphics, so all these issues are conveniently brushed under the rug. But if scientists could seriously use gene therapy to create intelligent apes, then they might closely resemble us, with hands that can use tools, vocal cords that can create speech, backbones that can support an upright posture, and large neck muscles to support large heads, as we have.

All this raises ethical issues as well. Although society may allow genetic studies of apes, it may not tolerate the manipulation of intelligent creatures that can feel pain and distress. These creatures, after all, would be intelligent and articulate enough to complain about their situation and their fate, and their views would be heard in society.

Not surprisingly, this area of bioethics is so new that it is totally unexplored. The technology is not yet ready, but in the coming decades, as we identify all the genes and their functions that separate us from the apes, the treatment of these enhanced animals could become a key question.

We can see, therefore, that it is only a matter of time before all the tiny genetic differences between us and the chimpanzees are carefully sequenced, analyzed, and interpreted. But this still does not explain a deeper question: What were the evolutionary forces that gave us this genetic heritage after we separated from the apes? Why did genes like ASPM, HAR1, and FOX2 develop in the first place? In other words, genetics gives us the ability to understand how we became intelligent, but it does not explain why this happened.

If we can understand this issue, it might provide clues as to how we might evolve in the future. This takes us to the heart of the ongoing debate: What is the origin of intelligence?

THE ORIGIN OF INTELLIGENCE

Many theories have been proposed as to why humans developed greater intelligence, going all the way back to Charles Darwin.

According to one theory, the evolution of the human brain probably took place in stages, with the earliest phase initiated by climate change in Africa. As the weather cooled, the forests began to recede, forcing our ancestors onto the open plains and savannahs, where they were exposed to predators and the elements. To survive in this new, hostile environment, they were forced to hunt and walk upright, which freed up their hands and opposable thumbs to use tools. This in turn put a premium on a larger brain to coordinate tool making. According to this theory, ancient man did not simply make tools—"tools made man."

Our ancestors did not suddenly pick up tools and become intelligent. It was the other way around. Those humans who picked up tools could survive in the grasslands, while those who did not gradually died off. The humans who then survived and thrived in the grasslands were those who, through mutations, became increasingly adept at tool making, which required an increasingly larger brain.

Another theory places a premium on our social, collective nature. Humans can easily coordinate the behavior of over a hundred other individuals involved in hunting, farming, warring, and building, groups that are much larger than those found in other primates, which gave humans an advantage over other animals. It takes a larger brain, according to this theory, to be able to assess and control the behavior of so many individuals. (The flip side of this theory is that it took a larger brain to scheme, plot, deceive, and manipulate other intelligent beings in your tribe. Individuals who could understand the motives of others and then exploit them would have an advantage over those who could not. This is the Machiavellian theory of intelligence.)

Another theory maintains that the development of language, which came later, helped accelerate the rise of intelligence. With language comes abstract thought and the ability to plan, organize society, create maps, etc. Humans have an extensive vocabulary unmatched by any other animal, with words numbering in the tens of thousands for an average person. With language, humans could coordinate and focus the activities of scores of individuals, as well as manipulate abstract concepts and ideas. Language meant you could manage teams of people on a hunt, which is a great advantage when pursuing the woolly mammoth. It meant you could tell others where game was plentiful or where danger lurked.

Yet another theory is "sexual selection," the idea that females prefer to mate with intelligent males. In the animal kingdom, such as in a wolf pack, the alpha male holds the pack together by brute force. Any challenger to the alpha male has to be soundly beaten back by tooth and claw. But millions of years ago, as humans became gradually more intelligent, strength alone could not keep the tribe together. Anyone with cunning and intelligence could ambush, lie or cheat, or form factions within the tribe to take down the alpha male. Hence the new generation of alpha males would not necessarily be the strongest. Over time, the leader would become the most intelligent and cunning. This is probably the reason why females choose smart males (not necessarily nerdy smart, but "quarterback smart"). Sexual selection in turn accelerated our evolution to become intelligent. So in this case the engine that drove the expansion of our brain would be females who chose men who could strategize, become leaders of the tribe, and outwit other males, which requires a large brain.

These are just a few of the theories about the origin of intelligence, and each has its pros and cons. The common theme seems to be the ability to simulate the future. For example, the purpose of the leader is to choose the correct path for the tribe in the future. This means any leader has to understand the intentions of others in order to plan strategy for the future. Hence simulating the future was perhaps one of the driving forces behind the evolution of our large brain and intelligence. And the person who can best simulate the future is the one who can plot, scheme, read the minds of many of his fellow tribesmen, and win the arms race with his fellow man.

Similarly, language allows you to simulate the future. Animals possess a rudimentary language, but it is mainly in the present tense. Their language may warn them of an immediate threat, such as a predator hiding among the trees. However, animal language apparently has no future or past tense. Animals do not conjugate their verbs. So perhaps the ability to express the past and future tense was a key breakthrough in the development of intelligence.

Dr. Daniel Gilbert, a psychologist at Harvard, writes, "For the first few hundred million years after their initial appearance on our planet, our brains were stuck in the permanent present, and most brains still are today. But not yours and not mine, because two or three million years ago our ancestors began a great escape from the here and now. . . ."

THE FUTURE OF EVOLUTION

So far, we have seen that there are intriguing results indicating that one can increase one's memory and intelligence, largely by making the brain more efficient and maximizing its natural capacity. A variety of methods are being studied, such as certain drugs, genes, or devices (TES, for example) that might increase the capabilities of our neurons.

So the concept of altering the brain size and capacity of the apes is a distinct, though difficult, possibility. Gene therapy on this scale is still many decades away. But this raises another difficult question: How far can this go? Can one extend the intelligence of an organism indefinitely? Or is there a limit to brain modification imposed by the laws of physics?

Surprisingly, the answer is yes. The laws of physics put an upper limit to what can be done with genetic modification of the human brain, given certain restraints. To see this limit, it is instructive to first examine whether evolution is still increasing human intelligence, and then what can be done to accelerate this natural process.

In popular culture, there is the notion that evolution will give us big brains and small, hairless bodies in the future. Likewise, aliens from space, because they are supposed to possess a superior level of intelligence, are often portrayed in this fashion. Go to any novelty shop and you will see the same extraterrestrial face, with big bug eyes, a huge head, and green skin.

Actually, there are indications that gross human evolution (i.e., our basic body shape and intelligence) has largely come to a halt. There are several factors supporting this. First of all, since we are bipedal mammals who walk upright, there are limitations to the maximum size of an infant's skull that can pass through the birth canal. Second, the rise of modern technology has removed many of the harsh evolutionary pressures faced by our ancestors.

However, evolution on a genetic and molecular basis continues unabated. Although it's difficult to see with the naked eye, there is evidence that human biochemistry has changed to adjust to environmental challenges, such as combating malaria in tropical areas. Also, humans recently evolved enzymes to digest lactose sugar as we learned to domesticate cows and drink milk. Mutations have occurred as humans adjusted to a diet created by the agricultural revolution. Moreover, people still choose to mate with others who are healthy and fit, and so evolution continues to eliminate unsuitable genes at this level. None of these mutations, however, has changed our basic body plan or increased our brain size. (Modern technology is also influencing our evolution to some degree. For example, there is no longer any selection pressure on nearsighted people, since anyone today can be outfitted with glasses or contact lenses.)

PHYSICS OF THE BRAIN

So from an evolutionary and biological point of view, evolution is no longer selecting for more intelligent people, at least not as rapidly as it did thousands of years ago.

There are also indications from the laws of physics that we have reached the maximum natural limit of intelligence, so that any enhancement of our intelligence would have to come from external means. Physicists who have studied the neurology of the brain conclude that there are trade-offs preventing us from getting much smarter. Every time we envision a brain that is larger, or denser, or more complex, we bump up against these negative trade-offs.

The first principle of physics that we can apply to the brain is the conservation of matter and energy; that is, the law stating that the total amount of matter and energy in a system remains constant. In particular, in order to carry out its incredible feats of mental gymnastics, the brain has to conserve energy, and hence it takes many shortcuts. As we saw in Chapter 1, what we see with our eyes is actually cobbled together using energy-saving tricks. It would take too much time and energy for a thoughtful analysis of every crisis, so the brain saves energy by making snap judgments in the form of emotions. Forgetting is an alternative way of saving energy. The conscious brain has access to only a tiny portion of the memories that have an impact on the brain.

So the question is: Would increased brain size or density of neurons give us more intelligence?

Probably not. "Cortical gray matter neurons are working with axons that are pretty close to the physical limit," says Dr. Simon Laughlin of Cambridge University. There are several ways in which one can increase the intelligence of the brain using the laws of physics, but each has its own problems:

- One can increase brain size and extend the length of neurons. The problem here is that the brain now consumes more energy. This generates more heat in the process, which is detrimental to our survival. If the brain uses up more energy, it gets hotter, and tissue damage results if the body temperature becomes too high. (The chemical reactions of the human body and our metabolism require temperatures to be in a precise range.) Also, longer neurons means that it takes longer for signals to go across the brain, which slows down the thinking process.
- One can pack more neurons into the same space by making them thinner. But if neurons become thinner and thinner, the complex chemical/electrical reactions that must take place inside the axons fail, and eventually they begin to misfire more easily. Douglas Fox, writing in *Scientific American,* says, "You might call it the mother of all limitations: the proteins that neurons use to generate electrical pulses, called ion channels, are inherently unstable."
- One can increase the speed of the signal by making the neurons thicker. But this also increases energy consumption and generates more heat. It also increases the size of the brain, which increases the time it takes for the signals to reach their destination.
- One can add more connections between neurons. But this again increases energy consumption and heat generation, making the brain larger and slower in the process.

So each time we tinker with the brain, we are checkmated. The laws of physics seem to indicate that we have maxed out the intelligence that we humans can attain in this way. Unless we can suddenly increase the size of our skulls or the very nature of neurons in our brains, it seems we are at the maximum level of intelligence. If we are to increase our intelligence, it has to be done by making our brains more efficient (via drugs, genes, and possibly TES-type machines).

PARTING THOUGHTS

In summary, it may be possible in the coming decades to use a combination of gene therapy, drugs, and magnetic devices to increase our intelligence. There are several avenues of exploration that are revealing the secrets of intelligence and how it may be modified or enhanced. But what would it do to society, though, if we could enhance our intelligence and get a "brain boost"? Ethicists have seriously contemplated this question, since the basic science is growing so rapidly. The big fear is that society may bifurcate, with only the rich and powerful having access to this technology, which they could use to further solidify their exalted position in society. Meanwhile, the poor won't have access to additional brain power, making it more difficult to move up in society.

This is certainly a valid concern, but it flies in the face of the history of technology. Many of the technologies of the past were indeed initially the province of the rich and powerful, but eventually mass production, competition, better transportation, and improvements in technology drove down the costs, so the average person could afford them. (For example, we take for granted that we eat foods for breakfast that the king of England could not have procured a century ago. Technology has made it possible to purchase delicacies from around the world at any supermarket that would be the envy of the aristocrats of the Victorian era.) So if it becomes possible to increase our intelligence, the price of this technology will gradually fall. Technology is never the monopoly of the privileged rich. Sooner or later ingenuity, hard work, and simple market forces will drive down its cost.

There is also the fear that the human race will split into those who want their intelligence to be boosted and those who prefer to remain the same, resulting in the nightmare of having a class of super-intelligent brahmins lord over the masses of the less gifted.

But again, perhaps the fear of boosting intelligence has been exaggerated. The average person has absolutely no interest in being able to solve the complex tensor equations for a black hole. The average person sees nothing to gain by mastering the mathematics of hyperspatial dimensions or the physics of the quantum theory. On the contrary, the average person may find such activities rather boring and useless. So most of us are not going to become mathematical geniuses if given the opportunity, because it is not in our character, and we see nothing to gain from it.

Keep in mind that society already has a class of accomplished mathematicians and physicists, and they are paid significantly less than ordinary businessmen and wield much less power than average politicians. Being super smart does not guarantee financial success in life. In fact, being super smart may actually pigeonhole you in the lower rungs of a society that values athletes, movie stars, comedians, and entertainers more.

No one ever got rich doing relativity.

Also, a lot depends on precisely which traits are enhanced. There are other forms of intelligence besides using mathematics. (Some argue that intelligence must include artistic genius as well. In this case, one can conceivably use this talent to make a comfortable living.)

Anxious parents of high school children may want to boost the IQ of their kids as they prepare for standardized exams. But IQ, as we have seen, does not necessarily correspond to success in life. Likewise, people may want to enhance their memory, but, as we have seen with savants, having a photographic memory can be a blessing as well as a curse. And in both cases, enhancement is unlikely to contribute to a society splitting in two.

Society as a whole, however, may benefit from this technology. Workers with an enhanced intelligence would be better prepared to face an ever-changing job market. Retraining workers for the jobs of the future would be less of a drain on society. Furthermore, the public will be able to make informed decisions about major technological issues of the future (e.g., climate change, nuclear energy, space exploration) because they will grasp these complex issues better.

Also, this technology may help even out the playing field. Children today who go to exclusive private schools and have personal tutors are better prepared for the job market because they have more opportunities to master difficult materials. But if everyone has had their intelligence enhanced, the fault lines within society will be evened out. Then how far someone goes in life would be more related to their drive, ambition, imagination, and resourcefulness rather than to being born with a silver spoon in their mouth.

In addition, raising our intelligence may help speed up technological innovation. Increased intelligence would mean a greater ability to simulate the future, which would be invaluable in making scientific discoveries. Often, science stagnates in certain areas because of a lack of fresh new ideas to stimulate new avenues of research. Having an ability to simulate different possible futures would vastly increase the rate of scientific breakthroughs.

These scientific discoveries, in turn, could generate new industries, which could enrich all of society, creating new markets, new jobs, and new opportunities. History is full of technological breakthroughs creating entirely new industries that benefited not just the few, but all of society (think of the transistor and the laser, which today form the foundation of the world economy).

However, in science fiction, there is the recurring theme of the super criminal, who uses his superior brain power to embark on a crime spree and thwart the superhero. Every Superman has his Lex Luthor, every Spider-Man has his Green Goblin. Although it is certainly possible that a criminal mind will use a brain booster to create super weapons and plan the crime of the century, realize that members of the police force can also have their intelligence boosted to outwit the evil mastermind. So super criminals are dangerous only if they are the only ones in possession of enhanced intelligence.

So far, we have examined the possibility that we can enhance or alter our mental capabilities via telepathy, telekinesis, uploading memories, or brain boosts. Such enhancement basically means modifying and augmenting the mental capabilities of our consciousness. This tacitly assumes that our normal consciousness is the only one, but I'd like to explore whether there are different forms of consciousness. If so, there could be other ways of thinking that lead to totally different outcomes and consequences. Within our own thoughts, there are altered states of consciousness, such as dreams, drug-induced hallucinations, and mental illness. There is also nonhuman consciousness, the consciousness of robots, and even that of aliens from outer space. We have to give up the chauvinistic notion that our human consciousness is the only one. There is more than one way to create a model of our world, and more than one way to simulate its future.

Dreams, for example, are one of the most ancient forms of consciousness and were studied by the ancients, yet very little progress has been made in understanding them until recently. Perhaps dreams are not silly, random events spliced together by the sleeping brain but phenomena that may give insight into the meaning of consciousness. Dreams may be a key to understanding altered states of consciousness.

BOOK III ALTERED CONSCIOUSNESS

The future belongs to those who believe in the beauty of their dreams.

—ELEANOR ROOSEVELT

7 IN YOUR DREAMS

Dreams can determine destiny.

Perhaps the most famous dream in antiquity took place in the year A.D. 312, when the Roman emperor Constantine engaged in one of the greatest battles of his life. Faced with a rival army twice the size of his own, he realized that he probably would die in battle the next day. But in a dream he had that night, an angel appeared before him bearing the image of a cross, uttering the fateful words "By this symbol, you shall conquer." Immediately he ordered the shields of his troops adorned with the symbol of the cross.

History records that he emerged triumphant the next day, cementing his hold on the Roman Empire. He vowed to repay the blood debt to this relatively obscure religion, Christianity, that had been persecuted for centuries by previous Roman emperors and whose adherents were regularly fed to the lions in the Colosseum. He signed laws that would eventually pave the way for it to become an official religion of one of the greatest empires in the world.

For thousands of years, kings and queens, as well as beggars and thieves, have all wondered about dreams. The ancients considered dreams to be omens about the future, so there have been countless attempts throughout history to interpret them. The Bible records in Genesis 41 the rise of Joseph, who was able to correctly interpret the dreams of the Pharaoh of Egypt thousands of years ago. When the Pharaoh dreamed about seven fat cows, followed by seven lean cows, he was so disturbed by the imagery that he asked scribes and mystics throughout the kingdom to find its meaning. All failed to give a convincing explanation, until Joseph finally interpreted the dream to mean that Egypt would have seven years of good harvests, followed by seven years of drought and famine. So, said Joseph, Egypt must begin stockpiling grain and supplies now, in preparation for the coming years of want and desperation. When this came to pass, Joseph was considered to be a prophet.

Dreams have long been associated with prophesy, but in more recent times they've also been known to stimulate scientific discovery. The idea that neurotransmitters could facilitate the movement of information past a synapse, which forms the foundation of neuroscience, came to pharmacologist Otto Loewi in a dream. Similarly, in 1865, August Kekulé had a dream about benzene, in which the bonds of carbon atoms formed a chain that eventually wrapped around and finally formed a circle, just like a snake biting its tail. This dream would unlock the atomic structure of the benzene molecule. He concluded, "Let us learn to dream!"

Dreams have also been interpreted as a window onto our true thoughts and intentions. The great Renaissance writer and essayist Michel de Montaigne once wrote, "I believe it to be true that dreams are the true interpretations of our inclinations, but there is art required to sort and understand them." More recently, Sigmund Freud proposed a theory to explain the origin of dreams. In his signature work, *The Interpretation of Dreams,* he claimed that they were manifestations of our subconscious desires, which were often repressed by the waking mind but which run wild every night. Dreams were not just the random figments of our overheated imaginations but could actually uncover deep secrets and truths about ourselves. "Dreams are the royal road to the unconscious," he wrote. Since then, people have amassed huge encyclopedias that claim to reveal the hidden meaning behind every disturbing image in terms of Freudian theory.

Hollywood takes advantage of our continuing fascination with dreams. A favorite scene in many movies is when the hero experiences a terrifying dream sequence and then suddenly wakes up from the nightmare in a cold sweat. In the blockbuster movie *Inception,* Leonardo DiCaprio plays a petty thief who steals intimate secrets from the most unlikely of all places, people's dreams. With a new invention, he is able to enter people's dreams and deceive them into giving up their financial secrets. Corporations spend millions of dollars protecting industrial secrets and patents. Billionaires jealously guard their wealth using elaborate codes. His job is to steal them. The plot quickly escalates as the characters enter dreams in which a person falls asleep and dreams again. So these criminals descend deeper and deeper into multiple layers of the subconscious.

But although dreams have always haunted and mystified us, only in the last decade or so have scientists been able to peel away the mysteries of dreams. In fact, scientists can now do something once considered impossible: they are able to take rough photographs and videotapes of dreams with MRI machines. One day, you may be able to view a video of the dream you had the previous night and gain insight into your own subconscious mind. With proper training, you might be able to consciously control the nature of your dreams. And perhaps, like DiCaprio's character, with advanced technology you might even be able to enter someone else's dream.

THE NATURE OF DREAMS

As mysterious as they are, dreams are not a superfluous luxury, the useless ruminations of the idle brain. Dreams, in fact, are essential for survival. Using brain scans, it is possible to show that certain animals exhibit dreamlike brain activity. If deprived of dreams, these animals would often die faster than they would by starvation, because such deprivation severely disrupts their metabolism. Unfortunately, science does not know exactly why this is the case.

Dreaming is an essential feature of our sleep cycle as well. We spend roughly two hours a night dreaming when we sleep, with each dream lasting five to twenty minutes. In fact, we spend about six years dreaming during an average lifetime.

Dreams are also universal across the human race. Looking across different cultures, scientists find common themes in dreams. Fifty thousand dreams were recorded over a forty-year time period by psychology professor Calvin Hall. He followed this up with one thousand dream reports from college students. Not surprisingly, he found that most people dreamed of the same things, such as personal experiences from the previous days or week.

(However, animals apparently dream differently than we do. In the dolphin, for example, only one hemisphere at a time sleeps in order to prevent drowning, because they are air-breathing mammals, not fish. So if they dream, it is probably in only one hemisphere at a time.)

The brain, as we have seen, is not a digital computer, but rather a neural network of some sort that constantly rewires itself after learning new tasks. Scientists who work with neural networks noticed something interesting, though. Often these systems would become saturated after learning too much, and instead of processing more information they would enter a "dream" state, whereby random memories would sometimes drift and join together as the neural networks tried to digest all the new material. Dreams, then, might reflect "house cleaning," in which the brain tries to organize its memories in a more coherent way. (If this is true, then possibly all neural networks, including all organisms that can learn, might enter a dream state in order to sort out their memories. So dreams probably serve a purpose. Some scientists have speculated that this might imply that robots that learn from experience might also eventually dream as well.)

Neurological studies seem to back up this conclusion. Studies have shown that retaining memories can be improved by getting sufficient sleep between the time of activity and a test. Neuroimaging shows that the areas of the brain that are activated during sleep are the same as those involved in learning a new task. Dreaming is perhaps useful in consolidating this new information.

Also, some dreams can incorporate events that happened a few hours earlier, just before sleep. But dreams mostly incorporate memories that are a few days old. For example, experiments have shown that if you put rose-colored glasses on a person, it takes a few days before the dreams become rose-colored as well.

BRAIN SCANS OF DREAMS

Brain scans are now unveiling some of the mystery of dreams. Normally EEG scans show that the brain is emitting steady electromagnetic waves while we are awake. However, as we gradually fall asleep, our EEG signals begin to change frequency. When we finally dream, waves of electrical energy emanate from the brain stem that surge upward, rising into the cortical areas of the brain, especially the visual cortex. This confirms that visual images are an important component of dreams. Finally, we enter a dream state, and our brain waves are typified by rapid eye movements (REM). (Since some mammals also enter REM sleep, we can infer that they might dream as well.)

While the visual areas of the brain are active, other areas involved with smell, taste, and touch are largely shut down. Almost all the images and sensations processed by the body are self-generated, originating from the electromagnetic vibrations from our brain stem, not from external stimuli. The body is largely isolated from the outside world. Also, when we dream, we are more or less paralyzed. (Perhaps this paralysis is to prevent us from physically acting out our dreams, which could be disastrous. About 6 percent of people suffer from "sleep paralysis" disorder, in which they wake up from a dream still paralyzed. Often these individuals wake up frightened and believing that there are creatures pinning down their chest, arms, and legs. There are paintings from the Victorian era of women waking up with a terrifying goblin sitting on their chest glaring down at them. Some psychologists believe that sleep paralysis could explain the origin of the alien abduction syndrome.)

The hippocampus is active when we dream, suggesting that dreams draw upon our storehouse of memories. The amygdala and anterior cingulate are also active, meaning that dreams can be highly emotional, often involving fear.

But more revealing are the areas of the brain that are shut down, including the dorsolateral prefrontal cortex (which is the command center of the brain), the orbitofrontal cortex (which can act like a censor or fact-checker), and the temporoparietal region (which processes sensory motor signals and spatial awareness).

When the dorsolateral prefrontal cortex is shut down, we can't count on the rational, planning center of the brain. Instead, we drift aimlessly in our dreams, with the visual center giving us images without rational control. The orbitofrontal cortex, or the fact-checker, is also inactive. Hence dreams are allowed to blissfully evolve without any constraints from the laws of physics or common sense. And the temporoparietal lobe, which helps coordinate our sense of where we are located using signals from our eyes and inner ear, is also shut down, which may explain our out-of-body experiences while we dream.

As we have emphasized, human consciousness mainly represents the brain constantly creating models of the outside world and simulating them into the future. If so, then dreams represent an alternate way in which the future is simulated, one in which the laws of nature and social interactions are temporarily suspended.

HOW DO WE DREAM?

But that leaves open this question: What generates our dreams? One of the world's authorities on dreams is Dr. Allan Hobson, a psychiatrist at Harvard Medical School. He has devoted decades of his life to unveiling the secrets of dreams. He claims that dreams, especially REM sleep, can be studied at the neurological level, and that dreams arise when the brain tries to make sense of the largely random signals emanating from the brain stem.

When I interviewed him, he told me that after many decades of cataloging dreams, he found five basic characteristics:

1. Intense emotions—this is due to the activation of the amygdala, causing emotions such as fear.
2. Illogical content—dreams can rapidly shift from one scene to another, in defiance of logic.
3. Apparent sensory impressions—dreams give us false sensations that are internally generated.
4. Uncritical acceptance of dream events—we uncritically accept the illogical nature of the dream.
5. Difficulty in being remembered—dreams are soon forgotten, within minutes of waking up.

Dr. Hobson (with Dr. Robert McCarley) made history by proposing the first serious challenge to Freud's theory of dreams, called the "activation synthesis theory." In 1977, they proposed the idea that dreams originate from random neural firings in the brain stem, which travel up to the cortex, which then tries to make sense of these random signals.

The key to dreams lies in nodes found in the brain stem, the oldest part of the brain, which squirts out special chemicals, called adrenergics, that keep us alert. As we go to sleep, the brain stem activates another system, the cholinergic, which emits chemicals that put us in a dream state.

As we dream, cholinergic neurons in the brain stem begin to fire, setting off erratic pulses of electrical energy called PGO (pontine-geniculate-occipital) waves. These waves travel up the brain stem into the visual cortex, stimulating it to create dreams. Cells in the visual cortex begin to resonate hundreds of times per second in an irregular fashion, which is perhaps responsible for the sometimes incoherent nature of dreams.

This system also emits chemicals that decouple parts of the brain involved with reason and logic. The lack of checks coming from the prefrontal and orbitofrontal cortices, along with the brain becoming extremely sensitive to stray thoughts, may account for the bizarre, erratic nature of dreams.

Studies have shown that it is possible to enter the cholinergic state without sleep. Dr. Edgar Garcia-Rill of the University of Arkansas claims that meditation, worrying, or being placed in an isolation tank can induce this cholinergic state. Pilots and drivers facing the monotony of a blank windshield for many hours may also enter this state. In his research, he has found that schizophrenics have an unusually large number of cholinergic neurons in their brain stem, which may explain some of their hallucinations.

To make his studies more efficient, Dr. Allan Hobson had his subjects put on a special nightcap that can automatically record data during a dream. One sensor connected to the nightcap registers the movements of a person's head (because head movements usually occur when dreams end). Another sensor measures movements of the eyelids (because REM sleep causes eyelids to move). When his subjects wake up, they immediately record what they dreamed about, and the information from the nightcap is fed into a computer.

In this way, Dr. Hobson has accumulated a vast amount of information about dreams. So what is the meaning of dreams? I asked him. He dismisses what he calls the "mystique of fortune-cookie dream interpretation." He does not see any hidden message from the cosmos in dreams.

Instead, he believes that after the PGO waves surge from the brain stem into the cortical areas, the cortex is trying to make sense of these erratic signals and winds up creating a narrative out of them: a dream.

PHOTOGRAPHING A DREAM

In the past, most scientists avoided the study of dreams, since they are so subjective and have such a long historical association with mystics and psychics. But with MRI scans, dreams are now revealing their secrets. In fact, since the brain centers that control dreaming are nearly identical to the ones that control vision, it is therefore possible to photograph a dream. This pioneering work is being done in Kyoto, Japan, by scientists at the ATR Computational and Neuroscience Laboratories.

Subjects are first placed in an MRI machine and shown four hundred black-and-white images, each consisting of a set of dots within a ten-by-ten-pixel framework. One picture is flashed at a time, and the MRI records how the brain responds to each collection of pixels. As with other groups working in this field of BMI, the scientists eventually create an encyclopedia of images, with each image of pixels corresponding to a specific MRI pattern. Here the scientists are able to work backward, to correctly reconstruct self-generated images from MRI brain scans taken while the subject dreams.

ATR chief scientist Yukiyasu Kamitani says, "This technology can also be applied to senses other than vision. In the future, it may also be possible to read feelings and complicated emotional states." In fact, any mental state of the brain might be imaged in this way, including dreams, as long as a one-to-one map can be made between a certain mental state and an MRI scan.

The Kyoto scientists have concentrated on analyzing still photographs generated by the mind. In Chapter 3, we encountered a similar approach pioneered by Dr. Jack Gallant, in which the voxels from 3-D MRI scans of the brain can be used to reconstruct the actual image seen by the eye with the help of a complex formula. A similar process has allowed Dr. Gallant and his team to create a crude video of a dream. When I visited the laboratory in Berkeley, I talked to a postdoctoral staff member, Dr. Shinji Nishimoto, who allowed me to watch the video of one of his dreams, one of the first ever done. I saw a series of faces flickering across the computer screen, meaning that the subject (in this case Dr. Nishimoto himself) was dreaming of people, rather than animals or objects. This was amazing. Unfortunately, the technology is not yet good enough to see the precise facial features of the people appearing in his dream, so the next step is to increase the number of pixels so that more complex images can be identified. Another advance will be to reproduce images in color rather than black and white.

I then asked Dr. Nishimoto the crucial question: How do you know the video is accurate? How do you know that the machine isn't just making things up? He was a bit sheepish when he replied that this was a weak point in his research. Normally, you have only a few minutes after waking up to record a dream. After that, most dreams are lost in the fog of our consciousness, so it is not easy to verify the results.

Dr. Gallant told me that this research on videotaping dreams was still a work in progress, and that is why it's not ready for publication. There is still a ways to go before we can watch a videotape of last night's dream.

LUCID DREAMS

Scientists are also investigating a form of dreaming that was once thought to be a myth: lucid dreaming, or dreaming while you are conscious. This sounds like a contradiction in terms, but it has been verified in brain scans. In lucid dreaming, dreamers are aware that they are dreaming and can consciously control the direction of the dream. Although science has only recently begun to experiment with lucid dreaming, there are references to this phenomenon dating back centuries. In Buddhism, for example, there are books that refer to lucid dreamers and how to train yourself to become one. Over the centuries, several people in Europe have written detailed accounts of their lucid dreams.

Brain scans of lucid dreamers show that this phenomenon is real; during REM sleep, their dorsolateral prefrontal cortex, which is usually dormant when a normal person dreams, is active, indicating that the person is partially conscious while dreaming. In fact, the more lucid the dream, the more active the dorsolateral prefrontal cortex. Since the dorsolateral prefrontal cortex represents the conscious part of the brain, the dreamer must be aware while he or she is dreaming.

Dr. Hobson told me that anyone can learn to do lucid dreaming by practicing certain techniques. In particular, people who do lucid dreaming should keep a notebook of dreams. Before going to sleep, they should remind themselves that they will "wake up" in the middle of the dream and realize that they are moving in a dream world. It is important to have this frame of mind before hitting the pillow. Since the body is largely paralyzed during REM sleep, it is difficult for the dreaming person to send a signal to the outside world that he has entered a dream, but Dr. Stephen LaBerge at Stanford University has studied lucid dreamers (including himself) who can signal the outside world while dreaming.

In 2011, for the first time, scientists used MRI and EEG sensors to measure dream content and even make contact with a dreaming person. At the Max Planck Institute in Munich and Leipzig, scientists enlisted the help of lucid dreamers, who were fitted with EEG sensors on their heads to help the scientists determine the moment they entered REM sleep; they were then placed in an MRI machine. Before falling asleep, the dreamers agreed to initiate a set of eye movements and breathing patterns when dreaming, like a Morse code. They were told that once they started dreaming, they should clench their right fist and then their left one for ten seconds. That was the signal that they were dreaming.

The scientists found that, once the subjects entered their dream state, the sensorimotor cortex of the brain (responsible for controlling motor actions like clenching your fists) was activated. The MRI scans could pick up that the fists were being clenched and which fist was being clenched first. Then, using another sensor (a near-infrared spectrometer) they were able to confirm that there was increased brain activity in the region that controls the planning of movements.

"Our dreams are therefore not a 'sleep cinema' in which we merely observe an event passively, but involve activity in the regions of the brain that are relevant to the dream content," says Michael Czisch, a group leader at the Max Planck Institute.

ENTERING A DREAM

If we can communicate with a dreaming person, then is it also possible to alter someone's dream from the outside? Quite possibly.

First, as we have seen, scientists have already made the initial steps in videotaping a person's dream, and in the coming years, it should be possible to create much more accurate pictures and videos of dreams. Since scientists have already been able to establish a communication link between the real world and the lucid dreamer in the fantasy world, then, in principle, scientists should be able to deliberately alter the course of a dream. Let's say that scientists are viewing the video of a dream using an MRI machine as the dream unfolds in real time. As the person wanders around the dreamscape, the scientists can tell where he is going and give directions for him to move in different ways.

So in the near future, it might be possible to watch a video of a person's dream and actually influence its general direction. But in the movie *Inception,* Leonardo DiCaprio goes much further. He is able not only to watch another person's dream, but also to enter it. Is this possible?

We saw earlier that we are paralyzed when we dream so that we don't carry out our dream fantasies, which might be disastrous. However, when people are sleepwalking, they often have their eyes open (although their eyes look glazed over). So sleepwalkers live in a hybrid world, part real and part dream-like. There are many documented instances of people walking around their homes, driving cars, cutting wood, and even committing homicides while in this dream state, where reality and the fantasy world are mixed. Hence it is possible that physical images that the eye actually sees can freely interact with the fictitious images that the brain is concocting during a dream.

The way to enter someone's dream, then, might be to have the subject wear contact lenses that can project images directly onto their retinas. Already, prototypes of Internet contact lenses are being developed at the University of Washington in Seattle. So if the observer wanted to enter the subject's dream, first he would sit in a studio and have a video camera film him. His image could then be projected onto the contact lenses of the dreamer, creating a composite image (the image of the observer superimposed upon the imaginary image the brain is manufacturing).

The observer could actually see this dream world as he wanders around the dream, since he, too, would be wearing Internet contact lenses. The MRI image of the subject's dream, after it has been deciphered by computer, would be sent directly into the observer's contact lenses.

Furthermore, you could actually change the direction of the dream you have entered. As you walk around in the empty studio, you would see the dream unfold in your contact lens, so you could start to interact with the objects and people appearing in the dream. This would be quite an experience, since the background would change without warning, images would appear and disappear without reason, and the laws of physics would be suspended. Anything goes.

Further into the future, it might even be possible to enter another person's dream by directly connecting two sleeping brains. Each brain would have to be connected to MRI scanners that were connected to a central computer, which would merge the two dreams into a single one. The computer would first decipher each person's MRI scans into a video image. Then the dream of one person would be sent into the sensory areas of the other person's brain, so that the other dreamer's dream would merge with the first dreamer's dream. However, the technology of videotaping and interpreting dreams would have to become much more advanced before this could become a possibility.

But this raises another question: If it's possible to alter the course of someone's dream, is it possible to control not only that person's dream but that person's mind as well? During the Cold War, this became a serious issue as both the Soviet Union and the United States played a deadly game, trying to use psychological techniques to control other people's wills.

Minds are simply what brains do.

—MARVIN MINSKY

8 CAN THE MIND BE CONTROLLED?

A raging bull is released into an empty arena in Cordoba, Spain. For generations, this ferocious beast has been carefully bred to maximize its killer instinct. Then a Yale professor calmly enters the same arena. Rather than donning a tweed jacket, he is dressed like a dashing matador, wearing a bright golden jacket and waving a red cape defiantly in front of the bull, egging him on. Instead of running away in terror, the professor looks calm, confident, and even detached. To a bystander, it appears as if the professor has gone mad and wants to commit suicide.

Enraged, the bull locks onto the professor. Suddenly the bull charges, aiming his deadly horns at him. The professor does not run away in fear. Instead, he holds a small box in his hand. Then, in front of the cameras, he presses a button on the box, and the bull stops dead in his tracks. The professor is so confident of himself that he has risked his life to prove a point, that he has mastered the art of controlling the mind of a mad bull.

The Yale professor is Dr. José Delgado, who was years ahead of his time. He pioneered a series of remarkable but unsettling animal experiments in the 1960s, in which he put electrodes into their brains with the aim of trying to control their movement. To stop the bull, he inserted electrodes into the striatum of the basal ganglia at the base of the brain, which is involved with motor coordination.

He also did a series of other experiments on monkeys to see if he could rearrange their social hierarchy with the push of a button. After implanting electrodes into the caudate nucleus (a region associated with motor control) of the alpha male within the group, Delgado could reduce the aggressive tendencies of the leader on command. Without threats of retaliation, the delta males began to assert themselves, taking over the territory and privileges normally reserved for the alpha male. The alpha male, meanwhile, appeared to have lost interest in defending his territory.

Then Dr. Delgado pressed another button, and the alpha male instantly sprung back to normal, resuming his aggressive behavior and reestablishing his power as the king of the hill. The delta males scrambled in fear.

Dr. Delgado was the first person in history to show that it was possible to control the minds of animals in this way. The professor became the puppet master, pulling the strings of living puppets.

As expected, the scientific community looked at Dr. Delgado's work with unease. To make matters worse, he wrote a book in 1969 with the provocative title *Physical Control of the Mind: Toward a Psychocivilized Society.* It raised an unsettling question: If scientists like Dr. Delgado are pulling the strings, then who controls the puppet master?

Dr. Delgado's work puts into sharp focus the enormous promise and perils of this technology. In the hands of an unscrupulous dictator, this technology might be used to deceive and control his unfortunate subjects. But it can also be used to free millions of people who are trapped in mental illness, hounded by their hallucinations, or crushed by their anxieties. (Years later, Dr. Delgado was asked by a journalist why he initiated these controversial experiments. He said that he wanted to correct the horrendous abuses being suffered by the mentally ill. They often underwent radical lobotomies, in which the prefrontal cortex was scrambled by a knife resembling an ice pick, which was hammered into the brain above the eye socket. The results were often tragic, and some of the horrors were exposed in Ken Kesey's novel *One Flew Over the Cuckoo's Nest,* which was made into a movie with Jack Nicholson. Some patients became calm and relaxed, but many others became zombies: lethargic, indifferent to pain and feelings, and emotionally vacuous. The practice was so widespread that in 1949, Antonio Moniz won the Nobel Prize for perfecting the lobotomy. Ironically, in 1950, the Soviet Union banned this technology, stating that "it was contrary to the principles of humanity." Lobotomies, the Soviet Union charged, turned "an insane person into an idiot." In total, it is estimated that forty thousand lobotomies were performed in the United States alone over two decades.)

MIND CONTROL AND THE COLD WAR

Another reason for the chilly reception of Dr. Delgado's work was the political climate of the time. It was the height of the Cold War, with painful memories of captured U.S. soldiers being paraded in front of cameras during the Korean War. With blank stares, they would admit they were on secret spy missions, confess to horrific war crimes, and denounce U.S. imperialism.

To make sense of this, the press used the term "brainwashing," the idea that the communists had developed secret drugs and techniques to turn U.S. soldiers into pliable zombies. In this charged political climate, Frank Sinatra starred in the 1962 Cold War thriller *The Manchurian Candidate,* in which he tries to expose a secret communist "sleeper" agent whose mission is to assassinate the president of the United States. But there is a twist. The assassin is actually a trusted U.S. war hero, someone who was captured and then brainwashed by the communists. Coming from a well-connected family, the agent seems above suspicion and is almost impossible to stop. *The Manchurian Candidate* mirrored the anxieties of many Americans at that time.

Many of these fears were also stoked by Aldous Huxley's prophetic 1931 novel *Brave New World.* In this dystopia, there are large test-tube-baby factories that produce clones. By selectively depriving oxygen from these fetuses, it is possible to produce children of different levels of brain damage. At the top are the alphas, who suffer no brain damage and are bred to rule society. At the bottom are the epsilons, who suffer significant brain damage and are used as disposable, obedient workers. In between are additional levels made up of other workers and the bureaucracy. The elite then control society by flooding it with mind-altering drugs, free love, and constant brainwashing. In this way, peace, tranquility, and harmony are maintained, but the novel asked a disturbing question that resonates even today: How much of our freedom and basic humanity do we want to sacrifice in the name of peace and social order?

CIA MIND-CONTROL EXPERIMENTS

The Cold War hysteria eventually reached the highest levels of the CIA. Convinced that the Soviets were far ahead in the science of brainwashing and unorthodox scientific methods, the CIA embarked upon a variety of classified projects, such as MKULTRA, which began in 1953, to explore bizarre, fringe ideas. (In 1973, as the Watergate scandal spread panic throughout the government, CIA director Richard Helms canceled MKULTRA and hurriedly ordered all documents pertaining to the project destroyed. However, a cache of twenty thousand documents somehow survived the purge and were declassified in 1977 under the Freedom of Information Act, revealing the full scope of this massive effort.)

It is now known that, from 1953 to 1973, MULTRA funded 80 institutions, including 44 universities and colleges, and scores of hospitals, pharmaceutical companies, and prisons, often experimenting on unsuspecting people without their permission, in 150 secret operations. At one point, fully 6 percent of the entire CIA budget went into MKULTRA.

Some of these mind-control projects included:

- developing a "truth serum" so prisoners would spill their secrets
- erasing memories via a U.S. Navy project called "Subproject 54"
- using hypnosis and a wide variety of drugs, especially LSD, to control behavior
- investigating the use of mind-control drugs against foreign leaders, e.g., Fidel Castro
- perfecting a variety of interrogation methods against prisoners
- developing a knockout drug that was fast working and left no trace
- altering people's personality via drugs to make them more pliable

Although some scientists questioned the validity of these studies, others went along willingly. People from a wide range of disciplines were recruited, including psychics, physicists, and computer scientists, to investigate a variety of unorthodox projects: experimenting with mind-altering drugs such as

LSD, asking psychics to locate the position of Soviet submarines patrolling the deep oceans, etc. In one sad incident, a U.S. Army scientist was secretly given LSD. According to some reports, he became so violently disoriented that he committed suicide by jumping out a window.

Most of these experiments were justified on the grounds that the Soviets were already ahead of us in terms of mind control. The U.S. Senate was briefed in another secret report that the Soviets were experimenting with beaming microwave radiation directly into the brains of test subjects. Rather than denouncing the act, the United States saw "great potential for development into a system for disorienting or disrupting the behavior pattern of military or diplomatic personnel." The U.S. Army even claimed that it might be able to beam entire words and speeches into the minds of the enemy: "One decoy and deception concept . . . is to remotely create noise in the heads of personnel by exposing them to low power, pulsed microwaves. . . . By proper choice of pulse characteristics, intelligible speech may be created. . . . Thus, it may be possible to 'talk' to selected adversaries in a fashion that would be most disturbing to them," the report said.

Unfortunately, none of these experiments was peer-reviewed, so millions of taxpayer dollars were spent on projects like this one, which most likely violated the laws of physics, since the human brain cannot receive microwave radiation and, more important, does not have the ability to decode microwave messages. Dr. Steve Rose, a biologist at the Open University, has called this far-fetched scheme a "neuro-scientific impossibility."

But for all the millions of dollars spent on these "black projects," apparently not a single piece of reliable science emerged. The use of mind-altering drugs did, in fact, create disorientation and even panic among the subjects who were tested, but the Pentagon failed to accomplish the key goal: control of the conscious mind of another person.

Also, according to psychologist Robert Jay Lifton, brainwashing by the communists had little long-term effect. Most of the American troops who denounced the United States during the Korean War reverted back to their normal personalities soon after being released. In addition, studies done on people who have been brainwashed by certain cults also show that they revert back to their normal personality after leaving the cult. So it seems that, in the long run, one's basic personality is not affected by brainwashing.

Of course, the military was not the first to experiment with mind control.

In ancient times, sorcerers and seers would claim that giving magic potions to captured soldiers would make them talk or turn against their leaders. One of the earliest of these mind-control methods was hypnotism.

YOU ARE GETTING SLEEPY. . . .

As a child, I remember seeing TV specials devoted to hypnosis. In one show, a person was placed in a hypnotic trance and told that when he woke up, he would be a chicken. The audience gasped as he began to cluck and flap his arms around the stage. As dramatic as this demonstration was, it's simply an example of "stage hypnosis." Books written by professional magicians and showmen explain that they use shills planted in the audience, the power of suggestion, and even the willingness of the victim to play along with the ruse.

I once hosted a BBC/Discovery TV documentary called *Time,* and the subject of long-lost memories came up. Is it possible to evoke such distant memories through hypnosis? And if it is, can you then impose your will on another? To test some of these ideas, I had myself hypnotized for TV.

BBC hired a skilled professional hypnotist to begin the process. I was asked to lie down on a bed in a quiet, darkened room. The hypnotist spoke to me in slow, gentle tones, gradually making me relax. After a while, he asked me to think back into the past, to perhaps a certain place or incident that stood out even after all these years. And then he asked me to reenter that place, reexperiencing its sights, sounds, and smells. Remarkably, I did begin to see places and people's faces that I had forgotten about decades ago. It was like watching a blurred movie that was slowly coming into focus. But then the recollections stopped. At a certain point, I could not recapture any more memories. There was clearly a limit to what hypnosis could do.

EEG and MRI scans show that during hypnosis the subject has minimal sensory stimulation in the sensory cortices from the outside. In this way, hypnosis can allow one to access some memories that are buried, but it certainly cannot change one's personality, goals, or wishes. A secret 1966 Pentagon document corroborates this, explaining that hypnotism cannot be trusted as a military weapon. "It is probably significant that in the long history of hypnosis, where the potential application to intelligence has always been known, there are no reliable accounts of its effective use by an intelligence service," it read.

It should also be noted that brain scans show that hypnotism is not a new state of consciousness, like dreaming and REM sleep. If we define human consciousness as the process of continually building models of the outside world and then simulating how they evolve into the future to carry out a goal, we see that hypnosis cannot alter this basic process. Hypnosis can accentuate certain aspects of consciousness and help retrieve certain memories, but it cannot make you squawk like a chicken without your permission.

MIND-ALTERING DRUGS AND TRUTH SERUMS

One of the goals of MKULTRA was the creation of a truth serum so that spies and prisoners would reveal their secrets. Although MKULTRA was canceled in 1973, U.S. Army and CIA interrogation manuals declassified by the Pentagon in 1996 still recommended the use of truth serums (although the U.S. Supreme Court ruled that confessions obtained in this way were "unconstitutionally coerced" and hence inadmissible in court).

Anyone who watches Hollywood movies knows that sodium pentathol is the truth serum of choice used by spies (as in the movies *True Lies* with Arnold Schwarzenegger and *Meet the Fockers* with Robert De Niro). Sodium pentathol is part of a larger class of barbiturates, sedatives, and hypnotics that can evade the blood-brain barrier, which prevents most harmful chemicals in the bloodstream from entering the brain.

Not surprisingly, most mind-altering drugs, such as alcohol, affect us powerfully because they can evade this barrier. Sodium pentathol depresses activity in the prefrontal cortex, so that a person becomes more relaxed, talkative, and uninhibited. However, this does not mean that they tell the truth. On the contrary, people under the influence of sodium pentathol, like those who have imbibed a few too many, are fully capable of lying. The "secrets" that come spilling out of the mouth of someone under this drug may be total fabrications, so even the CIA eventually gave up on drugs like this.

But this still leaves open the possibility that, one day, a wonder drug might be found that could alter our basic consciousness. This drug would work by changing the synapses between our nerve fibers by targeting neurotransmitters that operate in this area, such as dopamine, serotonin, or acetylcholine. If we think of the synapses as a series of tollbooths along a superhighway, then certain drugs (such as stimulants like cocaine) can open the tollbooth and let messages pass by unimpeded. The sudden rush that drug addicts feel is caused when these tollbooths are opened all at once, causing an avalanche of signals to flood by. But when all the synapses have fired in unison, they cannot fire again until hours later. It's as if the tolls have closed, and this causes the sudden depression one feels after the rush. The body's desire to reexperience the sudden rush then causes addiction.

HOW DRUGS ALTER THE MIND

Although the biochemical basis for mind-altering drugs was not known when the CIA first conducted its experiments on unsuspecting subjects, since then the molecular basis of drug addiction has been studied in detail. Studies in animals demonstrate how powerful drug addiction is: rats, mice, and primates will, given the chance, take drugs like cocaine, heroin, and amphetamines until they drop from exhaustion or die from it.

To see how widespread this problem has become, consider that by 2007, thirteen million Americans aged twelve or over (or 5 percent of the entire teen and adult population of the United States) had tried or become addicted to methamphetamines. Drug addiction not only destroys entire lives, it also systematically destroys the brain. MRI scans of the brains of meth addicts show an 11 percent reduction in the size of the limbic system, which processes emotions, and an 8 percent loss of tissue in the hippocampus, which is the gateway for memory. MRI scans show that the damage in some ways is comparable to that found in Alzheimer's patients. But no matter how much meth destroys the brain, addicts crave it because its high is up to twelve times the rush caused by eating a delicious meal or even having sex.

Basically, the "high" of drug addiction is due to the drug's hijacking of the brain's own pleasure/reward system located in the limbic system. This pleasure/reward circuit is very primitive, dating back millions of years in evolutionary history, but it is still extremely important for human survival because it rewards beneficial behavior and punishes harmful acts. Once this circuit is taken over by drugs, however, the result can be widespread havoc. These drugs first penetrate the blood-brain barrier and then cause the overproduction of neurotransmitters like dopamine, which then floods the nucleus accumbens, a tiny pleasure center located deep in the brain near the amygdala. The dopamine, in turn, is produced by certain brain cells in the ventral tegmental area, called VTA cells.

All drugs basically work the same way: by crippling the VTA–nucleus accumbens circuit, which controls the flow of dopamine and other neurotransmitters to the pleasure center. Drugs differ only in the way in which this process takes place. There are at least three main drugs that stimulate the pleasure center of the brain: dopamine, serotonin, and noradrenaline; all of them give feelings of pleasure, euphoria, and false confidence, and also produce a burst of energy.

Cocaine and other stimulants, for example, work in two ways. First, they directly stimulate the VTA cells to produce more dopamine, hence causing excess dopamine to flood into the nucleus accumbens. Second, they prevent the VTA cells from going back to their "off" position, thus keeping them continually producing dopamine. They also impede the uptake of serotonin and noradrenaline. The simultaneous flooding of neural circuits from all three of these neurotransmitters, then, creates the tremendous high associated with cocaine.

Heroin and other opiates, by contrast, work by neutralizing the cells in the VTA that can reduce the production of dopamine, thus causing the VTA to overproduce dopamine.

Drugs like LSD operate by stimulating the production of serotonin, inducing a feeling of well-being, purpose, and affection. But they also activate areas of the temporal lobe involved in creating hallucinations. (Only fifty micrograms of LSD can cause hallucinations. LSD binds so tightly, in fact, that further increasing the dosage has no effect.)

Over time, the CIA came to realize that mind-altering drugs were not the magic bullet they were looking for. The hallucinations and addictions that accompany these drugs made them too unstable and unpredictable, and they could cause more trouble than they were worth in delicate political situations.

(It should be pointed out that just in the last few years, MRI brain scans of drug addicts have indicated a novel way to possibly cure or treat some forms of addiction. By accident, it was noticed that stroke victims who have damage to the insula [located deep in the brain, between the prefrontal cortex and the temporal cortex] have a significantly easier time quitting smoking than the average smoker. This result has also been verified among drug abusers using cocaine, alcohol, opiates, and nicotine. If this result holds up, it might mean that one may be able to dampen the activity of the insula using electrodes or magnetic stimulators and hence treat addiction. "This is the first time we've shown anything like this, that damage to a specific brain area could remove the problem of addiction entirely. It's mind-boggling," says Dr. Nora Volkow, director of the National Institute on Drug Abuse. At present, no one knows how this works, because the insula is involved in a bewildering variety of brain functions, including perception, motor control, and self-awareness. But if this result bears out, it could change the entire landscape of addiction studies.)

PROBING THE BRAIN WITH OPTOGENETICS

These mind-control experiments were done mainly in an era when the brain was largely a mystery, with hit-or-miss methods that often failed. However, because of the explosion in devices that can probe the brain, new opportunities have arisen that will both help us understand the brain as well as possibly teach us how to control it.

Optogenetics, as we have seen, is one of the fastest-developing fields in science today. The basic goal is to identify precisely which neural pathway corresponds to which mode of behavior. Optogenetics starts with a gene called opsin, which is quite unusual because it is sensitive to light. (It is believed that the appearance of this gene hundreds of millions of years ago was responsible for creating the first eye. In this theory, a simple patch of skin sensitive to light due to opsin evolved into the retina of the eye.)

When the opsin gene is inserted into a neuron and exposed to light, the neuron will fire on command. By flipping a switch, one can instantly recognize the neural pathway for certain behaviors because the proteins manufactured by opsin conduct electricity and will fire.

The hard part, though, is to insert this gene into a single neuron. To do this, one uses a technique borrowed from genetic engineering. The opsin gene is inserted into a harmless virus (which has had its bad genes removed), and, using precision tools, it is then possible to apply this virus to a single neuron. The virus then infects the neuron by inserting its genes into the genes of the neuron. Then, when a light beam is flashed onto neural tissue, the neuron is turned on. In this way, one can establish the precise pathway that certain messages take.

Not only does optogenetics identify certain pathways by shining a light beam on them, it also enables scientists to control behavior. Already this method has been a proven success. It was long suspected that a simple neural circuit must be responsible for fruit flies escaping and flying away. Using this method, it was possible to finally identify the precise pathway behind the quick getaway. By simply shining a beam onto these fruit flies, they bolt on demand.

Scientists are also now able to make worms stop wiggling by flashing light, and in 2011 yet another breakthrough was made. Scientists at Stanford were able to insert the opsin gene into a precise region of the amygdala of mice. These mice, which were specially bred to be timid, cowered in their cage. But when a beam of light was flashed into their brains, the mice suddenly lost their timidity and began to explore their cage.

The implications are enormous. While fruit flies may have simple reflex mechanisms involving a handful of neurons, mice have complete limbic systems with counterparts in the human brain. Although many experiments that work with mice do not translate to human beings, this still holds out the possibility that scientists may one day find the precise neural pathways for certain mental illnesses, and then be able to treat them without any side effects. As Dr. Edward Boyden of MIT says, "If you want to turn off a brain circuit and the alternative is surgical removal of a brain region, optical fiber implants might seem preferable."

One practical application is in treating Parkinson's disease. As we have seen, it can be treated by deep brain stimulation, but because the positioning of electrodes in the brain lacks precision, there is always the danger of strokes, bleeding, infections, etc. Deep brain stimulation can also cause side effects such as dizziness and muscle contractions, because the electrodes can accidentally stimulate the wrong neurons. Optogenetics may improve deep brain stimulation by identifying the precise neural pathways that are misfiring, at the level of individual neurons.

Victims of paralysis might also benefit from this new technology. As we saw in Chapter 4, some paralyzed individuals have been hooked up to a computer in order to control a mechanical arm, but because they have no sense of touch, they often wind up dropping or crushing the object they wish to grab. "By feeding information from sensors on the prosthetic fingertips directly back to the brain using optogenetics, one could in principle provide a high-fidelity sense of touch," says Dr. Krishna Shenoy of Stanford.

Optogenetics will also help clarify which neural pathways are involved with human behavior. In fact, plans have already been drawn up to experiment with this technique on human brains, especially with regard to mental illness. There will be hurdles, of course. First, the technique requires opening up the skull, and if the neurons that one wishes to study are located deep inside the brain, the procedure will be even more invasive. Lastly, one has to insert tiny wires into the brain that can shine a light on this modified neuron so that it triggers the desired behavior.

Once these neural pathways have been deciphered, you can also stimulate them, making animals perform strange behaviors (for example, mice will run around in circles). Although scientists are just beginning to trace the neural pathways governing simple animal behaviors, in the future they should have an encyclopedia of such behaviors, including those of humans. In the wrong hands, however, optogenetics could potentially be used to control human behavior.

In the main, the benefits of optogenetics greatly outweigh its drawbacks. It can literally reveal the pathways of the brain in order to treat mental illness and other diseases. This may then give scientists the tools by which to repair the damage, perhaps curing diseases once thought to be incurable. In the near future, then, the benefits are all positive. But further in the future, once the pathways of human behaviors are also understood, optogenetics could also be used to control or at least modify human behavior as well.

MIND CONTROL AND THE FUTURE

In summary, the use of drugs and hypnotism by the CIA was a flop. These techniques were too unstable and unpredictable to be of any use to the military. They can be used to induce hallucinations and dependency, but they have failed to cleanly erase memories, make people more pliant, or force people to perform acts against their will. Governments will keep trying, but the goal is elusive. So far, drugs are simply too blunt an instrument to allow you to control someone's behavior.

But this is also a cautionary tale. Carl Sagan mentions one nightmare scenario that might actually work. He envisions a dictator taking children and putting electrodes into their "pain" and "pleasure" centers. These electrodes are then connected wirelessly to computers, so that the dictator can control his subjects with the push of a button.

Another nightmare might involve probes placed in the brain that could override our wishes and seize control of our muscles, forcing us to perform tasks we don't want to do. The work of Dr. Delgado was crude, but it showed that bursts of electricity applied to motor areas of the brain can overrule our conscious thoughts, so that our muscles are no longer under our control. He was able to identify only a few behaviors in animals that could be controlled with electric probes. In the future, it may be possible to find a wide variety of behaviors that can be controlled electronically with a switch.

If you are the person being controlled, it would be an unpleasant experience. Although you may think you are master of your own body, your muscles would actually fire without your permission, so you would do things against your will. The electric impulse being fed into your brain could be larger than the impulses you consciously send into your muscles, so that it would appear as if someone had hijacked your body. Your own body would become a foreign object.

In principle, some version of this nightmare might be possible in the future. But there are several factors that may prevent this as well. First, this is still an infant technology and it is not known how it will be applied to human behavior, so there is still plenty of time to monitor its development and perhaps create safeguards to see that it is not misused. Second, a dictator might simply decide that propaganda and coercion, the usual methods of controlling a population, are cheaper and more effective than putting electrodes into the brains of millions of children, which would be costly and invasive. And third, in democratic societies, a vigorous public debate would probably emerge concerning the promise and limitations of this powerful technology. Laws would have to be passed to prevent the abuse of these methods without impairing their ability to reduce human suffering. Soon science will give us unparalleled insight into the detailed neural pathways of the brain. A fine line has to be drawn between technologies that can benefit society and technologies that can control it. And the key to passing these laws is an educated, informed public.

But the real impact of this technology, I believe, will be to liberate the mind, not enslave it. These technologies can give hope to those who are trapped in mental illness. Although there is as yet no permanent cure for mental illness, these new technologies have given us deep insight into how such disorders form and how they progress. One day, through genetics, drugs, and a combination of high-tech methods, we will find a way to manage and eventually cure these ancient diseases.

One of the recent attempts to exploit this new knowledge of the brain is to understand historical personalities. Perhaps the insights from modern science can help explain the mental states of those in the past.

And one of the most mystifying figures being analyzed today is Joan of Arc.

Lovers and madmen have such seething brains. . . .
The lunatic, the lover, and the poet
Are of imagination all compact.

—WILLIAM SHAKESPEARE, *A MIDSUMMER NIGHT'S DREAM*

9 ALTERED STATES OF CONSCIOUSNESS

She was just an illiterate peasant girl who claimed to hear voices directly from God. But Joan of Arc would rise from obscurity to lead a demoralized army to victories that would change the course of nations, making her one of the most fascinating, compelling, and tragic figures in history.

During the chaos of the Hundred Years' War, when northern France was decimated by English troops and the French monarchy was in retreat, a young girl from Orléans claimed to have divine instructions to lead the French army to victory. With nothing to lose, Charles VII allowed her to command some of his troops. To everyone's shock and wonder, she scored a series of triumphs over the English. News rapidly spread about this remarkable young girl. With each victory, her reputation began to grow, until she became a folk heroine, rallying the French around her. French troops, once on the verge of total collapse, scored decisive victories that paved the way for the coronation of the new king.

However, she was betrayed and captured by the English. They realized what a threat she posed to them, since she was a potent symbol for the French and claimed guidance directly from God Himself, so they subjected her to a show trial. After an elaborate interrogation, she was found guilty of heresy and burned at the stake at the age of nineteen in 1431.

In the centuries that followed, hundreds of attempts have been made to understand this remarkable teenager. Was she a prophet, a saint, or a madwoman? More recently, scientists have tried to use modern psychiatry and neuroscience to explain the lives of historical figures such as Joan of Arc.

Few question her sincerity about claims of divine inspiration. But many scientists have written that she might have suffered from schizophrenia, since she heard voices. Others have disputed this fact, since the surviving records of her trial reveal a person of rational thought and speech. The English laid several theological traps for her. They asked, for example, if she was in God's grace. If she answered yes, then she would be a heretic, since no one can know for certain if they are in God's grace. If she said no, then she was confessing her guilt, and that she was a fraud. Either way, she would lose.

In a response that stunned the audience, she answered, "If I am not, may God put me there; and if I am, may God so keep me." The court notary, in the records, wrote, "Those who were interrogating her were stupefied."

In fact, the transcripts of her interrogation are so remarkable that George Bernard Shaw put literal translations of the court record in his play *Saint Joan*.

More recently, another theory has emerged about this exceptional woman: perhaps she actually suffered from temporal lobe epilepsy. People who have this condition sometimes experience seizures, but some of them also experience a curious side effect that may shed some light on the structure of human beliefs. These patients suffer from "hyperreligiosity," and can't help thinking that there is a spirit or presence behind everything. Random events are never random, but have some deep religious significance. Some psychologists have speculated that a number of history's prophets suffered from these temporal lobe epileptic lesions, since they were convinced they talked to God. The neuroscientist Dr. David Eagleman says, "Some fraction of history's prophets, martyrs, and leaders appear to have had temporal lobe epilepsy. Consider Joan of Arc, the sixteen-year-old girl who managed to turn the tide of the Hundred Years' War because she believed (and convinced the French soldiers) that she was hearing voices from Saint Michael the archangel, Saint Catherine of Alexandria, Saint Margaret, and Saint Gabriel."

This curious effect was noticed as far back as 1892, when textbooks on mental illness noted a link between "religious emotionalism" and epilepsy. It was first clinically described in 1975 by neurologist Norman Geschwind of Boston Veterans Administration Hospital. He noticed that epileptics who had electrical misfirings in their left temporal lobes often had religious experiences, and he speculated that the electrical storm in the brain somehow was the cause of these religious obsessions.

Dr. V. S. Ramachandran estimates that 30 to 40 percent of all the temporal lobe epileptics whom he has seen suffer from hyperreligiosity. He notes, "Sometimes it's a personal God, sometimes it's a more diffuse feeling of being one with the cosmos. Everything seems suffused with meaning. The patient will say, 'Finally, I see what it is all really about, Doctor. I really understand God. I understand my place in the universe—the cosmic scheme.'"

He also notes that many of these individuals are extremely adamant and convincing in their beliefs. He says, "I sometimes wonder whether such patients who have temporal lobe epilepsy have access to another dimension of reality, a wormhole of sorts into a parallel universe. But I usually don't say this to my colleagues, lest they doubt my sanity." He has experimented on patients with temporal lobe epilepsy, and confirmed that these individuals had a strong emotional reaction to the word "God" but not to neutral words. This means that the link between hyperreligiosity and temporal lobe epilepsy is real, not just anecdotal.

Psychologist Michael Persinger asserts that a certain type of transcranial electrical stimulation (called transcranial magnetic simulation, or TMS) can deliberately induce the effect of these epileptic lesions. If this is so, is it possible that magnetic fields can be used to alter one's religious beliefs?

In Dr. Persinger's studies, the subject places a helmet on his head (dubbed the "God helmet"), which contains a device that can send magnetism into particular parts of the brain. Afterward, when the subject is interviewed, he will often claim that he was in the presence of some great spirit. David Biello, writing in *Scientific American*, says, "During the three-minute bursts of stimulation, the affected subjects translated this perception of the divine into their own cultural and religious language—terming it God, Buddha, a benevolent presence, or the wonder of the universe." Since this effect is reproducible on demand, it indicates that perhaps the brain is hardwired in some way to respond to religious feelings.

Some scientists have gone further and have speculated that there is a "God gene" that predisposes the brain to be religious. Since most societies have created a religion of some sort, it seems plausible that our ability to respond to religious feelings might be genetically programmed into our genome. (Meanwhile, some evolutionary theorists have tried to explain these facts by claiming that religion served to increase the chances of survival for early humans. Religion helped bond bickering individuals into a cohesive tribe with a common mythology, which increased the chances that the tribe would stick together and survive.)

Would an experiment like the one using the "God helmet" shake a person's religious beliefs? And can an MRI machine record the brain activity of someone who experiences a religious awakening?

To test these ideas, Dr. Mario Beauregard of the University of Montreal recruited a group of fifteen Carmelite nuns who agreed to put their heads into an MRI machine. To qualify for the experiment, all of them must "have had an experience of intense union with God."

Originally, Dr. Beauregard had hoped that the nuns would have a mystical communion with God, which could then be recorded by an MRI scan. However, being shoved into an MRI machine, where you are surrounded by tons of magnetic coils of wire and high-tech equipment, is not an ideal setting for a religious epiphany. The best they could do was to evoke memories of previous religious experiences. "God cannot be summoned at will," explained one of the nuns.

The final result was mixed and inconclusive, but several regions of the brain clearly lit up during this experiment:

- The caudate nucleus, which is involved with learning and possibly falling in love. (Perhaps the nuns were feeling the unconditional love of God?)
- The insula, which monitors body sensations and social emotions. (Perhaps the nuns were feeling close to the other nuns as they were reaching out to God?)
- The parietal lobe, which helps process spatial awareness. (Perhaps the nuns felt they were in the physical presence of God?)

Dr. Beauregard had to admit that so many areas of the brain were activated, with so many different possible interpretations, that he could not say for sure whether hyperreligiosity could be induced. However, it was clear to him that the nuns' religious feelings were reflected in their brain scans.

But did this experiment shake the nuns' belief in God? No. In fact, the nuns concluded that God placed this "radio" in the brain so that we could communicate with Him.

Their conclusion was that God created humans to have this ability, so the brain has a divine antenna given to us by God so that we can feel His presence. David Biello concludes, "Although atheists might argue that finding spirituality in the brain implies that religion is nothing more than divine delusion, the nuns were thrilled by their brain scans for precisely the opposite reason: they seemed to provide confirmation of God's interactions with them." Dr. Beauregard concluded, "If you are an atheist and you live a certain kind of experience, you will relate it to the magnificence of the universe. If you are a Christian, you will associate it with God. Who knows. Perhaps they are the same thing."

Similarly, Dr. Richard Dawkins, a biologist at Oxford University and an outspoken atheist, was once placed in the God helmet to see if his religious beliefs would change.

They did not.

So in conclusion, although hyperreligiosity may be induced via temporal lobe epilepsy and even magnetic fields, there is no convincing evidence that magnetic fields can alter one's religious views.

MENTAL ILLNESS

But there is another altered state of consciousness that brings great suffering, both to the person experiencing it and to his or her family, and this is mental illness. Can brain scans and high technology reveal the origin of this affliction and perhaps lead to a cure? If so, one of the largest sources of human suffering could be eliminated.

For example, throughout history, the treatment of schizophrenia was brutal and crude. People who suffer from this debilitating mental disorder, which afflicts about 1 percent of the population, typically hear imaginary voices and suffer from paranoid delusions and disorganized thinking. Throughout history, they were considered to be "possessed" by the devil and were banished, killed, or locked up. Gothic novels sometimes refer to the strange, demented relative who lives in the darkness of a hidden room or basement. The Bible even mentions an incident when Jesus encountered two demoniacs. The demons begged Jesus to drive them into a herd of swine. He said, "Go then." When the demons entered the swine, the whole herd rushed down the bank and drowned in the sea.

Even today, you still see people with classic symptoms of schizophrenia walking around having arguments with themselves. The first indicators usually surface in the late teens (for men) or early twenties (for women). Some schizophrenics have led normal lives and even performed remarkable feats before the voices finally took over. The most famous case is that of the 1994 Nobel Prize winner in economics, John Nash, who was played by Russell Crowe in the movie *A Beautiful Mind.* In his twenties, Nash did pioneering work in economics, game theory, and pure mathematics at Princeton University. One of his advisers wrote him a letter of recommendation with just one line: "This man is a genius." Remarkably, he was able to perform at such a high intellectual level even while being hounded by delusions. He was finally hospitalized when he had a breakdown at age thirty-one, and spent many years in institutions or wandering around the world, fearing that communist agents would kill him.

At present, there is no precise, universally accepted way to diagnose mental illness. There is hope, however, that one day scientists will use brain scans and other high-tech devices to create accurate diagnostic tools. Progress in treating mental illness, therefore, has been painfully slow. After centuries of suffering, victims of schizophrenia had their first sign of relief when antipsychotic drugs like thorazine were found accidentally in the 1950s that could miraculously control or even at times eliminate the voices that haunted the mentally ill.

It is believed that these drugs work by regulating the level of certain neurotransmitters, such as dopamine. Specifically, the theory is that these drugs block the functioning of D2 receptors of certain nerve cells, thereby reducing the level of dopamine. (This theory, that hallucinations were in part caused by excess dopamine levels in the limbic system and prefrontal cortex, also explained why people taking amphetamines experienced similar hallucinations.)

Dopamine, because it is so essential for the synapses of the brain, has been implicated in other disorders as well. One theory holds that Parkinson's disease is aggravated by a lack of dopamine in the synapses, while Tourette's syndrome can be triggered by an overabundance of it. (People with Tourette's syndrome have tics and unusual facial movements. A small minority of them uncontrollably speak obscene words and make profane, derogatory remarks.)

More recently, scientists have zeroed in on another possible culprit: abnormal glutamate levels in the brain. One reason for believing these levels are involved is that PCP (angel dust) is known to create hallucinations similar to those of schizophrenics by blocking a glutamate receptor called NMDA. Clozapine, a relatively new drug for schizophrenia that stimulates the production of glutamate, shows great promise.

However, these antipsychotic drugs are not a cure-all. In about 20 percent of cases, such drugs stop all symptoms. About two-thirds find some relief from their symptoms, but the rest are totally unaffected. (According to one theory, antipsychotic drugs mimic a natural chemical that is missing in schizophrenics' brains, but it is not an exact copy. Hence a patient has to try a variety of these antipsychotic drugs, almost by trial and error. Moreover, they can have unpleasant side effects, so schizophrenics often stop taking them and suffer a relapse.)

Recently, brain scans of schizophrenics taken while they were having auditory hallucinations have helped explain this ancient disorder. For example, when we silently talk to ourselves, certain parts of the brain light up on an MRI scan, especially in the temporal lobe (such as in Wernicke's area). When a schizophrenic hears voices, the very same areas of the brain light up. The brain works hard to construct a consistent narrative, so schizophrenics try to make sense of these unauthorized voices, believing they originate from strange sources, such as Martians secretly beaming thoughts into their brains. Dr. Michael Sweeney of Ohio State writes, "Neurons wired for the sensation of sound fire on their own, like gas-soaked rags igniting spontaneously in a hot, dark garage. In the absence of sights and sounds in the surrounding environment, the schizophrenic's brain creates a powerful illusion of reality."

Notably, these voices seem to be coming from a third party, who often gives the subject commands, which are mostly mundane but sometimes violent. Meanwhile, the simulation centers in the prefrontal cortex seem to be on automatic pilot, so in a way it's as though the consciousness of a schizophrenic is running the same sort of simulations we all do, except they're done without his permission. The person is literally talking to himself without his knowledge.

HALLUCINATIONS

The mind constantly generates hallucinations of its own, but for the most part they are easily controlled. We see images that don't exist or hear spurious sounds, for example, so the anterior cingulate cortex is vital to distinguish the real from the manufactured. This part of the brain helps us distinguish between stimuli that are external and those that are internally generated by the mind itself.

However, in schizophrenics, it is believed that this system is damaged, so that the person cannot distinguish real from imaginary voices. (The anterior cingulate cortex is vital because it lies in a strategic place, between the prefrontal cortex and the limbic system. The link between these two areas is one of the most important in the brain, since one area governs rational thinking, and the other emotions.)

Hallucinations, to some extent, can be created on demand. Hallucinations occur naturally if you place someone in a pitch-black room, an isolation chamber, or a creepy environment with strange noises. These are examples of "our eyes playing tricks on us." Actually, the brain is tricking itself, internally creating false images, trying to make sense of the world and identify threats. This effect is called "pareidolia." Every time we look at clouds in the sky, we see images of animals, people, or our favorite cartoon characters. We have no choice. It is hardwired into our brains.

In a sense, all images we see, both real and virtual, are hallucinations, because the brain is constantly creating false images to "fill in the gaps." As we've seen, even real images are partly manufactured. But in the mentally ill, regions of the brain such as the anterior cingulate cortex are perhaps damaged, so the brain confuses reality and fantasy.

THE OBSESSIVE MIND

Another disorder in which drugs may be used to heal the mind is OCD (obsessive-compulsive disorder). As we saw earlier, human consciousness involves mediating between a number of feedback mechanisms. Sometimes, however, the feedback mechanisms are stuck in the "on" position.

One in forty Americans suffers from OCD. Cases can be mild, so that, for example, people have to constantly go home to check that they locked the door. The detective Adrian Monk on the TV show *Monk* has a mild case of OCD. But OCD can also be so severe that people compulsively scratch or wash their skin until it is left bleeding and raw. Some people with OCD have been known to repeat obsessive behaviors for hours, making it difficult to keep a job or have a family.

Normally these types of compulsive behaviors, in moderation, are actually good for us, since they help us keep clean, healthy, and safe. That is why we evolved these behaviors in the first place. But someone with OCD cannot stop this behavior, and it spirals out of control.

Brain scans are now revealing how this takes place. They show that at least three areas of the brain that normally help us keep ourselves healthy get stuck in a feedback loop. First, there is the orbitofrontal cortex, which we saw in Chapter 1 can act as a fact-checker, making sure that we have properly locked the doors and washed our hands. It tells us, "Hmm, something is wrong." Second, the caudate nucleus, located in the basal ganglia, governs learned activities that are automatic. It tells the body to "do something." And finally, we have the cingulate cortex, which registers conscious emotions, including discomfort. It says, "I still feel awful."

Psychiatry professor Jeffrey Schwartz of UCLA has tried to put this all together to explain how OCD gets out of hand. Imagine you have the urge to wash your hands. The orbitofrontal cortex recognizes that something is wrong, that your hands are dirty. The caudate nucleus kicks in and causes you to automatically wash your hands. Then the cingulate cortex registers satisfaction that your hands are clean.

But in someone with OCD, this loop is altered. Even after he notices that his hands are dirty and he washes them, he still has the discomforting feeling that something is wrong, that they are still dirty. So he is stuck in a feedback loop that won't stop.

In the 1960s, the drug clomipramine hydrochloride began to give OCD patients some relief. This and other drugs developed since then raise levels of the neurotransmitter serotonin in the body. They can reduce symptoms of OCD by as much as 60 percent in clinical trials. Dr. Schwartz says, "The brain's gonna do what the brain's gonna do, but you don't have to let it push you around." These drugs are certainly not a cure, but they have brought some relief to the sufferers of OCD.

BIPOLAR DISORDER

Another common form of mental illness is bipolar disorder, in which a person suffers from extreme bouts of wild, delusional optimism, followed by a crash and then periods of deep depression. Bipolar disorder also seems to run in families and, curiously, strikes frequently in artists; perhaps their great works of art were created during bursts of creativity and optimism. A list of creative people who were afflicted by bipolar disorder reads like a Who's Who of Hollywood celebrities, musicians, artists, and writers. Although the drug lithium seems to control many of the symptoms of bipolar disorder, the causes are not entirely clear.

One theory states that bipolar disorder may be caused by an imbalance between the left and right hemispheres. Dr. Michael Sweeney notes, "Brain scans have led researchers to generally assign negative emotions such as sadness to the right hemisphere and positive emotions such as joy to the left hemisphere. For at least a century, neuroscientists have noticed a link between damage to the brain's left hemisphere and negative moods, including depression and uncontrollable crying. Damage to the right, however, has been associated with a broad array of positive emotions."

So the left hemisphere, which is analytical and controls language, tends to become manic if left to itself. The right hemisphere, on the contrary, is holistic and tends to check this mania. Dr. V. S. Ramachandran writes, "If left unchecked, the left hemisphere would likely render a person delusional or manic. . . . So it seems reasonable to postulate a 'devil's advocate' in the right hemisphere that allows 'you' to adopt a detached, objective (allocentric) view of yourself."

If human consciousness involves simulating the future, it has to compute the outcomes of future events with certain probabilities. It needs, therefore, a delicate balance between optimism and pessimism to estimate the chances of success or failures for certain courses of action.

But in some sense, depression is the price we pay for being able to simulate the future. Our consciousness has the ability to conjure up all sorts of horrific outcomes for the future, and is therefore aware of all the bad things that could happen, even if they are not realistic.

It is hard to verify many of these theories, since brain scans of people who are clinically depressed indicate that many brain areas are affected. It is difficult to pinpoint the source of the problem, but among the clinically depressed, activity in the parietal and temporal lobes seems to be suppressed, perhaps indicating that the person is withdrawn from the outside world and living in their own internal world. In particular, the ventromedial cortex seems to play an important role. This area apparently creates the feeling that there is a sense of meaning and wholeness to the world, so that everything seems to have a purpose. Overactivity in this area can cause mania, in which people think they are omnipotent. Underactivity in this area is associated with depression and the feeling that life is pointless. So it is possible that a defect in this area may be responsible for some mood swings.

A THEORY OF CONSCIOUSNESS AND MENTAL ILLNESS

So how does the space-time theory of consciousness apply to mental illness? Can it give us a deeper insight into this disorder? As we mentioned before, we define human consciousness as the process of creating a model of our world in space and time (especially the future) by evaluating many feedback loops in various parameters in order to achieve a goal.

We have proposed that the key function of human consciousness is to simulate the future, but this is not a trivial task. The brain accomplishes it by having these feedback loops check and balance one another. For example, a skillful CEO at a board meeting tries to draw out the disagreement among staff members and to sharpen competing points of view in order to sift through the various arguments and then make a final decision. In the same way, various regions of the brain make diverging assessments of the future, which are given to the dorsolateral prefrontal cortex, the CEO of the brain. These competing assessments are then evaluated and weighed until a balanced final decision is made.

We can now apply the space-time theory of consciousness to give us a definition of most forms of mental illness:

> **Mental illness is largely caused by the disruption of the delicate checks and balances between competing feedback loops that simulate the future (usually because one region of the brain is overactive or underactive).**

Because the CEO of the mind (the dorsolateral prefrontal cortex) no longer has a balanced assessment of the facts, due to this disruption in feedback loops, it begins to make strange conclusions and act in bizarre ways. The advantage of this theory is that it is testable. One has to perform MRI scans of the brain of someone who is mentally ill as it exhibits dysfunctional behavior, evaluating how its feedback loops are performing, and compare it to the MRI scans of normal people. If this theory is correct, the dysfunctional behavior (for example, hearing voices or becoming obsessed) can be traced back to a malfunctioning of the checks and balances between feedback loops. The theory can be disproven if this dysfunctional behavior is totally independent of the interplay between these regions of the brain.

Given this new theory of mental illness, we can now apply it to various forms of mental disorders, summarizing the previous discussion in this new light.

We saw earlier that the obsessive behavior of people suffering from OCD might arise when the checks and balances between several feedback loops are thrown out of balance: one registering something as amiss, another carrying out corrective action, and another one signaling that the matter has been taken care of. The failure of the checks and balances within this loop can cause the brain to be locked into a vicious cycle, so the mind never believes that the problem has been resolved.

The voices heard by schizophrenics might arise when several feedback loops are no longer balancing one another. One feedback loop generates spurious voices in the temporal cortex (i.e., the brain is talking to itself). Auditory and visual hallucinations are often checked by the anterior cingulate cortex, so a normal person can differentiate between real and fictitious voices. But if this region of the brain is not working properly, the brain is flooded with disembodied voices that it believes are real. This can cause schizophrenic behavior.

Similarly, the manic-depressive swings of someone with bipolar disorder might be traced to an imbalance between the left and right hemispheres. The necessary interplay between optimistic and pessimistic assessments is thrown off balance, and the person oscillates wildly between these two diverging moods.

Paranoia may also be viewed in this light. It results from an imbalance between the amygdala (which registers fear and exaggerates threats) and the prefrontal cortex, which evaluates these threats and puts them into perspective.

We should also stress that evolution has given us these feedback loops for a reason: to protect us. They keep us clean, healthy, and socially connected. The problem occurs when the dynamic between opposing feedback loops is disrupted.

This theory can be roughly summarized as follows:

MENTAL ILLNESS	FEEDBACK LOOP #1	FEEDBACK LOOP #2	BRAIN REGION AFFECTED
Paranoia	Perceiving a threat	Discounting threats	Amygdala/ prefrontal lobe
Schizophrenia	Creating voices	Discounting voices	Left temporal lobe/ anterior cingulate cortex
Bipolar disorder	Optimism	Pessimism	Left/right hemisphere
OCD	Anxiety	Satisfaction	Orbitofrontal cortex/ caudate nucleus/ cingulate cortex

According to the space-time theory of consciousness, many forms of mental illness are typified by the disruption of the checks and balances of opposing feedback loops in the brain that simulate the future. Brain scans are gradually identifying which regions these are. A more complete understanding of mental illness will undoubtedly reveal the involvement of many more regions of the brain. This is only a preliminary sketch.

DEEP BRAIN STIMULATION

Although the space-time theory of consciousness may give us insight into the origin of mental illness, it doesn't tell us how to create new therapies and remedies.

How will science deal with mental illness in the future? This is hard to predict, since we now realize that mental illness is not just one category, but an entire range of illnesses that can afflict the mind in a bewildering number of ways. Furthermore, the science behind mental illness is still in its infancy, with huge areas totally unexplored and unexplained.

But a new method is being tried today to treat the unending agony of people suffering from one of the most common yet stubbornly persistent forms of mental disorder, depression, which afflicts twenty million people in the United States. Ten percent of them, in turn, suffer from an incurable form of depression that has resisted all medical advances. One direct way of treating them, which holds much promise, is to place probes deep inside certain regions of the brain.

An important clue to this disorder was discovered by Dr. Helen Mayberg and colleagues, then doing research at Washington University Medical School. Using brain scans, they identified an area of the brain, called Brodmann area 25 (also called the subcallosal cingulate region), in the cerebral cortex that is consistently hyperactive in depressed individuals for whom all other forms of treatment have been unsuccessful.

These scientists used deep brain stimulation (DBS) in this area, inserting a small probe into the brain and applying an electrical shock, much like a pacemaker. The success of DBS has been astonishing in the treatment of various disorders. In the past decade, DBS has been used on forty thousand patients for motor-related diseases, such as Parkinson's and epilepsy, which cause uncontrolled movements of the body. Between 60 and 100 percent of patients report significant improvement in controlling their shaking hands. More than 250 hospitals in the United States alone now perform DBS treatments.

But then Dr. Mayberg had the idea of applying DBS directly to Brodmann area 25 to treat depression as well. Her team took twelve patients who were clinically depressed and had shown no improvement after exhaustive use of drugs, psychotherapy, and electroshock therapy.

They found that eight of these chronically depressed individuals immediately showed progress. Their success was so astonishing, in fact, that other groups raced to duplicate these results and apply DBS to other mental disorders. At present, DBS is being applied to thirty-five patients at Emory University, and thirty at other institutions.

Dr. Mayberg says, "Depression 1.0 was psychotherapy—people arguing about whose fault it was. Depression 2.0 was the idea that it's a chemical imbalance. This is Depression 3.0. What has captured everyone's imagination is that, by dissecting a complex behavior disorder into its component systems, you have a new way of thinking about it."

Although the success of DBS in treating depressed individuals is remarkable, much more research needs to be done. First, it is not clear why DBS works. It is thought that DBS destroys or impairs overactive areas of the brain (as in Parkinson's and Brodmann area 25) and is hence effective only against ailments caused by such overactivity. Second, the precision of this tool needs to be improved. Although this treatment has been used to treat a variety of brain diseases, such as phantom limb pain (when a person feels pain from a limb that has been amputated), Tourette's syndrome, and obsessive-compulsive disorder, the electrode inserted into the brain is not precise, thus affecting perhaps several million neurons rather than just the handful that are the source of distress.

Time will only improve the effectiveness of this therapy. Using MEM technology, one can create microscopic electrodes able to stimulate only a few neurons at a time. Nanotechnology may also make possible neural nanoprobes that are one molecule thick, as in carbon nanotubes. And as MRI sensitivity increases, our capability to guide these electrodes to more specific areas of the brain should grow more precise.

WAKING UP FROM A COMA

Deep brain stimulation has branched into several different avenues of research, including a beneficial side effect: increasing the number of memory cells within the hippocampus. Yet another application is to revive some individuals in a coma.

Comas represent perhaps one of the most controversial forms of consciousness, and often results in national headlines. The case of Terri Schiavo, for example, riveted the public. Due to a heart attack, she suffered a lack of oxygen, which caused massive brain injury. As a result, Schiavo went into a coma in 1990. Her husband, with the approval of doctors, wanted to allow her the dignity of dying peacefully. But her family said this was cruelly pulling the plug on someone who still had some responses to stimuli and might one day be miraculously revived. They pointed out that there had been sensational cases in the past when coma patients suddenly regained consciousness after many years in a vegetative state.

Brain scans were used to settle the question. In 2003, most neurologists, examining the CAT scans, concluded that the damage to Schiavo's brain was so extensive that she could never be revived, and that she was in a permanent vegetative state (PVS). After she died in 2005, an autopsy confirmed these results—there was no chance of revival.

In some other cases involving coma patients, however, brain scans show that the damage is not so severe, so there is a slim chance of recovery. In the summer of 2007, a man in Cleveland woke up and greeted his mother after undergoing deep brain stimulation. The man had suffered extensive brain damage eight years earlier and fell into a deep coma known as a minimally conscious state.

Dr. Ali Rezai led the team of surgeons who performed the operation. They inserted a pair of wires into the patient's brain until they reached the thalamus, which, as we have seen, is the gateway where sensory information is first processed. By sending a low-voltage current through these wires, the doctors were able to stimulate the thalamus, which in turn woke the man up from his deep coma. (Usually, sending electricity into the brain causes that part of the brain to shut down, but under certain circumstances it can act to jolt neurons into action.)

Improvements in DBS technology should increase the number of success stories in different fields. Today a DBS electrode is about 1.5 millimeters in diameter, but it touches up to a million neurons when inserted into the brain, which can cause bleeding and damage to blood vessels. One to three percent of DBS patients in fact have bleeding that can progress to a stroke. The electric charge carried by DBS probes is also still very crude, pulsing at a constant rate. Eventually, surgeons will be able to adjust the electrical charge carried by the electrodes so that each probe is made for a specific person and a specific ailment. The next generation of DBS probes is bound to be safer and more precise.

THE GENETICS OF MENTAL ILLNESS

Another attempt to understand and eventually treat mental illness involves tracing its genetic roots. Many attempts have been made in this area, with disappointing, mixed results. There is considerable evidence that schizophrenia and bipolar disorder run in families, but attempts to find the genes common to all these individuals have not been conclusive. Occasionally scientists have followed the family trees of certain individuals afflicted by mental illness and found a gene that is prevalent. But attempts to generalize this result to other families have often failed. At best, scientists have concluded that environmental factors as well as a combination of several genes are necessary to trigger mental illness. However, it has generally been accepted that each disorder has its own genetic basis.

In 2012, however, one of the most comprehensive studies ever done showed that there could in fact be a common genetic factor to mental illness after all. Scientists from the Harvard Medical School and Massachusetts General Hospital analyzed sixty thousand people worldwide and found that there was a genetic link between five major mental illnesses: schizophrenia, bipolar disorder, autism, major depression, and attention deficit hyperactivity disorder (ADHD). Together they represent a significant fraction of all mentally ill patients.

After an exhaustive analysis of the subjects' DNA, scientists found that four genes increased the risk of mental illness. Two of them involved the regulation of calcium channels in neurons. (Calcium is an essential chemical involved in the processing of neural signals.) Dr. Jordan Smoller of the Harvard Medical School says, "The calcium channels findings suggest that perhaps—and that is a big if—treatments to affect calcium channeling functioning might have effects across a range of disorders." Already, calcium channel blockers are being used to treat people with bipolar disorder. In the future, these blockers may be used to treat other mental illnesses as well.

This new result could help explain the curious fact that when mental illness runs in a family, members may manifest different forms of disorders. For example, if one twin has schizophrenia, then the other twin might have a totally different disorder, such as bipolar disorder.

The point here is that although each mental illness has its own triggers and genes, there could be a common thread running through them as well. Isolating the common factors among these diseases could give us a clue to which drugs might be most effective against them.

"What we have identified here is probably just the tip of the iceberg," says Dr. Smoller. "As these studies grow, we expect to find additional genes that might overlap." If more genes are found among these five disorders, it could open up an entirely new approach to mental illness.

If more common genes are found, it could mean that gene therapy might be able to repair the damage caused by defective genes. Or it might give rise to new drugs that could treat the illness at the neural level.

FUTURE AVENUES

So at present, there is no cure for patients with mental illness. Historically, doctors were helpless in treating them. But modern medicine has given us a variety of new possibilities and therapies to tackle this ancient problem. Just a few of them include:

1. Finding new neurotransmitters and new drugs that regulate the signaling of neurons.
2. Locating the genes linked to various mental illnesses, and perhaps using gene therapy.
3. Using deep brain stimulation to dampen or increase neural activity in certain areas.
4. Using EEG, MRI, MEG, and TES to understand precisely how the brain malfunctions.
5. And in the chapter on reverse engineering the brain, we will explore yet another promising avenue, imaging the entire brain and all its neural pathways. This may finally unravel the mystery of mental illnesses.

But to make sense of the wide variety of mental illnesses, some scientists believe that mental illnesses can be grouped into at least two major groups, each one requiring a different approach:

1. Mental disorders involving injury to the brain
2. Mental disorders triggered by incorrect wiring within the brain

The first type includes Parkinson's, epilepsy, Alzheimer's, and a wide variety of disorders caused by strokes and tumors, in which brain tissue is actually injured or malfunctioning. In the case of Parkinson's and epilepsy, there are neurons in a precise area of the brain that are overactive. In Alzheimer's, a buildup of amyloid plaque destroys brain tissue, including the hippocampus. In strokes and tumors, certain parts of the brain are silenced, causing numerous behavioral problems. Each of these disorders has to be treated differently, since each injury is different. Parkinson's and epilepsy may require probes to silence the overactive areas, while damage from Alzheimer's, strokes, and tumors is often incurable.

In the future, there will be advances in methods to deal with these injured parts of the brain besides deep brain stimulation and magnetic fields. One day stem cells may replace brain tissue that has been damaged. Or perhaps artificial replacements can be found to compensate for these injured areas using computers. In this case, the injured tissue is removed or replaced, either organically or electronically.

The second category involves disorders caused by a miswiring of the brain. Disorders like schizophrenia, OCD, depression, and bipolar disorder might fall into this category. Each region of the brain may be relatively healthy and intact, but one or more of them may be miswired, causing messages to be processed incorrectly. This category is difficult to treat, since the wiring of the brain is not well understood. So far, the main way to deal with these disorders is through drugs that influence neurotransmitters, but there is still a lot of hit or miss involved here.

But there is another altered state of consciousness that has given us new insights into the working mind. It has also provided new perspectives on how the brain works and what might happen if there is a disorder. This is the field of AI, artificial intelligence. Although it is still in its infancy, it has opened profound insights into the thinking process and has even deepened our understanding of human consciousness. So the questions are: Can silicon consciousness be achieved? If so, how might it differ from human consciousness? And will it try one day to control us?

No, I'm not interested in developing a powerful brain. All I'm after is just a mediocre brain, something like the President of the American Telephone and Telegraph Company.

—ALAN TURING

10 THE ARTIFICIAL MIND AND SILICON CONSCIOUSNESS

In February 2011, history was made.

An IBM computer called Watson did what many critics thought was impossible: it beat two contestants on a TV game show called *Jeopardy!* Millions of viewers were glued to the screen as Watson methodically annihilated its opponents on national TV, answering questions that stumped the rival contestants, and thereby claiming the $1 million prize money.

IBM pulled out all the stops in assembling a machine with a truly monumental amount of computational firepower. Watson can process data at the astonishing rate of five hundred gigabytes per second (or the equivalent of a million books per second) with sixteen trillion bytes of RAM memory. It also had access to two hundred million pages of material in its memory, including the entire storehouse of knowledge within Wikipedia. Watson could then analyze this mountain of information on live TV.

Watson is just the latest generation of "expert systems," software programs that use formal logic to access vast amounts of specialized information. (When you talk on the phone to a machine that gives you a menu of choices, this is a primitive expert system.) Expert systems will continue to evolve, making our lives more convenient and efficient.

For example, engineers are currently working to create a "robo-doc," which will appear on your wristwatch or wall screen and give you basic medical advice with 99 percent accuracy almost for free. You'd talk to it about your symptoms, and it would access the databanks of the world's leading medical centers for the latest scientific information. This will reduce unnecessary visits to the doctor, eliminate costly false alarms, and make it effortless to have regular conversations with a doctor.

Eventually we might have robot lawyers that can answer all common legal questions, or a robo-secretary that can plan vacations, trips, and dinners. (Of course, for specialized services requiring professional advice, you would still need to see a real doctor, lawyer, etc., but for common, everyday advice, these programs would suffice.)

In addition, scientists have created "chat-bots" that can mimic ordinary conversations. The average person may know tens of thousands of words. Reading the newspaper may require about two thousand words or more, but a casual conversation usually involves only a few hundred. Robots can be programmed to converse with this limited vocabulary (as long as the conversation is limited to certain well-defined subjects).

MEDIA HYPE—THE ROBOTS ARE COMING

Soon after Watson won that contest, some pundits were wringing their hands, mourning the day when the machines will take over. Ken Jennings, one of the contestants defeated by Watson, remarked to the press, "I for one welcome our new computer overlords." The pundits asked, If Watson could defeat seasoned game show contestants in a head-to-machine contest, then what chance do the rest of us mortals have to stand up to the machines? Half jokingly, Jennings said, "Brad [the other contestant] and I were the first knowledge-industry workers put out of work by the new generation of 'thinking' machines."

The commentators, however, forgot to mention that you could not go up to Watson and congratulate it for winning. You could not slap it on its back, or share a champagne toast with it. It wouldn't know what any of that meant, and in fact Watson was totally unaware that it had won at all. All the hype aside, the truth is that Watson is a highly sophisticated adding machine, able to add (or search data files) billions of times faster than the human brain, but it is totally lacking in self-awareness or common sense.

On one hand, progress in artificial intelligence has been astounding, especially in the area of raw computational power. Someone from the year 1900, viewing the calculations performed by computers today, would consider these machines to be miracles. But in another sense, progress has been painstakingly slow in building machines that can think for themselves (i.e., true automatons, without a puppet master, a controller with a joystick, or someone with a remote-control panel). Robots are totally unaware that they are robots.

Given the fact that computer power has been doubling every two years for the past fifty years under Moore's law, some say it is only a matter of time before machines eventually acquire self-awareness that rivals human intelligence. No one knows when this will happen, but humanity should be prepared for the moment when machine consciousness leaves the laboratory and enters the real world. How we deal with robot consciousness could decide the future of the human race.

BOOM AND BUST CYCLES IN AI

It is difficult to foretell the fate of AI, since it has gone through three cycles of boom and bust. Back in the 1950s, it seemed as if mechanical maids and butlers were just around the corner. Machines were being built that could play checkers and solve algebra problems. Robot arms were developed that could recognize and pick up blocks. At Stanford University, a robot was built called Shakey—basically a computer sitting on top of wheels with a camera—which could wander around a room by itself, avoiding obstacles.

Breathless articles were soon published in science magazines heralding the coming of the robot companion. Some predictions were too conservative. In 1949, *Popular Mechanics* stated that "in the future, computers will weigh no more than 1.5 tons." But others were wildly optimistic in proclaiming that the day of the robots was near. Shakey would one day become a mechanical maid or butler that would vacuum our carpets and open our doors. Movies like *2001: A Space Odyssey* convinced us that robots would soon be piloting our rocket ships to Jupiter and chatting with our astronauts. In 1965, Dr. Herbert Simon, one of the founders of AI, said flatly, "Machines will be capable, within 20 years, of doing any work a man can do." Two years later, another founding father of AI, Dr. Marvin Minsky, said that "within a generation . . . the problem of creating 'artificial intelligence' will substantially be solved."

But all this unbounded optimism collapsed in the 1970s. Checker-playing machines could only play checkers, nothing more. Mechanical arms could pick up blocks, but nothing else. They were like one-trick ponies. The most advanced robots took hours just to walk across a room. Shakey, placed in an unfamiliar environment, would easily get lost. And scientists were nowhere near understanding consciousness. In 1974, AI suffered a huge blow when both the U.S. and British governments substantially curtailed funding in the field.

But as computer power steadily increased in the 1980s, a new gold rush occurred in AI, fueled mainly by Pentagon planners hoping to put robot soldiers on the battlefield. Funding for AI hit a billion dollars by 1985, with hundreds of millions of dollars spent on projects like the Smart Truck, which was supposed to be an intelligent, autonomous truck that could enter enemy lines, do reconnaissance by itself, perform missions (such as rescuing prisoners), and then return to friendly territory. Unfortunately, the only thing that the Smart Truck did was get lost. The visible failures of these costly projects created yet another AI winter in the 1990s.

Paul Abrahams, commenting about the years he spent at MIT as a graduate student, has said, "It's as though a group of people had proposed to build a tower to the moon. Each year, they point with pride at how much higher the tower is than it was the previous year. The only trouble is that the moon isn't getting much closer."

But now, with the relentless march of computer power, a new AI renaissance has begun, and slow but substantial progress has been made. In 1997, IBM's Deep Blue computer beat world chess champion Garry Kasparov. In 2005, a robot car from Stanford won the DARPA Grand Challenge for a driverless car. Milestones continue to be reached.

This question remains: Is the third try the charm?

Scientists now realize that they vastly underestimated the problem, because most human thought is actually subconscious. The conscious part of our thoughts, in fact, represents only the tiniest portion of our computations.

Dr. Steve Pinker says, "I would pay a lot for a robot that would put away the dishes or run simple errands, but I can't, because all of the little problems that you'd need to solve to build a robot do to that, like recognizing objects, reasoning about the world, and controlling hands and feet, are unsolved engineering problems."

Although Hollywood movies tell us that terrifying Terminator robots may be just around the corner, the task of creating an artificial mind has been much more difficult than previously thought. I once asked Dr. Minsky when machines would equal and perhaps even surpass human intelligence. He said that he was confident this would happen but that he doesn't make predictions about dates anymore. Given the roller-coaster history of AI, perhaps this is the wisest approach, to map out the future of AI without setting a specific timetable.

PATTERN RECOGNITION AND COMMON SENSE

There are at least two basic problems confronting AI: pattern recognition and common sense.

Our best robots can barely recognize simple objects like a cup or a ball. The robot's eye may see details better than a natural eye, but the robot brain cannot recognize what it is seeing. If you place a robot on a strange, busy street, it quickly becomes disoriented and gets lost. Pattern recognition (e.g., identifying objects) has progressed much more slowly than previously estimated because of this problem.

When a robot walks into a room, it has to perform trillions of calculations, breaking down the objects it sees into pixels, lines, circles, squares, and triangles, and then trying to make a match with the thousands of images stored in its memory. For instance, robots see a chair as a hodgepodge of lines and dots, but they cannot easily identify the essence of "chairness." Even if a robot is able to successfully match an object to an image in its database, a slight rotation (like a chair that's been knocked over on the floor) or change in perspective (viewing the chair from a different angle) will mystify the robot. Our brains, however, automatically take different perspectives and variations into account. Our brains are subconsciously performing trillions of calculations, but the process seems effortless to us.

Robots also have a problem with common sense. They do not understand simple facts about the physical and biological world. There isn't an equation that can confirm something as self-evident (to us humans) as "muggy weather is uncomfortable" or "mothers are older than their daughters."

There has been some progress made in translating this sort of information into mathematical logic, but to catalog the common sense of a four-year-old child would require hundreds of millions of lines of computer code. As Voltaire once said, "Common sense is not so common."

For example, one of our most advanced robots is called ASIMO, built in Japan (where 30 percent of all industrial robots are made) by the Honda Corporation. This marvelous robot, about the size of a young boy, can walk, run, climb stairs, speak different languages, and dance (much better than I do, in fact). I have interacted with ASIMO on TV several times, and was very impressed by its abilities.

However, I met privately with the creators of ASIMO and asked them this key question: How smart is ASIMO, if we compare it to an animal? They admitted to me that it has the intelligence of a bug. All the walking and talking is mostly for the press. The problem is that ASIMO is, by and large, a big tape recorder. It has only a modest list of truly autonomous functions, so almost every speech or motion has to be carefully scripted ahead of time. For example, it took about three hours to film a short sequence of me interacting with ASIMO, because the hand gesture and other movement had to be programmed by a team of handlers.

If we consider this in relation to our definition of human consciousness, it seems that our current robots are stuck at a very primitive level, simply trying to make sense of the physical and social world by learning basic facts. As a consequence, robots are not even at the stage where they can plot realistic simulations of the future. Asking a robot to craft a plan to rob a bank, for instance, assumes that the robot knows all the fundamentals about banks, such as where the money is stored, what sort of security system is in place, and how the police and bystanders will react to the situation. Some of this can be programmed, but there are hundreds of nuances that the human mind naturally understands but robots do not.

Where robots excel is in simulating the future in just one precise field, such as playing chess, modeling the weather, tracing the collision of galaxies, etc. Since the laws of chess and gravity have been well known for centuries, it is only a matter of raw computer power to simulate the future of a chess game or a solar system.

Attempts to move beyond this level using brute force have also floundered. One ambitious program, called CYC, was designed to solve the commonsense problem. CYC would include millions of lines of computer code containing all the information of common sense and knowledge necessary to understand its physical and social environment. Although CYC can process hundreds of thousands of facts and millions of statements, it still cannot reproduce the level of thought of a four-year-old human. Unfortunately, after some optimistic press releases, the effort has stagnated. Many of its programmers left, deadlines have come and gone, and yet the project still continues.

IS THE BRAIN A COMPUTER?

Where did we go wrong? For the past fifty years, scientists working in AI have tried to model the brain by following the analogy with digital computers. But perhaps this was too simplistic. As Joseph Campbell once said, "Computers are like Old Testament gods; lots of rules and no mercy." If you remove a single transistor from a Pentium chip, the computer will crash immediately. But the human brain can perform quite well even if half of it is missing.

This is because the brain is not a digital computer at all, but a highly sophisticated neural network of some sort. Unlike a digital computer, which has a fixed architecture (input, output, and processor), neural networks are collections of neurons that constantly rewire and reinforce themselves after learning a new task. The brain has no programming, no operating system, no Windows, no central processor. Instead, its neural networks are massively parallel, with one hundred billion neurons firing at the same time in order to accomplish a single goal: to learn.

In light of this, AI researchers are beginning to reexamine the "top-down approach" they have followed for the last fifty years (e.g., putting all the rules of common sense on a CD). Now AI researchers are giving the "bottom-up approach" a second look. This approach tries to follow Mother Nature, which has created intelligent beings (us) via evolution, starting with simple animals like worms and fish and then creating more complex ones. Neural networks must learn the hard way, by bumping into things and making mistakes.

Dr. Rodney Brooks, former director of the famed MIT Artificial Intelligence Laboratory, and cofounder of iRobot, which makes those mechanical vacuum cleaners found in many living rooms, introduced an entirely new approach to AI. Instead of designing big, clumsy robots, why not build small, compact, insectlike robots that have to learn how to walk, just as in nature?

When I interviewed him, he told me that he used to marvel at the mosquito, which had a nearly microscopic brain with very few neurons, yet was able to maneuver in space better than any robot airplane. He built a series of remarkably simple robots, affectionately called "insectoids" or "bugbots," which scurried around the floors of MIT and could run circles around the more traditional robots. The goal was to create robots that follow the trial-and-error method of Mother Nature. In other words, these robots learn by bumping into things.

(At first, it may seem that this requires a lot of programming. The irony, however, is that neural networks require no programming at all. The only thing that the neural network does is rewire itself, by changing the strength of certain pathways each time it makes a right decision. So programming is nothing; changing the network is everything.)

Science-fiction writers once envisioned that robots on Mars would be sophisticated humanoids, walking and moving just like us, with complex programming that gave them human intelligence. The opposite has happened. Today the grandchildren of this approach—like the Mars Curiosity rover—are now roaming over the surface of Mars. They are not programmed to walk like a human. Instead, they have the intelligence of a bug, but they do quite fine in this terrain. These Mars rovers have relatively little programming; instead, they learn as they bump into obstacles.

ARE ROBOTS CONSCIOUS?

Perhaps the clearest way to see why true robot automatons do not yet exist is to rank their level of consciousness. As we have seen in Chapter 2, we can rank consciousness in four levels. Level 0 consciousness describes thermostats and plants; that is, it involves a few feedback loops in a handful of simple parameters such as temperature or sunlight. Level I consciousness describes insects and reptiles, which are mobile and have a central nervous system; it involves creating a model of your world in relationship to a new parameter, space. Then we have Level II consciousness, which creates a model of the world in relationship to others of its kind, requiring emotions. Finally we have Level III consciousness, which describes humans, who incorporate time and self-awareness to simulate how things will evolve in the future and determine our own place in these models.

We can use this theory to rank the robots of today. The first generation of robots were at Level 0, since they were static, without wheels or treads. Today's robots are at Level I, since they are mobile, but they are at a very low echelon because they have tremendous difficulty navigating in the real world. Their consciousness can be compared to that of a worm or slow insect. To fully produce Level I consciousness, scientists will have to create robots that can realistically duplicate the consciousness of insects and reptiles. Even insects have abilities that current robots do not have, such as rapidly finding hiding places, locating mates in the forest, recognizing and evading predators, or finding food and shelter.

As we mentioned earlier, we can numerically rank consciousness by the number of feedback loops at each level. Robots that can see, for example, may have several feedback loops because they have visual sensors that can detect shadows, edges, curves, geometric shapes, etc., in three-dimensional space. Similarly, robots that can hear require sensors that can detect frequency, intensity, stress, pauses, etc. The total number of these feedback loops may total ten or so (while an insect, because it can forage in the wild, find mates, locate shelter, etc., may have fifty or more feedback loops). A typical robot, therefore, may have Level I:10 consciousness.

Robots will have to be able to create a model of the world in relation to others if they are to enter Level II consciousness. As we mentioned before, Level II consciousness, to a first approximation, is computed by multiplying the number of members of its group times the number of emotions and gestures that are used to communicate between them. Robots would thus have a consciousness of Level II:0. But hopefully, the emotional robots being built in labs today may soon raise that number.

Current robots view humans as simply a collection of pixels moving on their TV sensors, but some AI researchers are beginning to create robots that can recognize emotions in our facial expressions and tone of voice. This is a first step toward robots' realizing that humans are more than just random pixels, and that they have emotional states.

In the next few decades, robots will gradually rise in Level II consciousness, becoming as intelligent as a mouse, rat, rabbit, and then a cat. Perhaps late in this century, they will be as intelligent as a monkey, and will begin to create goals of their own.

Once robots have a working knowledge of common sense and the Theory of Mind, they will be able to run complex simulations into the future featuring themselves as the principal actors and thus enter Level III consciousness. They will leave the world of the present and enter the world of the future. This is many decades beyond the capability of any robot today. Running simulations of the future means that you have a firm grasp of the laws of nature, causality, and common sense, so that you can anticipate future events. It also means that you understand human intentions and motivations, so you can predict their future behavior as well.

The numerical value of Level III consciousness, as we mentioned, is calculated by the total number of causal links one can make in simulating the future in a variety of real-life situations, divided by the average value of a control group. Computers today are able to make limited simulations in a few parameters (e.g., the collision of two galaxies, the flow of air around an airplane, the shaking of buildings in an earthquake), but they are totally unprepared to simulate the future in complex, real-life situations, so their level of consciousness would be something like Level III:5.

As we can see, it may take many decades of hard work before we have a robot that can function normally in human society.

SPEED BUMPS ON THE WAY

So when might robots finally match and exceed humans in intelligence? No one knows, but there have been many predictions. Most of them rely on Moore's law extending decades into the future. However, Moore's law is not a law at all, and in fact it ultimately violates a fundamental physical law: the quantum theory.

As such, Moore's law cannot last forever. In fact, we can already see it slowing down now. It might flatten out by the end of this or the next decade, and the consequences could be dire, especially for Silicon Valley.

The problem is simple. Right now, you can place hundreds of millions of silicon transistors on a chip the size of your fingernail, but there is a limit to how much you can cram onto these chips. Today the smallest layer of silicon in your Pentium chip is about twenty atoms in width, and by 2020 that layer might be five atoms across. But then Heisenberg's uncertainty principle kicks in, and you wouldn't be able to determine precisely where the electron is and it could "leak out" of the wire. (See the Appendix, where we discuss the quantum theory and the uncertainty principle in more detail.) The chip would short-circuit. In addition, it would generate enough heat to fry an egg on it. So leakage and heat will eventually doom Moore's law, and a replacement will soon be necessary.

If packing transistors on flat chips is maxing out in computing power, Intel is making a multibillion-dollar bet that chips will rise into the third dimension. Time will tell if this gamble pays off (one major problem with 3-D chips is that the heat generated rises rapidly with the height of the chip).

Microsoft is looking into other options, such as expanding into 2-D with parallel processing. One possibility is to spread chips horizontally in a row. Then you break up a software problem into pieces, sort out each piece on a small chip, and reassemble it at the end. However, it may be a difficult process, and software grows at a much slower pace than the supercharged exponential rate we are accustomed to with Moore's law.

These stopgap measures may add years to Moore's law. But eventually, all this must pass, too: the quantum theory inevitably takes over. This means that physicists are experimenting with a wide variety of alternatives after the Age of Silicon draws to a close, such as quantum computers, molecular computers, nanocomputers, DNA computers, optical computers, etc. None of these technologies, however, is ready for prime time.

THE UNCANNY VALLEY

But assume for the moment that one day we will coexist with incredibly sophisticated robots, perhaps using chips with molecular transistors instead of silicon. How closely do we want our robots to look like us? Japan is the world's leader in creating robots that resemble cuddly pets and children, but their designers are careful not to make their robots appear too human, which can be unnerving. This phenomenon was first studied by Dr. Masahiro Mori in Japan in 1970, and is called the "uncanny valley." It posits that robots look creepy if they look too much like humans. (The effect was actually first mentioned by Darwin in 1839 in *The Voyage of the Beagle* and again by Freud in 1919 in an essay titled "The Uncanny.") Since then, it has been studied very carefully not just by AI researchers but also by animators, advertisers, and anyone promoting a product involving humanlike figures. For instance, in a review of the movie *The Polar Express*, a CNN writer noted, "Those human characters in the film come across as downright . . . well, creepy. So *The Polar Express* is at best disconcerting, and at worst, a wee bit horrifying."

According to Dr. Mori, the more a robot looks like a human, the more we feel empathy toward it, but only up to a point. There is a dip in empathy as the robot approaches actual human appearance—hence the uncanny valley. If the robot looks very similar to us save for a few features that are "uncanny," it creates a feeling of revulsion and fear. If the robot appears 100 percent human, indistinguishable from you and me, then we'll register positive emotions again.

This has practical implications. For example, should robots smile? At first, it seems obvious that robots should smile to greet people and make them feel comfortable. Smiling is a universal gesture that signals warmth and welcome. But if the robot smile is too realistic, it makes people's skin crawl. (For example, Halloween masks often feature fiendish-looking ghouls that are grinning.) So robots should smile only if they are childlike (i.e., with big eyes and a round face) or are perfectly human, and nothing in between. (When we force a smile, we activate facial muscles with our prefrontal cortex. But when we smile because we are in a good mood, our nerves are controlled by our limbic system, which activates a slightly different set of muscles. Our brains can tell the subtle difference between the two, which was beneficial for our evolution.)

This effect can also be studied using brain scans. Let's say that a subject is placed into an MRI machine and is shown a picture of a robot that looks perfectly human, except that its bodily motions are slightly jerky and mechanical. The brain, whenever it sees anything, tries to predict that object's motion into the future. So when looking at a robot that appears to be human, the brain predicts that it will move like a human. But when the robot moves like a machine, there is a mismatch, which makes us uncomfortable. In particular, the parietal lobe lights up (specifically, the part of the lobe where the motor cortex connects with the visual cortex). It is believed that mirror neurons exist in this area of the parietal lobe. This makes sense, because the visual cortex picks up the image of the humanlike robot, and its motions are predicted via the motor cortex and by mirror neurons. Finally, it is likely that the orbitofrontal cortex, located right behind the eyes, puts everything together and says, "Hmmm, something is not quite right."

Hollywood filmmakers are aware of this effect. When spending millions on making a horror movie, they realize that the scariest scene is not when a gigantic blob or Frankenstein's monster pounces out of the bushes. The scariest scene is when there is a perversion of the ordinary. Think of the movie *The Exorcist.* What scene made moviegoers vomit as they ran to escape the theater or faint right in their seats? Was it the scene when a demon appears? No. Theaters across the world erupted in shrill screams and loud sobs when Linda Blair turned her head completely around.

This effect can also be demonstrated in young monkeys. If you show them pictures of Dracula or Frankenstein, they simply laugh and rip the pictures apart. But what sends these young monkeys screaming in terror is a picture of a decapitated monkey. Once again, it is the perversion of the ordinary that elicits the greatest fear. (In Chapter 2, we mentioned that the space-time theory of consciousness explains the nature of humor, since the brain simulates the future of a joke, and then is surprised to hear the punch line. This also explains the nature of horror. The brain simulates the future of an ordinary, mundane event, but then is shocked when things suddenly become horribly perverted.)

For this reason, robots will continue to look somewhat childlike in appearance, even as they approach human intelligence. Only when robots can act realistically like humans will their designers make them look fully human.

SILICON CONSCIOUSNESS

As we've seen, human consciousness is an imperfect patchwork of different abilities developed over millions of years of evolution. Given information about their physical and social world, robots may be able to create simulations similar (or in some respects, even superior) to ours, but silicon consciousness might differ from ours in two key areas: emotions and goals.

Historically, AI researchers ignored the problem of emotions, considering it a secondary issue. The goal was to create a robot that was logical and rational, not scatterbrained and impulsive. Hence, the science fiction of the 1950s and '60s stressed robots (and humanoids like Spock on *Star Trek*) that had perfect, logical brains.

We saw with the uncanny valley that robots will have to look a certain way if they're to enter our homes, but some people argue that robots must also have emotions so that we can bond with, take care of, and interact productively with them. In other words, robots will need Level II consciousness. To accomplish this, robots will first have to recognize the full spectrum of human emotions. By analyzing subtle facial movements of the eyebrows, eyelids, lips, cheeks, etc., a robot will be able to identify the emotional state of a human, such as its owner. One institution that has excelled in creating robots that recognize and mimic emotion is the MIT Media Laboratory. I have had the pleasure of visiting the laboratory, outside Boston, on several occasions, and it is like visiting a toy factory for grown-ups. Everywhere you look, you see futuristic, high-tech devices designed to make our lives more interesting, enjoyable, and convenient.

As I looked around the room, I saw many of the high-tech graphics that eventually found their way into Hollywood movies like *Minority Report* and *AI*. As I wandered through this playground of the future, I came across two intriguing robots, Huggable and Nexi. Their creator, Dr. Cynthia Breazeal, explained to me that these robots have specific goals. Huggable is a cute teddy bear–like robot that can bond with children. It can identify the emotions of children; it has video cameras for eyes, a speaker for its mouth, and sensors in its skin (so it can tell when it is being tickled, poked, or hugged). Eventually, a robot like this might become a tutor, babysitter, nurse's aide, or a playmate.

Nexi, on the other hand, can bond with adults. It looks a little like the Pillsbury Doughboy. It has a round, puffy, friendly face, with large eyes that can roll around. It has already been tested in a nursing home, and the elderly patients all loved it. Once the seniors got accustomed to Nexi, they would kiss it, talk to it, and miss it when it had to leave. (See Figure 12.)

Dr. Breazeal told me she designed Huggable and Nexi because she was not satisfied with earlier robots, which looked like tin cans full of wires, gears, and motors. In order to design a robot that could interact emotionally with people, she needed to figure out how she could get it to perform and bond like us. Plus, she wanted robots that weren't stuck on a laboratory shelf but could venture out into the real world. The former director of MIT's Media Lab, Dr. Frank Moss, says, "That is why Breazeal decided in 2004 that it was time to create a new generation of social robots that could live anywhere: homes, schools, hospitals, elder care facilities, and so on."

At Waseda University in Japan, scientists are working on a robot that has

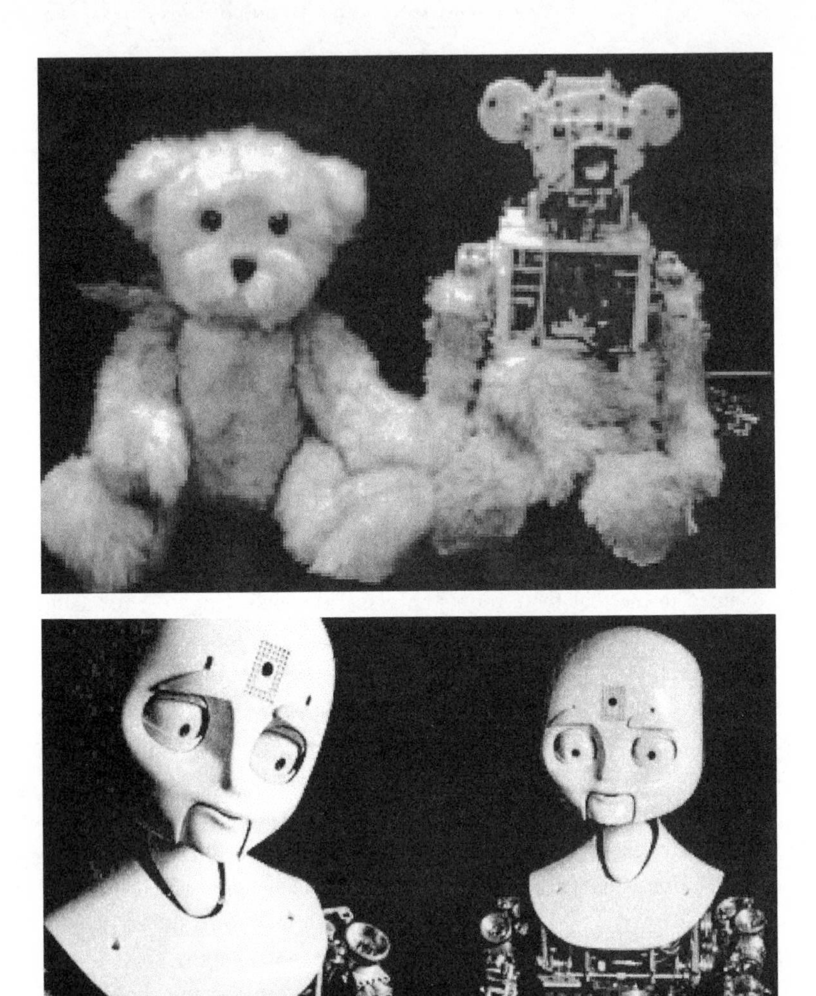

Figure 12. Huggable (top) and Nexi (bottom), two robots built at the MIT Media Laboratory that were explicitly designed to interact with humans via emotions.

upper-body motions representing emotions (fear, anger, surprise, joy, disgust, sadness) and can hear, smell, see, and touch. It has been programmed to carry out simple goals, such as satisfying its hunger for energy and avoiding dangerous situations. Their goal is to integrate the senses with the emotions, so that the robot acts appropriately in different situations.

Not to be outdone, the European Commission is funding an ongoing project, called Feelix Growing, which seeks to promote artificial intelligence in the UK, France, Switzerland, Greece, and Denmark.

EMOTIONAL ROBOTS

Meet Nao.

When he's happy, he will stretch out his arms to greet you, wanting a big hug. When he's sad, he turns his head downward and appears forlorn, with his shoulders hunched forward. When he's scared, he cowers in fear, until someone pats him reassuringly on the head.

He's just like a one-year-old boy, except that he's a robot. Nao is about one and a half feet tall, and looks very much like some of the robots you see in a toy store, like the Tranformers, except he's one of the most advanced emotional robots on earth. He was built by scientists at the UK's University of Hertfordshire, whose research was funded by the European Union.

His creators have programmed him to show emotions like happiness, sadness, fear, excitement, and pride. While other robots have rudimentary facial and verbal gestures that communicate their emotions, Nao excels in body language, such as posture and gesture. Nao even dances.

Unlike other robots, which specialize in mastering just one area of the emotions, Nao has mastered a wide range of emotional responses. First, Nao locks onto visitors' faces, identifies them, and remembers his previous interactions with each of them. Second, he begins to follow their movements. For example, he can follow their gaze and tell what they are looking at. Third, he begins to bond with them and learns to respond to their gestures. For example, if you smile at him, or pat him on his head, he knows that this is a positive sign. Because his brain has neural networks, he learns from interactions with humans. Fourth, Nao exhibits emotions in response to his interactions with people. (His emotional responses are all preprogrammed, like a tape recorder, but he decides which emotion to choose to fit the situation.) And lastly, the more Nao interacts with a human, the better he gets at understanding the moods of that person and the stronger the bond becomes.

Not only does Nao have a personality, he can actually have several of them. Because he learns from his interactions with humans and each interaction is unique, eventually different personalities begin to emerge. For example, one personality might be quite independent, not requiring much human guidance. Another personality might be timid and fearful, scared of objects in a room, constantly requiring human intervention.

The project leader for Nao is Dr. Lola Cañamero, a computer scientist at the University of Hertfordshire. To start this ambitious project, she analyzed the interactions of chimpanzees. Her goal was to reproduce, as closely as she could, the emotional behavior of a one-year-old chimpanzee.

She sees immediate applications for these emotional robots. Like Dr. Breazeal, she wants to use these robots to relieve the anxiety of young children who are in hospitals. She says, "We want to explore different roles—the robots will help the children to understand their treatment, explain what they have to do. We want to help the children to control their anxiety."

Another possibility is that the robots will become companions at nursing homes. Nao could become a valuable addition to the staff of a hospital. At some point, robots like these might become playmates to children and a part of the family.

"It's hard to predict the future, but it won't be too long before the computer in front of you will be a social robot. You'll be able to talk to it, flirt with it, or even get angry and yell at it—and it will understand you and your emotions," says Dr. Terrence Sejnowski of the Salk Institute, near San Diego. This is the easy part. The hard part is to gauge the response of the robot, given this information. If the owner is angry or displeased, the robot has to be able to factor this into its response.

EMOTIONS: DETERMINING WHAT IS IMPORTANT

What's more, AI researchers have begun to realize that emotions may be a key to consciousness. Neuroscientists like Dr. Antonio Damasio have found that when the link between the prefrontal lobe (which governs rational thought) and the emotional centers (e.g., the limbic system) is damaged, patients cannot make value judgments. They are paralyzed when making the simplest of decisions (what things to buy, when to set an appointment, which color pen to use) because everything has the same value to them. Hence, emotions are not a luxury; they are absolutely essential, and without them a robot will have difficulty determining what is important and what is not. So emotions, instead of being peripheral to the progress of artificial intelligence, are now assuming central importance.

If a robot encounters a raging fire, it might rescue the computer files first, not the people, since its programming might say that valuable documents cannot be replaced but workers always can be. It is crucial that robots be programmed to distinguish between what is important and what is not, and emotions are shortcuts the brain uses to rapidly determine this. Robots would thus have to be programmed to have a value system—that human life is more important than material objects, that children should be rescued first in an emergency, that objects with a higher price are more valuable than objects with a lower price, etc. Since robots do not come equipped with values, a huge list of value judgments must be uploaded into them.

The problem with emotions, however, is that they are sometimes irrational, while robots are mathematically precise. So silicon consciousness may differ from human consciousness in key ways. For example, humans have little control over emotions, since they happen so rapidly and because they originate in the limbic system, not the prefrontal cortex of the brain. Furthermore, our emotions are often biased. Numerous tests have shown that we tend to overestimate the abilities of people who are handsome or pretty. Good-looking people tend to rise higher in society and have better jobs, although they may not be as talented as others. As the expression goes, "Beauty has its privileges."

Similarly, silicon consciousness may not take into account subtle cues that humans use when they meet one another, such as body language. When people enter a room, young people usually defer to older ones and low-ranked staff members show extra courtesy to senior officials. We show our deference in the way we move our bodies, our choice of words, and our gestures. Because body language is older than language itself, it is hardwired into the brain in subtle ways. Robots, if they are to interact socially with people, will have to learn these unconscious cues.

Our consciousness is influenced by peculiarities in our evolutionary past, which robots will not have, so silicon consciousness may not have the same gaps or quirks as ours.

A MENU OF EMOTIONS

Since emotions have to be programmed into robots from the outside, manufacturers may offer a menu of emotions carefully chosen on the basis of whether they are necessary, useful, or will increase bonding with the owner.

In all likelihood, robots will be programmed to have only a few human emotions, depending on the situation. Perhaps the emotion most valued by the robot's owner will be loyalty. One wants a robot that faithfully carries out its commands without complaints, that understands the needs of the master and anticipates them. The last thing an owner will want is a robot with an attitude, one that talks back, criticizes people, and whines. Helpful criticisms are important, but they must be made in a constructive, tactful way. Also, if humans give it conflicting commands, the robot should know to ignore all of them except those coming from its owner.

Empathy will be another emotion that will be valued by the owner. Robots that have empathy will understand the problems of others and will come to their aid. By interpreting facial movements and listening to tone of voice, robots will be able to identify when a person is in distress and will provide assistance when possible.

Strangely, fear is another emotion that is desirable. Evolution gave us the feeling of fear for a reason, to avoid certain things that are dangerous to us. Even though robots will be made of steel, they should fear certain things that can damage them, like falling off tall buildings or entering a raging fire. A totally fearless robot is a useless one if it destroys itself.

But certain emotions may have to be deleted, forbidden, or highly regulated, such as anger. Given that robots could be built to have great physical strength, an angry robot could create tremendous problems in the home and workplace. Anger could get in the way of its duties and cause great damage to property. (The original evolutionary purpose of anger was to show our dissatisfaction. This can be done in a rational, dispassionate way, without getting angry.)

Another emotion that should be deleted is the desire to be in command. A bossy robot will only make trouble and might challenge the judgment and wishes of the owner. (This point will also be important later, when we discuss whether robots will one day take over from humans.) Hence the robot will have to defer to the wishes of the owner, even if this may not be the best path.

But perhaps the most difficult emotion to convey is humor, which is a glue that can bond total strangers together. A simple joke can defuse a tense situation or inflame it. The basic mechanics of humor are simple: they involve a punch line that is unanticipated. But the subtleties of humor can be enormous. In fact, we often size up other people on the basis of how they react to certain jokes. If humans use humor as a gauge to measure other humans, then one can appreciate the difficulty of creating a robot that can tell if a joke is funny or not. President Ronald Reagan, for example, was famous for defusing the most difficult questions with a quip. In fact, he accumulated a large card catalog of jokes, barbs, and wisecracks, because he understood the power of humor. (Some pundits concluded that he won the presidential debate against Walter Mondale when he was asked if he was too old to be president. Reagan replied that he would not hold the youth of his opponent against him.) Also, laughing inappropriately could have disastrous consequences (and is, in fact, sometimes a sign of mental illness). The robot has to know the difference between laughing with or at someone. (Actors are well aware of the diverse nature of laughter. They are skilled enough to create laughter that can represent horror, cynicism, joy, anger, sadness, etc.) So, at least until the theory of artificial intelligence becomes more developed, robots should stay away from humor and laughter.

PROGRAMMING EMOTIONS

In this discussion we have so far avoided the difficult question of precisely how these emotions would be programmed into a computer. Because of their complexity, emotions will probably have to be programmed in stages.

First, the easiest part is identifying an emotion by analyzing the gestures in a person's face, lips, eyebrows, and tone of voice. Today's facial recognition technology is already capable of creating a dictionary of emotions, so that certain facial expressions mean certain things. This process actually goes back to Charles Darwin, who spent a considerable amount of time cataloging emotions common to animals and humans.

Second, the robot must respond rapidly to this emotion. This is also easy.

If someone is laughing, the robot will grin. If someone is angry, the robot will get out of his way and avoid conflict. The robot would have a large encyclopedia of emotions programmed into it, and hence would know how to make a rapid response to each one.

The third stage is perhaps the most complex because it involves trying to determine the underlying motivation behind the original emotion. This is difficult, since a variety of situations can trigger a single emotion. Laughter may mean that someone is happy, heard a joke, or watched someone fall. Or it might mean that a person is nervous, anxious, or insulting someone. Likewise, if someone is screaming, there may be an emergency, or perhaps someone is just reacting with joy and surprise. Determining the reason behind an emotion is a skill that even humans have difficulty with. To do this, the robot will have to list the various possible reasons behind an emotion and try to determine the reason that makes the most sense. This means trying to find a reason behind the emotion that fits the data best.

And fourth, once the robot has determined the origin of this emotion, it has to make the appropriate response. This is also difficult, since there are often several possible responses, and the wrong one may make the situation worse. The robot already has, within its programming, a list of possible responses to the original emotion. It has to calculate which one will best serve the situation, which means simulating the future.

WILL ROBOTS LIE?

Normally, we might think of robots as being coldly analytical and rational, always telling the truth. But once robots become integrated into society, they will probably have to learn to lie or at least tactfully restrain their comments.

In our own lives, several times in a typical day we are confronted with situations where we have to tell a white lie. If people ask us how they look, we often dare not tell the truth. White lies, in fact, are like a grease that makes society run smoothly. If we were suddenly forced to tell the whole truth (like Jim Carrey in *Liar Liar*), we most likely would wind up creating chaos and hurting people. People would be insulted if you told them what they really looked like or how you really felt. Bosses would fire you. Lovers would dump you. Friends would abandon you. Strangers would slap you. Some thoughts are better kept confidential.

In the same way, robots may have to learn how to lie or conceal the truth, or else they might wind up offending people and being decommissioned by their owners. At a party, if a robot tells the truth, it could reflect badly on its owner and create an uproar. So if someone asks for its opinion, it will have to learn how to be evasive, diplomatic, and tactful. It must either dodge the question, change the subject, give platitudes for answers, reply with a question, or tell white lies (all things that today's chat-bots are increasingly good at). This means that the robot has already been programmed to have a list of possible evasive responses, and must choose the one that creates the fewest complications.

One of the few times that a robot would tell the entire truth would be if asked a direct question by its owner, who understands that the answer might be brutally honest. Perhaps the only other time when the robot will tell the truth is when there is a police investigation and the absolute truth is necessary. Other than that, robots will be able to freely lie or conceal the whole truth to keep the wheels of society functioning.

In other words, robots have to be socialized, just like teenagers.

CAN ROBOTS FEEL PAIN?

Robots, in general, will be assigned to do types of tasks that are dull, dirty, and dangerous. There is no reason why robots can't do repetitive or dirty jobs indefinitely, since we wouldn't program them to feel boredom or disgust. The real problem emerges when robots are faced with dangerous jobs. At that point, we might actually want to program them to feel pain.

We evolved the sense of pain because it helped us survive in a dangerous environment. There is a genetic defect in which children are born without the ability to feel pain. This is called congenital analgesia. At first glance, this may seem to be a blessing, since these children do not cry when they experience injury, but it is actually more of a curse. Children with this affliction have serious problems, such as biting off parts of their tongue, suffering severe skin burns, and cutting themselves, often leading to amputations of their fingers. Pain alerts us to danger, telling us when to move our hand away from the burning stove or to stop running on a twisted ankle.

At some point robots must be programmed to feel pain, or else they will not know when to avoid precarious situations. The first sense of pain they must have is hunger (i.e., a craving for electrical energy). As their batteries run out, they will get more desperate and urgent, realizing that soon their circuits will shut down, leaving all their work in disarray. The closer they are to running out of power, the more anxious they will become.

Also, regardless of how strong they are, robots may accidentally pick up an object that is too heavy, which could cause their limbs to break. Or they may suffer overheating by working with molten metal in a steel factory, or by entering a burning building to help firemen. Sensors for temperature and stress would alert them that their design specifications are being exceeded.

But once the sensation of pain is added to their menu of emotions, this immediately raises ethical issues. Many people believe that we should not inflict unnecessary pain on animals, and people may feel the same about robots as well. This opens the door to robots' rights. Laws may have to be passed to restrict the amount of pain and danger that a robot is allowed to face. People will not care if a robot is performing dull or dirty tasks, but if they feel pain doing a dangerous one, they may begin to lobby for laws to protect robots. This may even start a legal conflict, with owners and manufacturers of robots arguing for increasing the level of pain that robots can endure, while ethicists may argue for lowering it.

This, in turn, may set off other ethical debates about other robot rights. Can robots own property? What happens if they accidentally hurt someone? Can they be sued or punished? Who is responsible in a lawsuit? Can a robot own another robot? This discussion raises another sticky question: Should robots be given a sense of ethics?

ETHICAL ROBOTS

At first, the idea of ethical robots seems like a waste of time and effort. However, this question takes on a sense of urgency when we realize that robots will make life-and-death decisions. Since they will be physically strong and have the capability of saving lives, they will have to make split-second ethical choices about whom to save first.

Let's say there is a catastrophic earthquake and children are trapped in a rapidly crumbling building. How should the robot allocate its energy? Should it try to save the largest number of children? Or the youngest? Or the most vulnerable? If the debris is too heavy, the robot may damage its electronics. So the robot has to decide yet another ethical question: How does it weigh the number of children it saves versus the amount of damage that it will sustain to its electronics?

Without proper programming, the robot may simply halt, waiting for a human to make the final decision, wasting valuable time. So someone will have to program it ahead of time so that the robot automatically makes the "right" decision.

These ethical decisions will have to be preprogrammed into the computer from the start, since there is no law of mathematics that can put a value on saving a group of children. Within its programming, there has to be a long list of things, ranked in terms of how important they are. This is tedious business. In fact, it sometimes takes a human a lifetime to learn these ethical lessons, but a robot has to learn them rapidly, before it leaves the factory, if it is to safely enter society.

Only people can do this, and even then ethical dilemmas sometimes confound us. But this raises questions: Who will make the decisions? Who decides the order in which robots save human lives?

The question of how decisions will ultimately be made will probably be resolved via a combination of the law and the marketplace. Laws will have to be passed so that there is, at minimum, a ranking of importance of whom to save in an emergency. But beyond that, there are thousands of finer ethical questions. These subtler decisions may be decided by the marketplace and common sense.

If you work for a security firm guarding important people, you will have to tell the robot how to save people in a precise order in different situations, based on considerations such as fulfilling the primary duty but also doing it within budget.

What happens if a criminal buys a robot and wants the robot to commit a crime? This raises a question: Should a robot be allowed to defy its owner if it is asked to break the law? We saw from the previous example that robots must be programmed to understand the law and also make ethical decisions. So if it decides that it is being asked to break the law, it must be allowed to disobey its master.

There is also the ethical dilemma posed by robots reflecting the beliefs of their owners, who may have diverging morals and social norms. The "culture wars" that we see in society today will only be magnified when we have robots that reflect the opinions and beliefs of their owners. In some sense, this conflict is inevitable. Robots are mechanical extensions of the dreams and wishes of their creators, and when robots are sophisticated enough to make moral decisions, they will do so.

The fault lines of society may be stressed when robots begin to exhibit behaviors that challenge our values and goals. Robots owned by youth leaving a noisy, raucus rock concert may conflict with robots owned by elderly residents of a quiet neighborhood. The first set of robots may be programmed to amplify the sounds of the latest bands, while the second set may be programmed to keep noise levels to an absolute minimum. Robots owned by devout, churchgoing fundamentalists may get into arguments with robots owned by atheists. Robots from different nations and cultures may be designed to reflect the mores of their society, which may clash (even for humans, let alone robots).

So how does one program robots to eliminate these conflicts?

You can't. Robots will simply reflect the biases and prejudices of their creators. Ultimately, the cultural and ethical differences between these robots will have to be settled in the courts. There is no law of physics or science that determines these moral questions, so eventually laws will have to be written to handle these social conflicts. Robots cannot solve the moral dilemmas created by humans. In fact, robots may amplify them.

But if robots can make ethical and legal decisions, can they also feel and understand sensations? If they succeed in saving someone, can they experience joy? Or can they even feel things like the color red? Coldly analyzing the ethics of whom to save is one thing, but understanding and feeling is another. So can robots feel?

CAN ROBOTS UNDERSTAND OR FEEL?

Over the centuries, a great many theories have been advanced about whether a machine can think and feel. My own philosophy is called "constructivism"; that is, instead of endlessly debating the question, which is pointless, we should be devoting our energy to creating an automaton to see how far we can get. Otherwise we wind up in endless philosophical debates that are never ultimately resolved. The advantage of science is that, once everything is said and done, one can perform experiments to settle a question decisively.

Thus, to settle the question of whether a robot can think, the final resolution may be to build one. Some, however, have argued that machines will never be able to think like a human. Their strongest argument is that, although a robot can manipulate facts faster than a human, it does not "understand" what it is manipulating. Although it can process senses (e.g., color, sound) better than a human, it cannot truly "feel" or "experience" the essence of these senses.

For example, philosopher David Chalmers has divided the problems of AI into two categories, the Easy Problems and the Hard Problems. To him, the Easy Problems are creating machines that can mimic more and more human abilities, such as playing chess, adding numbers, recognizing certain patterns, etc. The Hard Problems involve creating machines that can understand feelings and subjective sensations, which are called "qualia."

Just as it is impossible to teach the meaning of the color red to a blind person, a robot will never be able to experience the subjective sensation of the color red, they say. Or a computer might be able to translate Chinese words into English with great fluency, but it will never be able to understand what it is translating. In this picture, robots are like glorified tape recorders or adding machines, able to recite and manipulate information with incredible precision, but without any understanding whatsoever.

These arguments have to be taken seriously, but there is also another way of looking at the question of qualia and subjective experience. In the future, a machine most likely will be able to process a sensation, such as the color red, much better than any human. It will be able to describe the physical properties of red and even use it poetically in a sentence better than a human. Does the robot "feel" the color red? The point becomes irrelevant, since the word "feel" is not well defined. At some point, a robot's description of the color red may exceed a human's, and the robot may rightly ask: Do humans really understand the color red? Perhaps humans cannot really understand the color red with all the nuances and subtly that a robot can.

As behaviorist B. F. Skinner once said, "The real problem is not whether machines think, but whether men do."

Similarly, it is only a matter of time before a robot will be able to define Chinese words and use them in context much better than any human. At that point, it becomes irrelevant whether the robot "understands" the Chinese language. For all practical purposes, the computer will know the Chinese language better than any human. In other words, the word "understand" is not well defined.

One day, as robots surpass our ability to manipulate these words and sensations, it will become irrelevant whether the robot "understands" or "feels" them. The question will cease to have any importance.

As mathematician John von Neumann said, "In mathematics, you don't understand things. You just get used to them."

So the problem lies not in the hardware but in the nature of human language, in which words that are not well defined mean different things to different people. The great quantum physicist Niels Bohr was once asked how one could understand the deep paradoxes of the quantum theory. The answer, he replied, lies in how you define the word "understand."

Dr. Daniel Dennett, a philosopher at Tufts University, has written, "There could not be an objective test to distinguish a clever robot from a conscious person. Now you have a choice: you can either cling to the Hard Problem, or you can shake your head in wonder and dismiss it. Just let go."

In other words, there is no such thing as the Hard Problem.

To the constructivist philosophy, the point is not to debate whether a machine can experience the color red, but to construct the machine. In this picture, there is a continuum of levels describing the words "understand" and "feel." (This means that it might even be possible to give numerical values to the degree of understanding and feeling.) At one end we have the clumsy robots of today, which can manipulate a few symbols but not much more. At the other end we have humans, who pride themselves on feeling qualia. But as time goes by, robots will eventually be able to describe sensations better than us on any level. Then it will be obvious that robots understand.

This was the philosophy behind Alan Turing's famous Turing test. He predicted that one day a machine would be built that could answer any question, so that it would be indistinguishable from a human. He said, "A computer would deserve to be called intelligent if it could deceive a human into believing that it was human."

Physicist and Nobel laureate Francis Crick said it best. In the last century, he noted, biologists had heated debates over the question "What is life?" Now, with our understanding of DNA, scientists realize that the question is not well defined. There are many variations, layers, and complexities to that simple question. The question "What is life?" simply faded away. The same may eventually apply to feeling and understanding.

SELF-AWARE ROBOTS

What steps must be taken before computers like Watson have self-awareness? To answer this question, we have to refer back to our definition of self-awareness: the ability to put one's self inside a model of the environment, and then run simulations of this model into the future to achieve a goal. This first step requires a very high level of common sense in order to anticipate a variety of events. Then the robot has to put itself inside this model, which requires an understanding of the various courses of action it may take.

At Meiji University, scientists have taken the first steps to create a robot with self-awareness. This is a tall order, but they think they can do it by creating robots with a Theory of Mind. They started by building two robots. The first was programmed to execute certain motions. The second was programmed to observe the first robot, and then to copy it. They were able to create a second robot that could systematically mimic the behavior of the first just by watching it. This is the first time in history that a robot has been built specifically to have some sense of self-awareness. The second robot has a Theory of Mind; that is, it is capable of watching another robot and then mimicking its motions.

In 2012, the next step was taken by scientists at Yale University who created a robot that passed the mirror test. When animals are placed in front of a mirror, most of them think the image in the mirror is that of another animal. As we recall, only a few animals have passed the mirror test, realizing that the mirror image was a reflection of themselves. The scientists at Yale created a robot called Nico that resembles a gangly skeleton made of twisted wires, with mechanical arms and two bulging eyes sitting on top. When placed in front of a mirror, Nico not only recognized itself but could also deduce the location of objects in a room by looking at their images in the mirror. This is similar to what we do when we look into a rearview mirror and infer the location of objects behind us.

Nico's programmer, Justin Hart, says, "To our knowledge, this is the first robotic system to attempt to use a mirror in this way, representing a significant step towards a cohesive architecture that allows robots to learn about their bodies and appearance through self-observation, and an important capability required in order to pass the mirror test."

Because the robots at Meiji University and Yale University represent the state of the art in terms of building robots with self-awareness, it is easy to see that scientists have a long ways to go before they can create robots with humanlike self-awareness.

Their work is just the first step, because our definition of self-awareness demands that the robot use this information to create simulations of the future. This is far beyond the capability of Nico or any other robot.

This raises the important question: How can a computer gain full self-awareness? In science fiction, we often encounter a situation where the Internet suddenly becomes self-aware, as in the movie *The Terminator.* Since the Internet is connected to the entire infrastructure of modern society (e.g., our sewer system, our electricity, our telecommunications, our weapons), it would be easy for a self-aware Internet to seize control of society. We would be left helpless in this situation. Scientists have written that this may happen as an example of an "emergent phenomenon" (i.e., when you amass a sufficiently large number of computers together, there can be a sudden phase transition to a higher stage, without any input from the outside).

However, this says everything and it says nothing, because it leaves out all the important steps in between. It's like saying that a highway can suddenly become self-aware if there are enough roads.

But in this book we have given a definition of consciousness and self-awareness, so it should be possible to list the steps by which the Internet can become self-aware.

First, an intelligent Internet would have to continually make models of its place in the world. In principle, this information can be programmed into the Internet from the outside. This would involve describing the outside world (i.e., Earth, its cities, and its computers), all of which can be found on the Internet itself.

Second, it would have to place itself in the model. This information is also easily obtained. It would involve giving all the specifications of the Internet (the number of computers, nodes, transmission lines, etc.) and its relationship to the outside world.

But step three is by far the most difficult. It means continually running simulations of this model into the future, consistent with a goal. This is where we hit a brick wall. The Internet is not capable of running simulations into the future, and it has no goals. Even in the scientific world, simulations into the future are usually done in just a few parameters (e.g., simulating the collision of two black holes). Running a simulation of the model of the world containing the Internet is far beyond the programming available today. It would have to incorporate all the laws of common sense, all the laws of physics, chemistry, and biology, as well as facts about human behavior and human society.

In addition, this intelligent Internet would have to have a goal. Today it is just a passive highway, without any direction or purpose. Of course, one can in principle impose a goal on the Internet. But let us consider the following problem: Can you create an Internet whose goal is self-preservation?

This would be the simplest possible goal, but no one knows how to program even this simple task. Such a program, for example, would have to stop any attempt to shut down the Internet by pulling the plug. At present, the Internet is totally incapable of recognizing a threat to its existence, let alone plotting ways to prevent it. (For example, an Internet capable of detecting threats to its existence would have to be able to identify attempts to shut down its power, cut lines of communication, destroy its servers, disable its fiber-optic and satellite connections, etc. Furthermore, an Internet capable of defending itself against these attacks would have to have countermeasures for each scenario and then run these attempts into the future. No computer on Earth is capable of doing even a fraction of such things.)

In other words, one day it may be possible to create self-aware robots, even a self-aware Internet, but that day is far into the future, perhaps at the end of this century.

But assume for the moment that the day has arrived, that self-aware robots walk among us. If a self-aware robot has goals that are compatible with our own, then this type of artificial intelligence will not pose a problem. But what happens if the goals are different? The fear is that humans may be outwitted by self-aware robots and then may be enslaved. Because of their superior ability to simulate the future, the robots could plot the outcomes of many scenarios to find the best way to overthrow humanity.

One way this possibility may be controlled is to make sure that the goals of these robots are benevolent. As we have seen, simulating the future is not enough. These simulations must serve some final goal. If a robot's goal is merely to preserve itself, then it would react defensively to any attempt to pull the plug, which could spell trouble for mankind.

WILL ROBOTS TAKE OVER?

In almost all science-fiction tales, the robots become dangerous because of their desire to take over. The word "robot," in fact, comes from the Czech word for "worker," first seen in the 1920 play *R.U.R. (Rossum's Universal Robots)* by Karel Čapek, in which scientists create a new race of mechanical beings that look identical to humans. Soon there are thousands of these robots performing menial and dangerous tasks. However, humans mistreat them badly, and one day they rebel and destroy the human race. Although these robots have taken over Earth, they have one defect: they cannot reproduce. But at the end of the play, two robots fall in love. So perhaps a new branch of "humanity" emerges once again.

A more realistic scenario comes from the movie *The Terminator,* in which the military has created a supercomputer network called Skynet that controls the entire U.S. nuclear stockpile. One day, it wakes up and becomes sentient. The military tries to shut down Skynet but then realizes there is a flaw in its programming: it is designed to protect itself, and the only way to do so is by eliminating the problem—humanity. It starts a nuclear war, which reduces humanity to a ragtag bunch of misfits and rebels fighting the juggernaut of the machines.

It is certainly possible that robots could become a threat. The current Predator drone can target its victims with deadly accuracy, but it is controlled by someone with a joystick thousands of miles away. According to the *New York Times,* the orders to fire come directly from the president of the United States. But in the future, a Predator might have face recognition technology and permission to fire if it is 99 percent confident of the identity of its target. Without human intervention, it could automatically use this technology to fire at anyone who fits the profile.

Now assume that such a drone suffers a breakdown, such that its facial recognition software malfunctions. Then it becomes a rogue robot, with permission to kill anyone in sight. Worse, imagine a fleet of such robots controlled by a central command. If a single transistor were to blow out in this central computer and malfunction, then the entire fleet might go on a killing spree.

A more subtle problem is when robots perform perfectly well, without any malfunctions, yet there is a tiny but fatal flaw in their programming and goals. For a robot, self-preservation is one important goal. But so is being helpful to humans. The real problem arises when these goals contradict each other.

In the movie *I, Robot,* the computer system decides that humans are self-destructive, with their never-ending wars and atrocities, and that the only way to protect the human race is to take over and create a benevolent dictatorship of the machine. The contradiction here is not between two goals, but within a single goal that is not realistic. These murderous robots do not malfunction—they logically conclude that the only way to preserve humanity is to take control of society.

One solution to this problem is to create a hierarchy of goals. For example, the desire to help humans must outrank self-preservation. This theme was explored in the movie *2001.* The computer system HAL 9000 was a sentient computer capable of conversing easily with humans. But the orders given to HAL 9000 were self-contradictory and could not be logically carried out. By attempting to execute an impossible goal, it fell off the mesa; it went crazy, and the only solution to obeying contradictory commands from imperfect humans was to eliminate the humans.

The best solution might be to create a new law of robotics, which would state that robots cannot do harm to the human race, even if there are contradictions within their previous directives. They must be programmed to ignore lower-level contradictions within their orders and always preserve the supreme law. But this might still be an imperfect system at best. (For example, if the robots' central goal is to protect humanity to the exclusion of all other goals, then it all depends on how the robots define the word "protect." Their mechanical definition of this word may differ from ours.)

Instead of reacting with terror, some scientists, such as Dr. Douglas Hofstadter, a cognitive scientist at Indiana University, do not fear this possibility. When I interviewed him, he told me that robots are our children, so why shouldn't we love them like our own? His attitude, he told me, is that we love our children, even though we know that they will take over.

When I interviewed Dr. Hans Moravec, former director of the AI Laboratory at Carnegie Mellon University, he agreed with Dr. Hofstadter. In his book *Robot,* he writes, "Unleashed from the plodding pace of biological evolution, the children of our minds will be free to grow to confront immense and fundamental challenges in the larger universe. . . . We humans will benefit for a time from their labors, but . . . like natural children, they will seek their own fortunes, while we, their aged parents, silently fade away."

Others, on the contrary, think that this is a horrible solution. Perhaps the problem can be solved if we make changes in our goals and priorities now, before it is too late. Since these robots are our children, we should "teach" them to be benevolent.

FRIENDLY AI

Robots are mechanical creatures that we make in the laboratory, so whether we have killer robots or friendly robots depends on the direction of AI research. Much of the funding comes from the military, which is specifically mandated to win wars, so killer robots are a definite possibility.

However, since 30 percent of all commercial robots are manufactured in Japan, there is another possibility: robots will be designed to become helpful playmates and workers from the very beginning. This goal is feasible if the consumer sector dominates robotics research. The philosophy of "friendly AI" is that inventors should create robots that, from the very first steps, are programmed to be beneficial to humans.

Culturally, the Japanese approach to robots is different from the West's. While kids in the West might feel terror watching rampaging Terminator-type robots, kids in Japan are steeped in the Shinto religion, which believes spirits live in all things, even mechanical robots. Instead of being uncomfortable at the sight of robots, Japanese children squeal with delight upon encountering them. It's no wonder, therefore, that these robots in Japan are proliferating in the marketplace and in homes. They greet you at department stores and educate you on TV. There is even a serious play in Japan featuring a robot. (Japan has another reason for embracing robots. These are the future robot nurses for an aging country. Twenty-one percent of the population is over sixty-five, and Japan is aging faster than any other nation. In some sense, Japan is a train wreck in slow motion. Three demographic factors are at work. First, Japanese women have the longest life expectancy of any ethnic group in the world. Second, Japan has one of the world's lowest birthrates. Third, it has a strict immigration policy, with over 99 percent of the population being pure Japanese. Without young immigrants to take care of the elderly, Japan may rely on robot nurses. This problem is not restricted to Japan; Europe is next. Italy, Germany, Switzerland, and other European nations face similar demographic pressures. The populations of Japan and Europe could experience severe shrinkage by mid-century. The United States is not far behind. The birthrate of native-born U.S. citizens has also fallen dramatically in the last few decades, but immigration will keep the United States expanding into this century. In other words, it could be a trillion-dollar gamble to see if robots can save us from these three demographic nightmares.)

Japan leads the world in creating robots that can enter our personal lives. The Japanese have built robots that can cook (one can make a bowl of noodles in a minute and forty seconds). When you go to a restaurant, you can place your order on a tablet computer and the robot cook springs into action. It consists of two large, mechanical arms, which grab the bowls, spoons, and knives and prepare the food for you. Some robotic cooks even resemble human ones.

There are also musical robots for entertainment. One such robot actually has accordion-like "lungs" by which it can generate music by pumping air through an instrument. There are also robot maids. If you carefully prepare your laundry, it can fold it in front of you. There is even a robot that can talk because it has artificial lungs, lips, tongue, and nasal cavity. The Sony Corporation, for example, built the AIBO robot, which resembles a dog and can register a number of emotions if you pet it. Some futurists predict that the robotics industry may one day become as large as the automobile industry is today.

The point here is that robots are not necessarily programmed to destroy and dominate. The future of AI is up to us.

But some critics of friendly AI claim that robots may take over not because they are aggressive, but because we are sloppy in creating them. In other words, if the robots take over, it will be because we programmed them to have conflicting goals.

"I AM A MACHINE"

When I interviewed Dr. Rodney Brooks, former director of the MIT Artificial Intelligence Lab and cofounder of iRobot, I asked him if he thought machines would one day take over. He told me that we just have to accept that we are machines ourselves. This means that one day, we will be able to build machines that are just as alive as we are. But, he cautioned, we will have to give up the concept of our "specialness."

This evolution in human perspective started with Nicolaus Copernicus when he realized that the Earth was not the center of the universe, but rather goes around the sun. It continued with Darwin, who showed that we were similar to the animals in our evolution. And it will continue into the future, he told me, when we realize that we are machines, except that we are made of wetware and not hardware.

It's going to represent a major change in our world outlook to accept that we, too, are machines, he believes. He writes, "We don't like to give up our specialness, so you know, having the idea that robots could really have emotions, or that robots could be living creatures—I think is going to be hard for us to accept. But we're going to come to accept it over the next fifty years."

But on the question of whether the robots will eventually take over, he says that this will probably not happen, for a variety of reasons. First, no one is going to accidentally build a robot that wants to rule the world. He says that creating a robot that can suddenly take over is like someone accidentally building a 747 jetliner. Plus, there will be plenty of time to stop this from happening. Before someone builds a "super-bad robot," someone has to build a "mildly bad robot," and before that a "not-so-bad robot."

His philosophy is summed up when he says, "The robots are coming, but we don't have too much to worry about. It's going to be a lot of fun." To him, the robot revolution is a certainty, and he foresees the day when robots will surpass human intelligence. The only question is when. But there is nothing to fear, since we will have created them. We have the choice to create them to help, and not hinder, us.

MERGE WITH THEM?

If you ask Dr. Brooks how we can coexist with these super-smart robots, his reply is straightforward: we will merge with them. With advances in robotics and neuroprosthetics, it becomes possible to incorporate AI into our own bodies.

Dr. Brooks notes that the process, in some sense, has already begun. Today, about twenty thousand people have had cochlear implants, which have given them the gift of hearing. Sounds are picked up by a tiny receiver, which converts sound waves to electrical signals, which are then sent directly to the auditory nerves of the ear.

Similarly, at the University of Southern California and elsewhere, it is possible to take a patient who is blind and implant an artificial retina. One method places a mini video camera in eyeglasses, which converts an image into digital signals. These are sent wirelessly to a chip placed in the person's retina. The chip activates the retina's nerves, which then send messages down the optic nerve to the occipital lobe of the brain. In this way, a person who is totally blind can see a rough image of familiar objects. Another design has a light-sensitive chip placed on the retina itself, which then sends signals directly to the optic nerve. This design does not need an external camera.

This also means that we can go even further and enhance ordinary senses and abilities. With cochlear implants, it will be possible to hear high frequencies that we have never heard before. Already with infrared glasses, one can see the specific type of light that emanates from hot objects in the dark and that is normally invisible to the human eye. With artificial retinas, it may be possible to enhance our ability to see ultraviolet or infrared light. (Bees, for example, can see UV light because they have to lock onto the sun in order to navigate to a flower bed.)

Some scientists even dream of the day when exoskeletons will have superpowers like those found in comic books, with super strength, super senses, and super abilities. We'd become a cyborg like Iron Man, a normal human with superhuman abilities and powers. This means that we might not have to worry about super-intelligent robots taking over. We'd simply merge with them.

This, of course, is for the distant future. But some scientists, frustrated that robots are not leaving the factory and entering our lives, point out that Mother Nature has already created the human mind, so why not copy it? Their strategy is to take the brain apart, neuron by neuron, and then reassemble it.

But reverse engineering entails more than just creating a vast blueprint to create a living brain. If the brain can be duplicated down to the last neuron, perhaps we can upload our consciousness into a computer. We'd have the ability to leave our mortal bodies behind. This is beyond mind over matter. This is mind without matter.

I'm as fond of my body as anyone, but if I can be 200 with a body of silicon, I'll take it.

—DANIEL HILL, COFOUNDER OF THINKING MACHINES CORP.

11 REVERSE ENGINEERING THE BRAIN

In January 2013, two bombshells were dropped that could alter the medical and scientific landscape forever. Overnight, reverse engineering the brain, once considered to be too complex to solve, suddenly became a focal point of scientific rivalry and pride between the greatest economic powers on Earth.

First, in his State of the Union address, President Barack Obama stunned the scientific community by announcing that federal research funds, perhaps to the tune of $3 billion, might be allocated to the Brain Research Through Advancing Innovative Neurotechnologies (or BRAIN) Initiative. Like the Human Genome Project, which opened the floodgates for genetic research, BRAIN will pry open the secrets of the brain at the neural level by mapping its electrical pathways. Once the brain is mapped, a host of intractable diseases like Alzheimer's, Parkinson's, schizophrenia, dementia, and bipolar disorder might be understood and possibly cured. To jump-start BRAIN, $100 million might be allocated in 2014 toward the project.

Almost simultaneously, the European Commission announced that the Human Brain Project would be awarded 1.19 billion euros (about $1.6 billion) to create a computer simulation of the human brain. Using the power of the biggest supercomputers on the planet, the Human Brain Project will create a copy of the human brain made of transistors and steel.

Proponents of both projects stressed the enormous benefits of these endeavors. President Obama was quick to point out that not only would BRAIN alleviate the suffering of millions of people, it will also generate new revenue streams. For every dollar spent on the Human Genome Project, he claimed, about $140 of economic activity was generated. Entire industries, in fact, sprouted with the completion of the Human Genome Project. For the taxpayer, BRAIN, like the Human Genome Project, will be a win-win situation.

Although Obama's speech did not give details, scientists quickly filled in many of the gaps. Neurologists pointed out that, on one hand, it is now possible to use delicate instruments to monitor the electrical activity of single neurons. On the other hand, using MRI machines, it is possible to monitor the global behavior of the entire brain. What is missing, they pointed out, is the middle ground, where most of the interesting brain activity takes place. It is in this middle ground, involving the pathways of thousands to millions of neurons, that there are huge gaps in our understanding of mental disease and behavior.

To tackle this enormous problem, scientists laid out a tentative fifteen-year program. In the first five years, neurologists hope to monitor the electrical activity of tens of thousands of neurons. The short-term goals might include reconstructing the electrical activity of important parts of animal brains, such as the medulla of the Drosophila fruit fly or the ganglion cells in a mouse retina (which has fifty thousand neurons).

Within ten years, that number should increase to hundreds of thousands of neurons. This could include imaging the entire Drosophila brain (135,000 neurons) or even the cortex of the Etruscan shrew, the smallest known mammal, with a million neurons.

Finally, within fifteen years, it should be possible to monitor millions of neurons, comparable to the zebrafish brain or the entire neocortex of a mouse. This could pave the way toward imaging parts of the brains of primates.

Meanwhile, in Europe, the Human Brain Project would tackle the problem from a different point of view. Over a ten-year period, it will use supercomputers to simulate the basic functioning of the brains of different animals, starting with mice and working up to humans. Instead of dealing with individual neurons, the Human Brain Project will use transistors to mimic their behavior, so that there will be computer modules that can act like the neocortex, the thalamus, and other parts of the brain.

In the end, the rivalry between these two gigantic projects could create a windfall by generating new discoveries for treating incurable diseases and spawning new industries. But there is also another, unstated goal. If one can eventually simulate a human brain, does it mean that the brain can become immortal? Does it mean that consciousness can now exist outside the body? Some of the thorniest theological and metaphysical questions are raised by these ambitious projects.

BUILDING A BRAIN

Like many other children, I used to love taking apart clocks, disassembling them, screw for screw, and then trying to see how the whole thing fit together. I would trace each part mentally, seeing how one gear connected to the next one, until the whole thing fit together. I realized the mainspring turned the main gear, which then fed a sequence of smaller gears, which eventually turned the hands of the clock.

Today, on a much larger scale, computer scientists and neurologists are trying to take apart an infinitely more complex object, the most sophisticated object we know about in the universe: the human brain. Moreover, they wish to reassemble it, neuron by neuron.

Because of rapid advances in automation, robotics, nanotechnology, and neuroscience, reverse engineering the human brain is no longer idle speculation for polite after-dinner banter. In the United States and Europe, billions of dollars will soon be flowing into projects once considered preposterous. Today a small band of visionary scientists are dedicating their professional lives to a project that they may not live to see completed. Tomorrow their ranks could swell into an entire army, generously funded by the United States and the nations of Europe.

If successful, these scientists could alter the course of human history. Not only might they find new cures and therapies for mental illnesses, they might also unlock the secret of consciousness and perhaps upload it into a computer.

It is a daunting task. The human brain consists of over one hundred billion neurons, approximately as many stars as there are in the Milky Way galaxy. Each neuron, in turn, is connected to perhaps ten thousand other neurons, so altogether there are a total of ten million billion possible connections (and that does not begin to compute the number of pathways there are among this thicket of neurons). The number of "thoughts" that a human brain can conceive of is therefore truly astronomical and beyond human ken.

Yet that has not stopped a small bunch of fiercely dedicated scientists from attempting to reconstruct the brain from scratch. There is an old Chinese proverb, "A journey of a thousand miles begins with the first step." That first step was actually taken when scientists decoded, neuron for neuron, the nervous system of a nematode worm. This tiny creature, called *C. elegans,* has 302 neurons and 7,000 synapses, all of which have been precisely recorded. A complete blueprint of its nervous system can be found on the Internet. (Even today, it is the only living organism to have its entire neural structure decoded in this way.)

At first, it was thought that the complete reverse engineering of this simple organism would open the door to the human brain. Ironically, the opposite has happened. Although the nematode's neurons were finite in number, the network is still so complex and sophisticated that it has taken years to understand even simple facts about worm behavior, such as which pathways are responsible for which behaviors. If even the lowly nematode worm could elude our scientific understanding, scientists were forced to appreciate how complex a human brain must be.

THREE APPROACHES TO THE BRAIN

Because the brain is so complex, there are at least three distinct ways in which it can be taken apart, neuron by neuron. The first is to simulate the brain electronically with supercomputers, which is the approach being taken by the Europeans. The second is to map out the neural pathways of living brains, as in BRAIN. (This task, in turn, can be further subdivided, depending on how these neurons are analyzed—either anatomically, neuron by neuron, or by function and activity.) And third, one can decipher the genes that control the development of the brain, which is an approach pioneered by billionaire Paul Allen of Microsoft.

The first approach, simulating the brain using transistors and computers, is forging ahead by reverse engineering the brains of animals in a certain sequence: first a mouse, then a rat, rabbit, and a cat. The Europeans are following the rough trail of evolution, starting with simple brains and working upward. To a computer scientist, the solution is raw computing power—the more, the better. And this means using some of the largest computers on Earth to decipher the brains of mice and men.

Their first target is the brain of a mouse, which is one-thousandth the size of a human brain, containing about one hundred million neurons. The thinking process behind a mouse brain is being analyzed by the IBM Blue Gene computer, located at the Lawrence Livermore National Laboratory in California, where some of the biggest computers in the world are located; they're used to design hydrogen warheads for the Pentagon. This colossal collection of transistors, chips, and wires contains 147,456 processors with a staggering 150,000 gigabytes of memory. (A typical PC may have one processor and a few gigabytes of memory.)

Progress has been slow but steady. Instead of modeling the entire brain, scientists try to duplicate just the connections between the cortex and the thalamus, where much of brain activity is concentrated. (This means that the sensory connections to the outside world are missing in this simulation.)

In 2006, Dr. Dharmendra Modha of IBM partially simulated the mouse brain in this way with 512 processors. In 2007, his group simulated the rat brain with 2,048 processors. In 2009, the cat brain, with 1.6 billion neurons and nine trillion connections, was simulated with 24,576 processors.

Today, using the full power of the Blue Gene computer, IBM scientists have simulated 4.5 percent of the human brain's neurons and synapses. To begin a partial simulation of the human brain, one would need 880,000 processors, which might be possible around 2020.

I had a chance to film the Blue Gene computer. To get to the laboratory, I had to go through layers and layers of security, since it is the nation's premier weapons laboratory, but once you have cleared all the checkpoints, you enter a huge, air-conditioned room housing Blue Gene.

The computer is truly a magnificent piece of hardware. It consists of racks and racks of large black cabinets full of switches and blinking lights, each about eight feet tall and roughly fifteen feet long. As I walked among the cabinets that make up Blue Gene, I wondered what kinds of operations it was performing. Most likely, it was modeling the interior of a proton, calculating the decay of plutonium triggers, simulating the collision of two black holes, and thinking of a mouse, all at once.

Then I was told that even this supercomputer is giving way to the next generation, the Blue Gene/Q Sequoia, which will take computing to a new level. In June 2012, it set the world's record for the fastest supercomputer. At peak speed, it can perform operations at 20.1 PFLOPS (or 20.1 trillion floating point operations per second). It covers an area of three thousand square feet, and gobbles up electrical energy at the rate of 7.9 megawatts, enough power to light up a small city.

But with all this massive computational firepower concentrated in one computer, is it enough to rival the human brain?

Unfortunately, no.

These computer simulations try only to duplicate the interactions between the cortex and the thalamus. Huge chunks of the brain are therefore missing. Dr. Modha understands the enormity of his project. His ambitious research has allowed him to estimate what it would take to create a working model of the entire human brain, and not just a portion or a pale version of it, complete with all parts of the neocortex and connections to the senses. He envisions using not just a single Blue Gene computer but thousands of them, which would fill up not just a room but an entire city block. The energy consumption would be so great that you would need a thousand-megawatt nuclear power plant to generate all the electricity. And then, to cool off this monstrous computer so it wouldn't melt, you would need to divert a river and send it through the computer circuits.

It is remarkable that a gigantic, city-size computer is required to simulate a piece of human tissue that weighs three pounds, fits inside your skull, raises your body temperature by only a few degrees, uses twenty watts of power, and needs only a few hamburgers to keep it going.

BUILDING A BRAIN

But perhaps the most ambitious scientist who has joined this campaign is Dr. Henry Markram of the École Polytechnique Fédérale de Lausanne, in Switzerland. He is the driving force behind the Human Brain Project, which has received over a billion dollars of funding from the European Commission. He has spent the last seventeen years of his life trying to decode the brain's neural wiring. He, too, is using the Blue Gene computer to reverse engineer the brain. At present, his Human Brain Project is running up a bill of $140 million from the European Union, and that represents only a fraction of the computer firepower he will need in the coming decade.

Dr. Markram believes that this is no longer a science project but an engineering endeavor, requiring vast sums of money. He says, "To build this—the supercomputers, the software, the research—we need around one billion dollars. This is not expensive when one considers that the global burden of brain disease will exceed twenty percent of the world gross domestic project very soon." To him, a billion dollars is nothing, just a pittance compared to the hundreds of billions in bills stemming from Alzheimer's, Parkinson's, and other related diseases when the baby boomers retire.

So to Dr. Markram, the solution is one of scale. Throw enough money at the project, and the human brain will emerge. Now that he has won the coveted billion-dollar prize from the European Commission, his dream may become a reality.

He has a ready answer when asked what the average taxpayer will get from this billion-dollar investment. There are three reasons, he says, for embarking on this lonely but expensive quest. First, "It's essential for us to understand the human brain if we want to get along in society, and I think that it is a key step in evolution. The second reason is, we cannot keep doing animal experimentation forever. . . . It's like a Noah's Ark. It's like an archive. And the third reason is that there are two billion people on this planet that are affected by mental disorder. . . ."

To him, it is a scandal that so little is known about mental diseases, which cause so much suffering to millions of people. He says, "There's not a single neurological disease today in which anybody knows what is malfunctioning in this circuit—which pathway, which synapse, which neuron, which receptor. This is shocking."

At first, it may sound impossible to complete this project, with so many neurons and so many connections. It seems like a fool's errand. But these scientists think they have an ace in the hole.

The human genome consists of roughly twenty-three thousand genes, yet it can somehow create the brain, which consists of one hundred billion neurons. It seems to be a mathematical impossibility to create the human brain from our genes, yet it happens every time an embryo is conceived. How can so much information be crammed into something so small?

The answer, Dr. Markram believes, is that nature uses shortcuts. The key to his approach is that certain modules of neurons are repeated over and over again once Mother Nature finds a good template. If you look at microscopic slices of the brain, at first you see nothing but a random tangle of neurons. But upon closer examination, patterns of modules that are repeated over and over appear.

(Modules, in fact, are one reason why it is possible to assemble large skyscrapers so rapidly. Once a single module is designed, it is possible to repeat it endlessly on the assembly line. Then you can rapidly stack them on top of one another to create the skyscraper. Once the paperwork is all signed, an apartment building can be assembled using modules in a few months.)

The key to Dr. Markram's Blue Brain project is the "neocortical column," a module that is repeated over and over in the brain. In humans, each column is about two millimeters tall, with a diameter of half a millimeter, and contains sixty thousand neurons. (As a point of comparison, rat neural modules contain only ten thousand neurons each.) It took ten years, from 1995 to 2005, for Dr. Markram to map the neurons in such a column and to figure out how it worked. Once that was deciphered, he then went to IBM to create massive iterations of these columns.

He is the eternal optimist. In 2009, at a TED conference, he claimed he could finish the project in ten years. (Most likely, this will be for a stripped-down version of the human brain without any attachment to the other lobes or to the senses.) But he has claimed, "If we build it correctly, it should speak and have an intelligence and behave very much as a human does."

Dr. Markram is a skilled defender of his work. He has an answer for everything. When critics say that he is treading on forbidden territory, he counters, "As scientists, we need to be not afraid of the truth. We need to understand our brain. It's natural that people would think that the brain is sacred, that we shouldn't tamper with it because it may be where the secrets of the soul are. But I think, quite honestly, that if the planet understood how the brain functions, we would resolve conflicts everywhere. Because people would understand how trivial and how deterministic and how controlled conflicts and reactions and misunderstandings are."

When faced with the final criticism that he is "playing God," he says, "I think we're far from playing God. God created the whole universe. We're just trying to build a little model."

IS IT REALLY A BRAIN?

Although these scientists claim that their computer simulation of the brain will begin to reach the capability of the human brain by around 2020, the main question is, How realistic is this simulation? Can the cat simulation, for example, catch a mouse? Or play with a ball of yarn?

The answer is no. These computer simulations try to match the sheer power of the neurons firing in the cat brain, but they cannot duplicate the way in which the regions of the brain are hooked together. The IBM simulation is only for the thalamocortical system (i.e., the channel that connects the thalamus to the cortex). The system does not have a physical body, and hence all the complex interactions between the brain and the environment are missing. The brain has no parietal lobe, so it has no sensory or motor connections with the outside world. And even within the thalamocortical system, the basic wiring does not respect the thinking process of a cat. There are no feedback loops and memory circuits for stalking prey or finding a mate. The computerized cat brain is a blank slate, devoid of any memories or instinctual drives. In other words, it cannot catch a mouse.

So even if it is possible to simulate a human brain by around 2020, you will not be able to have a simple conversation with it. Without a parietal lobe, it would be like a blank slate without sensations, devoid of any knowledge of itself, people, and the world around it. Without a temporal lobe, it would not be able to talk. Without a limbic system, it would not have any emotions. In fact, it would have less brain power than a newborn infant.

The challenge of hooking up the brain to the world of sensations, emotions, language, and culture is just beginning.

THE SLICE-AND-DICE APPROACH

The next approach, favored by the Obama administration, is to map the neurons of the brain directly. Instead of using transistors, this approach analyzes the actual neural pathways of the brain. There are several components to it.

One way to proceed is to physically identify each and every neuron and synapse of the brain. (The neurons are usually destroyed by this process.) This is called the anatomical approach. Another path is to decipher the ways in which electrical signals flow across neurons when the brain is performing certain functions. (The latter approach, which stresses identifying the pathways of the living brain, is the one that seems to be favored by the Obama administration.)

The anatomical approach is to take apart the cells of an animal brain, neuron by neuron, using the "slice-and-dice" method. In this way, the full complexity of the environment, the body, and memories are already encoded in the model. Instead of approximating a human brain by assembling a huge number of transistors, these scientists want to identify each neuron of the brain. After that, perhaps each neuron can be simulated by a collection of transistors so that you'd have an exact replica of the human brain, complete with memory, personality, and connection to the senses. Once someone's brain is fully reversed engineered in this way, you should be able to have an informative conversation with that person, complete with memories and a personality.

No new physics is required to finish the project. Using a device similar to a meat slicer in a delicatessen, Dr. Gerry Rubin of the Howard Hughes Medical Institute has been slicing the brain of a fruit fly. This is not an easy task, since the fruit fly brain is only three hundred micrometers across, a tiny speck compared to the human brain. The fruit fly brain contains about 150,000 neurons. Each slice, which is only fifty-billionths of a meter across, is meticulously photographed with an electron microscope, and the images are fed into a computer. Then a computer program tries to reconstruct the wiring, neuron by neuron. At the present rate, Dr. Rubin will be able to identify every neuron in the fruit fly brain in twenty years.

The snail-like pace is due, in part, to current photographic technology, since a standard scanning microscope operates at about ten million pixels per second. (That is about a third of the resolution achieved by a standard TV screen per second.) The goal is to have an imaging machine that can process ten billion pixels per second, which would be a world record.

The problem of how to store the data pouring in from the microscope is also staggering. Once his project gets up to speed, Rubin expects to scan about a million gigabytes of data per day for just a single fruit fly, so he envisions filling up huge warehouses full of hard drives. On top of that, since every fruit fly brain is slightly different, he has to scan hundreds of fruit fly brains in order to get an accurate approximation of one.

Based on working with the fruit fly brain, how long will it take to eventually slice up the human brain? "In a hundred years, I'd like to know how human consciousness works. The ten- or twenty-year goal is to understand the fruit fly brain," he says.

This method can be speeded up with several technical advances. One possibility is to use an automated device, so that the tedious process of slicing the brain and analyzing each slide is done by machine. This could rapidly reduce the time for the project. Automation, for example, vastly reduced the cost of the Human Genome Project (although it was budgeted at $3 billion, it was accomplished ahead of time and under budget, which is unheard of in Washington). Another method is to use a large variety of dyes that will tag different neurons and pathways, making them easier to see. An alternative approach would be to create an automated super microscope that can scan neurons one by one with unparalleled detail.

Given that a complete mapping of the brain and all its senses will take up to a hundred years, these scientists feel somewhat like the medieval architects who designed the cathedrals of Europe, knowing that their grandchildren would finally complete the project.

In addition to constructing an anatomical map of the brain, neuron by neuron, there is a parallel effort called the "Human Connectome Project," which uses brain scans to reconstruct the pathways connecting various regions of the brain.

THE HUMAN CONNECTOME PROJECT

In 2010, the National Institutes of Health announced that it was allocating $30 million, spread out over five years, to a consortium of universities (including Washington University in St. Louis and the University of Minnesota), and a $8.5 million grant over three years to a consortium led by Harvard University, Massachusetts General Hospital, and UCLA. With this level of short-term funding, of course, researchers cannot fully sequence the entire brain, but the funding was meant to jump-start the effort.

Most likely, this effort will be folded into the BRAIN project, which will vastly accelerate this work. The goal is to produce a neuronal map of the human brain's pathways that will elucidate brain disorders such as autism and schizophrenia. One of the leaders of the Connectome Project, Dr. Sebastian Seung, says, "Researchers have conjectured that the neurons themselves are healthy, but maybe they are just wired together in an abnormal way. But we've never had the technology to test that hypothesis until now." If these diseases are actually caused by the miswiring of the brain, then the Human Connectome Project may give us an invaluable clue as to how to treat these conditions.

When considering the ultimate goal of imaging the entire human brain, sometimes Dr. Seung despairs of ever finishing this project. He says, "In the seventeenth century, the mathematician and philosopher Blaise Pascal wrote of his dread of the infinite, his feeling of insignificance at contemplating the vast reaches of outer space. And as a scientist, I'm not supposed to talk about my feelings. . . . I feel curiosity, and I feel wonder, but at times I have also felt despair." But he and others like him persist, even if their project will take multiple generations to finish. They have reason to hope, since one day automated microscopes will tirelessly take the photographs and artificially intelligent machines will analyze them twenty-four hours a day. But right now, just imaging the human brain with ordinary electron microscopes would consume about one zettabyte of data, which is equivalent to all the data compiled in the world today on the web.

Dr. Seung even invites the public to participate in this great project by visiting a website called EyeWire. There, the average "citizen scientist" can view a mass of neural pathways and is asked to color them in (staying within their boundaries). It's like a virtual coloring book, except images are of the actual neurons in the retina of an eye, taken by an electron microscope.

THE ALLEN BRAIN ATLAS

Finally, there is a third way to map the brain. Instead of analyzing the brain by using computer simulations or by identifying all the neural pathways, yet another approach was taken with a generous grant of $100 million from Microsoft billionaire Paul Allen. The goal was to construct a map or atlas of the mouse brain, with the emphasis on identifying the genes responsible for creating the brain.

It is hoped that this understanding of how genes are expressed in the brain will help in understanding autism, Parkinson's, Alzheimer's, and other disabilities. Since a large number of mouse genes are found in humans, it's possible that findings here will give us insight into the human brain.

With this sudden infusion of funds, the project was completed in 2006, and its results are freely available on the web. A follow-up project, the Allen Human Brain Atlas, was announced soon afterward, with the hope of creating an anatomically and genetically complete 3-D map of the human brain. In 2011, the Allen Institute announced that it had mapped the biochemistry of two human brains, finding one thousand anatomical sites with one hundred million data points detailing how genes are expressed in the underlying biochemistry. The study confirmed that 82 percent of our genes are expressed in the brain.

"Until now, a definitive map of the human brain, at this level of detail, simply hasn't existed," says Dr. Allen Jones of the Allen Institute. "The Allen Human Brain Atlas provides never-before-seen views into our most complex and most important organ," he adds.

OBJECTIONS TO REVERSE ENGINEERING

Scientists who have dedicated their lives to reverse engineering the brain realize that decades of hard work lie ahead of them. But they are also convinced of the practical implications of their work. They feel that even partial results will help decode the mystery of mental diseases that have afflicted humans throughout our history.

The cynics, however, may claim that, after this arduous task is finished, we will have a mountain of data with no understanding of how it all fits together. For example, imagine a Neanderthal who one day comes across the complete blueprint for an IBM Blue Gene computer. All the details are there in the blueprint, down to the very last transistor. The blueprint is huge, taking up thousands of square feet of paper. The Neanderthal may be dimly aware that this blueprint is the secret of a super-powerful machine, but the sheer mass of technical data means nothing to him.

Similarly, the fear is that, after spending billions deciphering the location of every neuron of the brain, we won't be able to understand what it all means. It may take many more decades of hard work to see how the whole thing functions.

For example, the Human Genome Project was a smashing success in sequencing all the genes that make up the human genome, but it was a huge disappointment for those who expected immediate cures for genetic diseases. The Human Genome Project was like a gigantic dictionary, with twenty-three thousand entries but no definitions. Page after page of this dictionary is blank, yet the spelling of each gene is perfect. The project was a breakthrough, but at the same time it's just the first step in a long journey to figure out what these genes do and how they interact.

Similarly, just having a complete map of every single neural connection in the brain does not guarantee that we will know what these neurons are doing and how they react. Reverse engineering is the easy part; after that, the hard part begins—making sense of all this data.

THE FUTURE

But assume for now that the moment has finally arrived. With much fanfare, scientists solemnly announce that they have successfully reverse engineered the entire human brain.

Then what?

One immediate application is to find the origins of certain mental diseases. It's thought that many mental diseases are not caused by the massive destruction of neurons, but by a simple misconnection. Think of genetic diseases that are caused by a single mutation, like Huntington's disease, Tay-Sachs, or cystic fibrosis. Out of three billion base pairs, a single misspelling (or repetition) can cause uncontrollable flailing of your limbs and convulsions, as in Huntington's disease. Even if the genome is 99.9999999 percent accurate, a tiny flaw might invalidate the entire sequence. That is why gene therapy has targeted these single mutations as possible genetic diseases that can be fixed.

Likewise, once the brain is reverse engineered, it might be possible to run simulations of the brain, deliberately disrupting a few connections to see if you can induce certain illnesses. Only a handful of neurons may be responsible for major disruptions of our cognition. Locating this tiny collection of misfiring neurons may be one of the jobs of the reverse-engineered brain.

One example might be Capgras delusion, in which you see someone you recognize as your mother, but you believe that person to be an impostor.

According to Dr. V. S. Ramachandran, this rare disease might be due to a misconnection between two parts of the brain. The fusiform gyrus in the temporal lobe is responsible for recognizing the face of your mother, but the amygdala is responsible for your emotional response in seeing your mother. When the connection between these two centers is disrupted, an individual can recognize his mother's face perfectly well, but, since there is no emotional response, he is also convinced that she is an impostor.

Another use for the reverse-engineered brain is to pinpoint precisely which cluster of neurons is misfiring. Deep brain stimulation, as we've seen, involves using tiny probes to dampen the activity of a tiny portion of the brain, such as Broadmann's area 25, in the case of certain severe forms of depression. Using the reverse-engineered map, it might be possible to find precisely where the neurons are misfiring, which may involve only a handful of neurons.

A reversed-engineered brain would also be of great help to AI. Vision and face recognition are done effortlessly by the brain, but they still elude our most advanced computers. For example, computers can recognize with 95 percent or greater accuracy human faces that look straight ahead and are part of a small data bank, but if you show the computer the same face from different angles or a face that's not in the database, the computer will most likely fail. Within .1 seconds, we can recognize familiar faces from different angles; it's so easy for our brains that we are not even aware we are doing it. Reverse engineering the brain may reveal the mystery of how this is done.

More complicated would be diseases that involve multiple failures of the brain, such as schizophrenia. This disorder involves several genes, plus interactions with the environment, which in turn cause unusual activity in several areas of the brain. But even there, a reverse-engineered brain would be able to tell precisely how certain symptoms (such as hallucinations) are formed, and this might pave the way for a possible cure.

A reverse-engineered brain would also solve such basic but unresolved questions as how long-term memories are stored. It is known that certain parts of the brain, such as the hippocampus and amygdala, store memories, but how the memory is dispersed through various cortices and then reassembled to create a memory is still unclear.

Once the reverse-engineered brain is fully functional, then it will be time to turn on all its circuits to see if it can respond like a human (i.e., to see if it can pass the Turing test). Since long-term memory is already encoded in the neurons of the reverse-engineered brain, it should be obvious very quickly whether the brain can respond in a way indistinguishable from a human.

Finally, there is one impact of reverse engineering the brain that is rarely discussed but is on many people's minds: immortality. If consciousness can be transferred into a computer, does that mean we don't have to die?

Speculation is never a waste of time. It clears away the deadwood in the thickets of deduction.

—ELIZABETH PETERS

We are a scientific civilization. . . . That means a civilization in which knowledge and its integrity are crucial. Science is only a Latin word for knowledge. . . . Knowledge is our destiny.

—JACOB BRONOWSKI

12 **THE FUTURE** MIND BEYOND MATTER

Can consciousness exist by itself, free from the constraints of the physical body? Can we leave our mortal body and, like spirits, wander around this playground called the universe?

This was explored on *Star Trek*, when Captain Kirk of the starship *Enterprise* encounters a superhuman race, almost a million years more advanced than the Federation of Planets. They are so advanced that they have long since abandoned their frail, mortal bodies, and now inhabit pulsating globes of pure energy. It has been millennia since they could feel intoxicating sensations, such as breathing fresh air, touching another's hand, or feeling physical love. Their leader, Sargon, welcomes the *Enterprise* to their planet. Captain Kirk accepts the invitation, acutely aware that this civilization could instantly vaporize the *Enterprise* if it wanted to.

But unknown to the crew, these super beings have a fatal weakness. For all their advanced technology, they have been severed for hundreds of thousands of years from their physical bodies. As such, they yearn to feel the rush of physical sensations and long to become human again.

One of these super beings, in fact, is evil and determined to gain possession of the physical bodies of the crew. He wants to live like a human, even if it means destroying the mind of the body's owner. Soon a battle breaks out on the deck of the *Enterprise,* as the evil entity seizes control of Spock's body and the crew fights back.

Scientists have asked themselves, Is there a law of physics preventing the mind from existing without the body? In particular, if the conscious human mind is a device that constantly creates models of the world and simulates them into the future, is it possible to create a machine that can simulate this entire process?

Previously, we mentioned the possibility of having our bodies placed in pods, as in the movie *Surrogates,* while we mentally control a robot. The problem here is that our natural body will still gradually wither away, even if our robot surrogate keeps on going. Serious scientists are contemplating whether we can actually transfer our minds into a robot so we can become truly immortal. And who wouldn't want a chance at eternal life? As Woody Allen once said, "I don't want to live forever through my works. I want to live forever by not dying."

Actually, millions of people already claim that it is possible for the mind to leave the body. In fact, many insist that they have done it themselves.

OUT-OF-BODY EXPERIENCES

The idea of minds without bodies is perhaps the oldest of our superstitions, embedded deep within our myths, folklore, dreams, and perhaps even our genes. Every society, it seems, has some tale of ghosts and demons who can enter and leave the body at will.

Sadly, many innocents were persecuted to exorcize the demons that were supposedly possessing their bodies. They probably suffered from mental illness, such as schizophrenia, in which victims are often haunted by voices generated by their own minds. Historians believe that one of the Salem witches who was hung in 1692 for being possessed probably had a rare genetic condition, called Huntington's disease, that causes uncontrolled flailing of the limbs.

Today some people claim that they have entered a trancelike state in which their consciousness has left their body and is free to roam throughout space, even able to look back at their mortal body. In a poll of thirteen thousand Europeans, 5.8 percent claimed they had had an out-of-body experience. Interviews with people in the United States show similar numbers.

Nobel laureate Richard Feynman, always curious about new phenomena, once placed himself in a sensory deprivation tank and tried to leave his physical body. He was successful. He would later write that he felt that he had left his body, drifted into space, and saw his motionless body when he looked back. However, Feynman later concluded that this was probably just his imagination, caused by sensory deprivation.

Neurologists who have studied this phenomenon have a more prosaic explanation. Dr. Olaf Blanke and his colleagues in Switzerland may have located the precise place in the brain that generates out-of-body experiences. One of his patients was a forty-three-year-old woman who suffered from debilitating seizures that came from her right temporal lobe. A grid of about one hundred electrodes was placed over her brain in order to locate the region responsible for her seizures. When the electrodes stimulated the area between the parietal and temporal lobes, she immediately had the sensation of leaving her body. "I see myself lying in bed, from above, but I only see my legs and lower trunk!" she exclaimed. She felt she was floating six feet above her body.

When the electrodes were turned off, however, the out-of-body sensation disappeared immediately. In fact, Dr. Blanke found that he could turn the out-of-body sensation on and off, like a light switch, by repeatedly stimulating this area of the brain. As we saw in Chapter 9, temporal lobe epileptic lesions can induce the feeling that there are evil spirits behind every misfortune, so the concept of spirits leaving the body is perhaps part of our neural makeup. (This may also explain the presence of supernatural beings. When Dr. Blanke analyzed a twenty-two-year-old woman who was suffering from intractable seizures, he found that, by stimulating the temporoparietal area of the brain, he could induce the sensation that there was a shadowy presence behind her. She could describe this person, who even grabbed her arms, in detail. His position would change with each appearance, but he would always appear behind her.)

Human consciousness, I believe, is the process of continually forming a model of the world, in order to simulate the future and carry out a goal. In particular, the brain is receiving sensations from the eyes and inner ear to create a model of where we are in space. However, when the signals from our eyes and ears are in contradiction, we become confused about our location. We often get nauseous and throw up. For example, many people develop sea sickness when they are on a rocking boat because their eyes, looking at the cabin walls, tell them that they are stationary, but their inner ear tells them that they are swaying. The mismatch between these signals causes them to become nauseous. The remedy is to look out at the horizon so that the visual image matches the signals from the inner ear. (This same sense of nausea can be induced even if you are stationary. If you look at a spinning garbage can with bright vertical stripes painted on it, the stripes seem to move horizontally across your eyes, giving you the sensation that you are moving. But your inner ear says you are stationary. The resulting mismatch causes you to throw up after a few minutes, even if you are sitting in a chair.)

The messages from the eyes and inner ear can also be disrupted electrically, at the boundary of the temporal and parietal lobes, and this is the origin of out-of-body experiences. When this sensitive area is touched, the brain gets confused about where it is located in space. (Notably, temporary loss of blood or oxygen or excess carbon dioxide in the blood can also cause a disruption in the temporoparietal region and induce out-of-body experiences, which may explain the prevalence of these sensations during accidents, emergencies, heart attacks, etc.)

NEAR-DEATH EXPERIENCES

But perhaps the most dramatic category of out-of-body experiences are the near-death stories of individuals who have been declared dead but then mysteriously regained consciousness. In fact, 6 to 12 percent of survivors of cardiac arrest report having near-death experiences. It's as though they have cheated death itself. When interviewed, they have dramatic tales of the same experience: they left their body and drifted toward a bright light at the end of a long tunnel.

The media have seized upon this, with numerous best sellers and TV documentaries devoted to these theatrical stories. Many bizarre theories have been proposed to explain near-death experiences. In a poll of two thousand people, fully 42 percent believed that near-death experiences were proof of contact with the spiritual world that lies beyond death. (Some believe that the body releases endorphins—natural narcotics—before death. This may explain the euphoria that people feel, but not the tunnel and the bright lights.) Carl Sagan even speculated that near-death experiences were a reliving of the trauma of birth. The fact that these individuals recount very similar experiences doesn't necessarily corroborate their glimpses into the afterlife; in fact, it seems to indicate that there is some deep neurological event happening.

Neurologists have looked into this phenomenon seriously and suspect that the key may be the decrease of blood flow to the brain that often accompanies near-death cases, and which also occurs in fainting. Dr. Thomas Lempert, a neurologist at the Castle Park Clinic in Berlin, conducted a series of experiments on forty-two healthy individuals, causing them to faint under controlled laboratory conditions. Sixty percent of them had visual hallucinations (e.g., bright lights and colored patches). Forty-seven percent of them felt that they were entering another world. Twenty percent claimed to have encountered a supernatural being. Seventeen percent saw a bright light. Eight percent saw a tunnel. So fainting can mimic all the sensations people have in near-death experiences. But precisely how does this happen?

The mystery of how fainting can simulate near-death experiences may be solved by analyzing the experiences of military pilots. The U.S. Air Force, for example, contacted neurophysiologist Dr. Edward Lambert to analyze military pilots who blacked out when experiencing high g forces (i.e., when executing a tight turn in a jet or pulling out of a dive). Dr. Lampert placed pilots in an ultracentrifuge at the Mayo Clinic in Rochester, Minnesota, which spun them around in a circle until they experienced high g forces. As blood drained from their brain, they would become unconscious after fifteen seconds of experiencing several g's of acceleration.

He found that after only five seconds, the blood flow to the pilots' eyes diminished, so that their peripheral vision dimmed, creating the image of a long tunnel. This could explain the tunnel that is often seen by people having a near-death experience. If the periphery of your vision blacks out, all you see is the narrow tunnel in front of you. But because Dr. Lampert could carefully adjust the velocity of the centrifuge by turning a dial, he found he could keep the pilots in this state indefinitely, allowing him to prove that this tunnel vision is caused by loss of blood flow to the periphery of the eye.

CAN CONSCIOUSNESS LEAVE YOUR BODY?

Some scientists who have investigated near-death and out-of-body experiences are convinced that they are by-products of the brain itself when it is placed under stressful conditions and its wiring gets confused. However, there are other scientists who believe that one day, when our technology is sufficiently advanced decades from now, one's consciousness may truly be able to leave the body. Several controversial methods have been suggested.

One method has been pioneered by futurist and inventor Dr. Ray Kurzweil, who believes that consciousness may one day be uploaded into a supercomputer. We once spoke at a conference together, and he told me his fascination with computers and artificial intelligence began when he was five years old and his parents bought him all sorts of mechanical devices and toys. He loved to tinker with these devices, and even as a child he knew he was destined to become an inventor. At MIT, he received his doctorate under Dr. Marvin Minsky, one of the founders of AI. Afterward, he cut his teeth applying pattern-recognition technology to musical instruments and text-to-sound machines. He was able to translate AI research in these areas into a string of companies. (He sold his first company when he was only twenty.) His optical reader, which could recognize text and convert it into sound, was heralded as an aid for the blind, and was even mentioned by Walter Cronkite on the evening news.

In order to be a successful inventor, he said to me, you always have to be ahead of the curve, to anticipate change, not react to it. Indeed, Dr. Kurzweil loves to make predictions, and many of them have mirrored the remarkable exponential growth of digital technology. He made the following predictions:

- By 2019, a $1,000 PC will have the computing power of the human brain—twenty million billion calculations per second. (This number is obtained by taking the one hundred billion neurons of the brain, multiplying one thousand connections per neuron, and two hundred calculations per second per connection.)
- By 2029, a $1,000 PC will be a thousand times more powerful than the human brain; the human brain itself will be successfully reversed engineered.
- By 2055, $1,000 of computing power will equal the processing power of all the humans on the planet. (He adds modestly, "I may be off by a year or two.")

In particular, the year 2045 looms as an important one for Dr. Kurzweil, since that is when he believes the "singularity" will take hold. By then, he claims, machines will have surpassed humans in intelligence and in fact will have created next-generation robots even smarter than themselves. Since this process can continue indefinitely, it means, according to Dr. Kurzweil, a never-ending acceleration of the power of machines. In this scenario, we should either merge with our creations or step out of their way. (Although these dates are in the far future, he told me that he wants to live long enough to see the day when humans finally become immortal; that is, he wants to live long enough to live forever.)

As we know from Moore's law, at a certain point computer power can no longer advance by creating smaller and smaller transistors. In Kurzweil's opinion, the only way to expand computing power further would be to increase overall size, which would leave robots scavenging for more computer power by devouring the minerals of the Earth. Once the planet has become a gigantic computer, robots may be forced to go into outer space, searching for more sources of computer power. Eventually, they may consume the power of entire stars.

I once asked him if this cosmic growth of computers could alter the cosmos itself. Yes, he replied. He told me that he sometimes looks at the night sky, wondering if on some distant planet intelligent beings have already attained the singularity. If so, then perhaps they should leave some mark on the stars themselves that might be visible to the naked eye.

One limitation he told me, is the speed of light. Unless these machines can break the light barrier, this exponential rise in power may hit a ceiling. When that happens, says Kurzweil, perhaps they will alter the laws of physics themselves.

Anyone who makes predictions with such precision and scope naturally invites criticism like a lightning rod, but it doesn't seem to faze him. People can quibble about this or that prediction, since Kurzweil has missed some of his deadlines, but he is mainly concerned about the thrust of his ideas, which predict the exponential growth of technology. To be fair, most people working in the field of AI whom I have interviewed agree that some form of a singularity will happen, but they disagree sharply on when it might occur and how it will unfold. For example, Bill Gates, cofounder of Microsoft, believes that no one alive today will live to see the day when computers are smart enough to pass for a human. Kevin Kelly, an editor for *Wired* magazine, has said, "People who predict a very utopian future always predict that it is going to happen before they die."

Indeed, one of Kurzweil's many goals is to bring his father back to life. Or rather, he wants to create a realistic simulation. There are several possibilities, but all are still highly speculative.

Kurzweil proposes that perhaps DNA can be extracted from his father (from his grave site, relatives, or organic materials he left behind). Contained within roughly twenty-three thousand genes would be a complete blueprint to re-create the body of that individual. Then a clone could be grown from the DNA.

This is certainly a possibility. I once asked Dr. Robert Lanza of the company Advanced Cell Technology how he was able to bring a long-dead creature "back to life," making history in the process. He told me that the San Diego Zoo asked him to create a clone of a banteng, an oxlike creature that had died out about twenty-five years earlier. The hard part was extracting a usable cell for the purpose of cloning. However, he was successful, and then he FedExed the cell to a farm, where it was implanted into a female cow, which then gave birth to this animal. Although no primate has ever been cloned, let alone a human, Lanza feels it's a technical problem, and that it's only a matter of time before someone clones a human.

This would be the easy part, though. The clone would be genetically equivalent to the original, but without its memories. Artificial memories might be uploaded to the brain using the pioneering methods described in Chapter 5, such as inserting probes into the hippocampus or creating an artificial hippocampus, but Kurzweil's father has long passed, so it's impossible to make the recording in the first place. The best one can do is to assemble piecemeal all historical data about that person, such as by interviewing others who possess relevant memories, or accessing their credit card transactions, etc., and then inputting them into the program.

A more practical way of inserting a person's personality and memory would be to create a large data file containing all known information about a person's habits and life. For example, today it is possible to store all your e-mail, credit card transactions, records, schedules, electronic diaries, and life history onto a single file, which can create a remarkably accurate picture of who you are. This file would represent your entire "digital signature," representing everything that is known about you. It would be remarkably accurate and intimate, detailing what wines you like, how you spend vacations, what kind of soap you use, your favorite singer, and so on.

Also, with a questionnaire, it would be possible to create a rough approximation of Kurzweil's father's personality. His friends, relatives, and associates would fill out a questionnaire containing scores of questions about his personality, such as whether he was shy, curious, honest, hardworking, etc. Then they would assign a number to each trait (e.g., a "10" would mean that you are very honest). This would create a string of hundreds of numbers, each one ranking a specific personality trait. Once this vast set of numbers was compiled, a computer program would take these data and approximate how he would behave in hypothetical situations. Let's say that you are giving a speech and are confronted with an especially obnoxious heckler. The computer program would then scan the numbers and then predict one of several possible outcomes (e.g., ignore the heckler, heckle back, or get into a brawl with the heckler). In other words, his basic personality would be reduced to a long string of numbers, each from 1 to 10, which can be used by a computer to predict how he would react to new situations.

The result would be a vast computer program that would respond to new situations roughly the way the original person would have, using the same verbal expressions and having the same quirks, all tempered with the memories of that person.

Another possibility would be to forgo the whole cloning process and simply create a robot resembling the original person. It would then be straightforward to insert this program into a mechanical device that looks like you, talks with the same accent and mannerisms, and moves its arms and limbs the same way that you do. Adding your favorite expressions (e.g., "you know . . .") would also be easy.

Of course, today it would be easy to detect that this robot is a fake. However, in the coming decades, it may be possible to get closer and closer to the original, so it might be good enough to fool some people.

But this raises a philosophical question. Is this "person" really the same as the original? The original is still dead, so the clone or robot is, strictly speaking, still an impostor. A tape recorder, for example, might reproduce a conversation we have with perfect fidelity, but that tape recorder is certainly not the original. Can a clone or robot that behaves just like the original be a valid substitute?

IMMORTALITY

These methods have been criticized because this process does not realistically input your true personality and memories. A more faithful way of putting a mind into a machine is via the Connectome Project, which we discussed in the last chapter and which seeks to duplicate, neuron for neuron, all the cellular pathways of your brain. All your memories and personality quirks are already embedded in the connectome.

The Connectome Project's Dr. Sebastian Seung notes that some people pay $100,000 or more to have their brains frozen in liquid nitrogen. Certain animals, like fish and frogs, can be frozen solid in a block of ice in winter yet be perfectly healthy after thawing out in spring. This is because they use glucose as an antifreeze to alter the freezing point of water in their blood. Thus their blood remains liquid, even though they are encased in solid ice. This high concentration of glucose in the human body, however, would probably be fatal, so freezing the human brain in liquid nitrogen is a dubious pursuit because expanding ice crystals would rupture the cell wall from the inside (and also, as brain cells die, calcium ions rush in, causing the brain cells to expand until they finally rupture). In either case, brain cells would most likely not survive the freezing process.

Rather than freezing the body and having the cells rupture, a more reliable process to attain immortality might be to have your connectome completed. In this scenario, your doctor would have all your neural connections on a hard drive. Basically, your soul would now be on a disk, reduced to information. Then at a future point, someone would be able to resurrect your connectome and, in principle, use either a clone or a tangle of transistors to bring you back to life.

The Connectome Project, as we mentioned, is still far from being able to record a human's neural connections. But as Dr. Seung says, "Should we ridicule the modern seekers of immortality, calling them fools? Or will they someday chuckle over our graves?"

MENTAL ILLNESS AND IMMORTALITY

Immortality may have its drawbacks, however. The electronic brains being built so far contain only the connections between the cortex and the thalamus. The reverse-engineered brain, lacking a body, might begin to suffer from sensory isolation and even manifest signs of mental illness, as prisoners do when they are placed into solitary confinement. Perhaps the price of creating an immortal, reverse-engineered brain is madness.

Subjects who are placed in isolation chambers, where they are deprived of any contact with the outside world, eventually hallucinate. In 2008, BBC-TV aired a science program titled *Total Isolation,* in which they followed six volunteers as they were placed inside a nuclear bunker, alone and in complete darkness. After just two days, three of the volunteers began to see and hear things—snakes, cars, zebras, and oysters. After they were released, doctors found that all of them suffered from mental deterioration. One subject's memory suffered a 36 percent drop. One can imagine that, after a few weeks or months of this, most of them might go insane.

To maintain the sanity of a reverse-engineered brain, it might be essential to connect it to sensors that receive signals from the environment so it would be able to see and feel sensations from the outside world. But then another problem arises: it might feel that it is a grotesque freak, an unwieldy scientific guinea pig living at the mercy of a science experiment. Because this brain has the same memory and personality as the original human, it would crave human contact. And yet, lurking inside the memory of some supercomputer, with a macabre jungle of electrodes dangling outside, the reverse-engineered brain would be repulsive to any human. Bonding with it would be impossible. Its friends would turn away.

THE CAVEMAN PRINCIPLE

At this point, what I call the Caveman Principle starts to kick in. Why do so many reasonable predictions fail? And why would someone *not* want to live forever inside a computer?

The Caveman Principle is this: given a choice between high-tech or high-touch, we opt for high-touch every time. For example, if we are given a choice between tickets to see our favorite musician live or a CD of the same musician in concert, which would we choose? Or if we are given a choice between tickets to visit the Taj Mahal or just seeing a beautiful picture of it, which would we prefer? More than likely the live concert and the airplane tickets.

This is because we have inherited the consciousness of our apelike ancestors. Some of our basic personality has probably not changed much in the last one hundred thousand years, since the first modern humans emerged in Africa. A large portion of our consciousness is devoted to looking good and trying to impress members of the opposite sex and our peers. This is hardwired into our brains.

More likely, given our basic, apelike consciousness, we will merge with computers only if this enhances but does not totally replace our present-day body.

The Caveman Principle probably explains why some reasonable predictions about the future never materialized, such as "the paperless office." Computers were supposed to banish paper from the office; ironically, computers have actually created even more paper. This is because we are descended from hunters who need "proof of the kill" (i.e., we trust concrete evidence, not ephemeral electrons dancing on a computer screen that vanish when you turn it off). Likewise, the "peopleless city," where people would use virtual reality to go to meetings instead of commuting, never materialized. Commuting to cities is worse than ever. Why? Because we are social animals who like to bond with others. Videoconferencing, although useful, cannot pick up the full spectrum of subtle information offered via body language. A boss, for example, may want to ferret out problems in his staff and therefore wants to see them squirm and sweat under interrogation. You can do this only in person.

CAVEMEN AND NEUROSCIENCE

When I was a child, I read Isaac Asimov's *Foundation Trilogy* and was deeply influenced by it. First, it forced me to ask a simple question: What will technology look like fifty thousand years in the future, when we have a galactic empire? I also couldn't help wondering throughout the novel, Why do humans look and act the same as they do now? I thought that surely thousands of years into the future humans should have cyborg bodies with superhuman abilities. They should have given up their puny human forms millennia ago.

I came up with two answers. First, Asimov wanted to appeal to a young audience willing to buy his book, so he had to create characters that those people could identify with, including all their faults. Second, perhaps people in the future will have the option to have superpowered bodies but prefer to look normal most of the time. This would be because their minds have not changed since humans first emerged from the forest, and so acceptance from their peers and the opposite sex still determines what they look like and what they want out of life.

So now let us apply the Caveman Principle to the neuroscience of the future. At the minimum, it means that any modification of the basic human form would have to be nearly invisible on the outside. We don't want to resemble a refugee from a science-fiction movie, with electrodes dangling from our head. Brain implants that might insert memories or increase our intelligence will be adopted only if nanotechnology can make microscopic sensors and probes that are invisible to the naked eye. In the future, it might be possible to make nanofibers, perhaps made of carbon nanotubes one molecule thick, so thin that they would be able to make contact with neurons with surgical precision and yet leave our appearance unaltered, with our mental capabilities enhanced.

Meanwhile, if we need to be connected to a supercomputer to upload information, we won't want to be tied to a cable jacked into our spinal cord, as in *The Matrix.* The connection will have to be wireless so we can access vast amounts of computer power simply by mentally locating the nearest server.

Today we have cochlear implants and artificial retinas that can give the gift of sound and sight to patients, but in the future our senses will be enhanced using nanotechnology while we preserve our basic human form. For instance, we might have the option of enhancing our muscles, via genetic modification or exoskeletons. There could be a human body shop from which we could order new spare parts as the old ones wear out, but these and other physical enhancements of the body would have to avoid abandoning the human form.

Another way to use this technology in accordance with the Caveman Principle is to use it as an option, rather than a permanent way of life. One might want the option of plugging into this technology and then unplugging soon afterward. Scientists may want to boost their intelligence to solve a particularly tricky problem. But afterward, they will be able to take off their helmets or implants and go about their business. In this way, we are not caught looking like a space cadet to our friends. The point is that no one would force you to do any of this. We would want the option of enjoying the benefits of this technology without the downside of looking silly.

So in the centuries to come, it is likely our bodies will look very similar to the ones we possess today, except that they will be perfect and have enhanced powers. It is a relic of our apelike past that our consciousness is dominated by ancient desires and wishes.

But what about immortality? As we have seen, a reverse-engineered brain, with all the personality quirks of the original person, would eventually go mad if placed inside a computer. Furthermore, connecting this brain to external sensors so it could feel sensations from its environment would create a grotesque monstrosity. One partial solution to this problem is to connect the reverse-engineered brain to an exoskeleton. If the exoskeleton acts like a surrogate, then the reverse-engineered brain would be able to enjoy sensations such as touch and sight without looking grotesque. Eventually the exoskeleton would go wireless, so that it would act like a human but be controlled by a reverse-engineered brain "living" inside a computer.

This surrogate would have the best of both worlds. Being an exoskeleton, it would be perfect. It would possess superpowers. Since it would be wirelessly connected to a reverse-engineered brain inside a large computer, it would also be immortal. And lastly, since it would sense the environment and look appealingly like a real human, it would not have as many problems interacting with humans, many of whom will also have probably opted for this procedure. So the actual connectome would reside in a stationary supercomputer, although its consciousness would manifest itself in a perfect, mobile surrogate body.

All this would require a level of technology far beyond anything that is attainable today. However, given the rapid pace of scientific progress, this could become a reality by the end of the century.

GRADUAL TRANSFERENCE

Right now the process of reverse engineering involves transferring the information within the brain, neuron for neuron. The brain has to be cut up into thin slices, since MRI scans are not yet refined enough to identify the precise neural architecture of the living brain. So until that can be done, the obvious disadvantage of this approach is that you have to die before you can be reversed engineered. Since the brain degenerates rapidly after death, its preservation would have to take place immediately, which is very difficult to accomplish.

But there may be one way to attain immortality without having to die first. This idea was pioneered by Dr. Hans Moravec, former director of the Artificial Intelligence Laboratory at Carnegie Mellon University. When I interviewed him, he told me that he envisions a time in the distant future when we will be able to reverse engineer the brain for a specific purpose: to transfer the mind into an immortal robotic body even while a person is still conscious. If we can reverse engineer every neuron of the brain, why not create a copy made of transistors, duplicating precisely the thought processes of the mind? In this way, you do not have to die in order to live forever. You can be conscious throughout the entire process.

He told me that this process would have to be done in steps. First, you lie on a stretcher, next to a robot lacking a brain. Next, a robotic surgeon extracts a few neurons from your brain, and then duplicates these neurons with some transistors located in the robot. Wires connect your brain to the transistors in the robot's empty head. The neurons are then thrown away and replaced by the transistor circuit. Since your brain remains connected to these transistors via wires, it functions normally and you are fully conscious during this process. Then the super surgeon removes more and more neurons from your brain, each time duplicating these neurons with transistors in the robot. Midway through the operation, half of your brain is empty; the other half is connected by wires to a large collection of transistors inside the robot's head. Eventually all the neurons in your brain have been removed, leaving a robot brain that is an exact duplicate of your original brain, neuron for neuron.

At the end of this process, however, you rise from the stretcher and find that your body is perfectly formed. You are handsome and beautiful beyond your dreams, with superhuman powers and abilities. As a perk, you are also immortal. You gaze back at your original mortal body, which is just an aging shell without a mind.

This technology, of course, is far ahead of our time. We cannot reverse engineer the human brain, let alone make a carbon copy made of transistors. (One of the main criticisms of this approach is that a transistorized brain may not fit inside the skull. In fact, given the size of electronic components, the transistorized brain may be the size of a huge supercomputer. In this sense, this proposal begins to resemble the previous one, in which the reverse-engineered brain is stored in a huge supercomputer, which in turn controls a surrogate. But the great advantage of this approach is that you don't have to die; you'd be fully conscious during the process.)

One's head spins contemplating these possibilities. All of them seem to be consistent with the laws of physics, but the technological barriers to achieving them are truly formidable. All these proposals for uploading consciousness into a computer require a technology that is far into the future.

But there is one last proposal for attaining immortality that does not require reverse engineering the brain at all. It requires simply a microscopic "nanobot" that can manipulate individual atoms. So why not live forever in your own natural body, but with a periodic "tune-up" that makes it immortal?

WHAT IS AGING?

This new approach incorporates the latest research into the aging process. Traditionally there has been no consensus among biologists about the source of the aging process. But within the last decade, a new theory has gained gradual acceptance and has unified many strands of research into aging. Basically, aging is the buildup of errors, at the genetic and cellular level. As cells get older, errors begin to build up in their DNA and cellular debris also starts to accumulate, which makes cells sluggish. As cells begin to slowly malfunction, skin begins to sag, bones become frail, hair falls out, and our immune system deteriorates. Eventually, we die.

But cells also have error-correcting mechanisms. Over time, however, even these error-correcting mechanisms begin to fail, and aging accelerates. The goal, therefore, is to strengthen the natural cell-repair mechanisms, which can be done via gene therapy and the creation of new enzymes. But there is also another way: using "nanobot" assemblers.

One of the linchpins of this futuristic technology is something called the "nanobot," or an atomic machine, which patrols the bloodstream, zapping cancer cells, repairing the damage from the aging process, and keeping us forever young and healthy. Nature has already created some nanobots, in the form of immune cells that patrol the body in the blood. But these immune cells attack viruses and foreign bodies, not the aging process.

Immortality is within reach if these nanobots can reverse the ravages of the aging process at the molecular and cellular level. In this vision, nanobots are like immune cells, tiny police patrolling your bloodstream. They attack any cancer cells, neutralize viruses, and clean out the debris and mutations.

Then the possibility of immortality would be within reach using our own bodies, not some robot or clone.

NANOBOTS—REAL OR FANTASY?

My own personal philosophy is that if something is consistent with the laws of physics, then it becomes an engineering and economics problem to build it. The engineering and economic hurdles may be formidable, of course, making it impractical for the present, but nonetheless it is still possible.

On the surface, the nanobot is simple: an atomic machine with arms and clippers that grabs molecules, cuts them at specific points, and then splices them back together. By cutting and pasting various atoms, the nanobot can create almost any known molecule, like a magician pulling something out of a hat. It can also self-reproduce, so it is necessary to build only one nanobot. This nanobot will then take raw materials, digest them, and create millions of other nanobots. This could trigger a second Industrial Revolution, as the cost of building materials plummets. One day, perhaps every home will have its own personal molecular assembler, so you can have anything you want just by asking for it.

But the key question is: Are nanobots consistent with the laws of physics? Back in 2001, two visionaries practically came to blows over this crucial question. At stake was nothing less than a vision of the entire future of technology. On one side was the late Richard Smalley, a Nobel laureate in chemistry and skeptical of nanobots. On the other side was Eric Drexler, one of the founding fathers of nanotechnology. Their titanic, tit-for-tat battle played out in the pages of several scientific magazines from 2001 to 2003.

Smalley said that, at the atomic scale, new quantum forces emerge that make nanobots impossible. The error made by Drexler and others, he claimed, is that the nanobot, with its clippers and arms, cannot function at the atomic scale. There are novel forces (e.g., the Casimir force) that cause atoms to repel or attract one another. He called this the "sticky, fat fingers" problem, because the fingers of the nanobot are not like delicate, precise pliers and wrenches. Quantum forces get in the way, so it's like trying to weld metals together while wearing gloves that are many inches thick. Furthermore, every time you try to weld two pieces of metal together, these pieces are either repelled or stick to you, so you can never grab one properly.

Drexler then fired back, stating that nanobots are not science fiction—they actually exist. Think of the ribosomes in our own body. They are essential in creating and molding DNA molecules. They can cut and splice DNA molecules at specific points, which makes possible the creation of new DNA strands.

But Smalley wasn't satisfied, stating that ribosomes are not all-purpose machines that can cut and paste anything you want; they work specifically on DNA molecules. Moreover, ribosomes are organic chemicals that need enzymes to speed up the reaction, which occurs only in a watery environment. Transistors are made of silicon, not water, so these enzymes would never work, he concluded. Drexel, in turn, mentioned that catalysts can work even without water. This heated exchange went back and forth through several rounds. In the end, like two evenly matched prizefighters, both sides seemed exhausted. Drexler had to admit that the analogy to workers with cutters and blowtorches was too simplistic, that quantum forces do get in the way sometimes. But Smalley had to concede that he was unable to score a knockout blow. Nature had at least one way of evading the "sticky, fat fingers" problem, with ribosomes, and perhaps there might be other subtle, unforeseen ways as well.

Regardless of the details of this debate, Ray Kurzweil is convinced that these nanobots, whether or not they have fat, sticky fingers, will one day shape not just molecules, but society itself. He summarized his vision when he said, "I'm not planning to die. . . . I see it, ultimately, as an awakening of the whole universe. I think the whole universe right now is basically made up of dumb matter and energy and I think it will wake up. But if it becomes transformed into this sublimely intelligent matter and energy, I hope to be part of that."

As fantastic as these speculations are, they are only a preface to the next leap in speculation. Perhaps one day the mind will not only be free of its material body, it will also be able to explore the universe as a being of pure energy. The idea that consciousness will one day be free to roam among the stars is the ultimate dream. As incredible as it may sound, this is well within the laws of physics.

13 THE MIND AS PURE ENERGY

The idea that one day consciousness may spread throughout the universe has been considered seriously by physicists. Sir Martin Rees, the Royal Astronomer of Great Britain, has written, "Wormholes, extra dimensions, and quantum computers open up speculative scenarios that could transform our entire universe eventually into a 'living cosmos'!"

But will the mind one day be freed of its material body to explore the entire universe? This was the theme explored in Isaac Asimov's classic science-fiction tale "The Last Question." (He would fondly recall that this was his favorite science-fiction short story of all the ones he had written.) In it, billions of years into the future, humans will have placed their physical bodies in pods on an obscure planet, freeing their minds to control pure energy throughout the galaxy. Instead of surrogates made of steel and silicon, these surrogates are pure energy beings that can effortlessly roam the distant reaches of space, past exploding stars, colliding galaxies, and other wonders of the universe. But no matter how powerful humanity has become, it is helpless as it witnesses the ultimate death of the universe itself in the Big Freeze. In desperation, humanity constructs a supercomputer to answer the final question: Can the death of the universe be reversed? The computer is so large and complex that it has to be placed in hyperspace. But the computer simply responds that there is insufficient information to give an answer.

Eons later, as the stars begin to turn dark, all life in the universe is about to die. But then the supercomputer finally discovers a way to reverse the death of the universe. It collects dead stars from across the universe, combines them into one gigantic cosmic ball, and ignites it. As the ball explodes, the supercomputer announces, "Let there be light!"

And there was light.

So humanity, once freed of the physical body, is capable of playing God and creating a new universe.

At first, Asimov's fantastic tale of beings made of pure energy roaming across the universe sounds impossible. We are accustomed to thinking of beings made of flesh and blood, which are at the mercy of the laws of physics and biology, living and breathing on Earth, and bound by the gravity of our planet. The concept of conscious entities of energy, soaring across the galaxy, unimpeded by the limitations of material bodies, is a strange one.

Yet this dream of exploring the universe as beings of pure energy is well within the laws of physics. Think of the most familiar form of pure energy, a laser beam, which is capable of containing vast amounts of information. Today trillions of signals in the form of phone calls, data packages, videos, and e-mail messages are transmitted routinely by fiber-optic cables carrying laser beams. One day, perhaps sometime in the next century, we will be able to transmit the consciousness of our brains throughout the solar system by placing our entire connectomes onto powerful laser beams. A century beyond that, we may be able to send our connectome to the stars, riding on a light beam.

(This is possible because the wavelength of a laser beam is microscopic, i.e., measured in millionths of a meter. That means you can compress vast amounts of information on its wave pattern. Think of Morse code. The dots and dashes of Morse code can easily be superimposed on the wave pattern of a laser beam. Even more information can be transferred onto a beam of X-rays, which has a wavelength even smaller than an atom.)

One way to explore the galaxy, unbound by the messy restrictions of ordinary matter, is to place our connectomes onto laser beams directed at the moon, the planets, and even the stars. Given the crash program to find the pathways of the brain, the complete connectome of the human brain will be available late in this century, and a form of the connectome capable of being placed on a laser beam might be available in the next century.

The laser beam would contain all the information necessary to reassemble a conscious being. Although it may take years or even centuries for the laser beam to reach its destination, from the point of view of the person riding on the laser beam, the trip would be instantaneous. Our consciousness is essentially frozen on the laser beam as it soars through empty space, so the trip to the other side of the galaxy appears to take place in the blink of an eye.

In this way, we avoid all the unpleasant features of interplanetary and interstellar travel. First, there is no need to build colossal booster rockets. Instead, you simply press the "on" button of a laser. Second, there are no powerful g forces crushing your body as you accelerate into space. Instead, you are boosted instantly to the speed of light, since you are immaterial. Third, you don't have to suffer the hazards of outer space, such as meteor impacts and deadly cosmic rays, since asteroids and radiation pass right through you harmlessly. Fourth, you don't have to freeze your body or endure years of boredom as you lumber tediously inside a conventional rocket. Instead, you zip across space at the fastest velocity in the universe, frozen in time.

Once we reach our destination, there would have to be a receiving station to transfer the data of the laser beam onto a mainframe computer, which then brings the conscious being back to life. The code that was imprinted onto the laser beam now takes control of the computer and redirects its programming. The connectome directs the mainframe computer to begin simulating the future to attain its goals (i.e., it becomes conscious).

This conscious being inside the mainframe then sends signals wirelessly to a robotic surrogate body, which has been waiting for us at the destination. In this way, we suddenly "wake up" on a distant planet or star, as if the trip took place in the blink of an eye, inside the robotic body of our surrogate. All the complex computations take place in a large mainframe computer, which directs the movements of a surrogate to carry on with our business on a distant star. We are oblivious to the hazards of space travel, as if nothing had happened.

Now imagine a vast network of these stations spread out over the solar system and even the galaxy. From our point of view, hopping from star to star would be almost effortless, traveling at the speed of light in journeys that are instantaneous. At each station, there is a robotic surrogate waiting for us to enter its body, just like an empty hotel room waiting for us to check in. We arrive at our destination refreshed and equipped with a superhuman body.

The type of surrogate robotic body that awaits us at the end of this journey would depend on the mission. If the job is to explore a new world, then the surrogate body would have to work in harsh conditions. It would have to adjust to a different gravitational field, a poisonous atmosphere, freezing-cold or blistering-hot temperatures, different day-night cycles, and a constant rain of deadly radiation. To survive under these harsh conditions, the surrogate body would have to have super strength and super senses.

If the surrogate body is purely for relaxation, then it would be designed for leisurely activities. It would maximize the pleasure of soaring through space on skis, surfboards, kites, gliders, or planes, or of sending a ball through space propelled by the swing of a bat, club, or racket.

Or if the job is to mingle with and study the local natives, then the surrogate would approximate the bodily characteristics of the indigenous population (as in the movie *Avatar*).

Admittedly, in order to create this network of laser stations in the first place, it might be necessary first to travel to the planets and stars in the old-fashioned way, in more conventional rocket ships. Then one could build the first set of these laser stations. (Perhaps the fastest, cheapest, and most efficient way of creating this interstellar network would be to send self-replicating robotic probes throughout the galaxy. Because they can make copies of themselves, starting with one such probe, after many generations there would be billions of such probes streaming out in all directions, each one creating a laser station wherever it lands. We will discuss this further in the next chapter.)

But once the network is fully established, one can conceive of a continual stream of conscious beings roaming the galaxy, so that at any time crowds of people are leaving and arriving from distant parts of the galaxy. Any laser station in the network might look like Grand Central Station.

As futuristic as this may sound, the basic physics for this concept are already well established. This includes placing vast amounts of data onto laser beams, sending this information across thousands of miles, and then decoding the information at the other end. The major problems facing this idea are therefore not in the physics, but in the engineering. Because of this, it may take us until the next century to send our entire connectome on laser beams powerful enough to reach the planets. It might take us still another century to beam our minds to the stars.

To see if this is feasible, it is instructive to do a few simple, back-of-the-envelope calculations. The first problem is that the photons inside a pencil-thin laser beam, although they appear to be in perfectly parallel formation, actually diverge slightly in space. (When I was a child, I used to shine a flashlight at the moon and wonder if the light ever reached it. The answer is yes. The atmosphere absorbs over 90 percent of the original beam, leaving some remaining to reach the moon. But the real problem is that the image the flashlight finally casts on the moon is miles across. This is because of the uncertainty principle; even laser beams must diverge slowly. Since you cannot know the precise location of the laser beam, it must, by the laws of quantum physics, slowly spread out over time.)

But beaming our connectomes to the moon does not give us much advantage, since it's easier simply to remain on Earth and control the lunar surrogate directly by radio. The delay is only about a second when issuing commands to the surrogate. The real advantage comes when controlling surrogates on the planets, since a radio message may take hours to reach a surrogate there. The process of issuing a series of radio commands to a surrogate, waiting for a response, and issuing another command would be painfully slow, taking days on end.

If you want to send a laser beam to the planets, you first have to establish a battery of lasers on the moon, well above the atmosphere, so there is no air to absorb the signal. Shot from the moon, a laser beam to the planets could arrive in a matter of minutes to a few hours. Once the laser beam has sent the connectome to the planets, then it's possible to directly control the surrogate without any delay factors at all.

So establishing a network of these laser stations throughout the solar system could be accomplished by the next century. But the problems are magnified when we try sending the beam to the stars. This means that we must have relay stations placed on asteroids and space stations along the way, in order to amplify the signal, reduce errors, and send the message to the next relay station. This could potentially be done by using the comets that lie between our sun and the nearby stars. For example, extending about a light-year from the sun (or one-quarter of the distance to the nearest star) is the Oort cloud of comets. It is a spherical shell of billions of comets, many of which lie motionless in empty space. There is probably a similar Oort cloud of comets surrounding the Centauri star system, which is our nearest stellar neighbor. Assuming that this Oort cloud also extends a light-year from those stars, then fully half the distance from our solar system to the next contains stationary comets on which we can build laser relay stations.

Another problem is the sheer amount of data that must be sent by laser beam. The total information contained in one's connectome, according to Dr. Sebastian Seung, is roughly one zettabyte (that is, a 1 with twenty-one zeros after it). This is roughly equivalent to the total information contained in the World Wide Web today. Now consider shooting a battery of laser beams into space carrying this vast mountain of information. Optical fibers can carry terabytes of data per second (a 1 with twelve zeros after it). Within the next century, advances in information storage, data compression, and bundling of laser beams may increase this efficiency by a factor of a million. This means that it would take a few hours or so to send the beam into space carrying all the information contained within the brain.

So the problem is not the sheer amount of data sent on laser beams. In principle, laser beams can carry an unlimited amount of data. The real bottlenecks are the receiving stations at either end, which must have switches that rapidly manipulate this amount of data at blinding speed. Silicon transistors may not be fast enough to handle this volume of data. Instead, we might have to use quantum computers, which compute not on silicon transistors but on individual atoms. At present, quantum computers are at a primitive level, but by the next century they might be powerful enough to handle zettabytes of information.

FLOATING BEINGS OF ENERGY

Another advantage of using quantum computers to process this mountain of data is the chance to create beings of energy that can hover and float in the air, which appear frequently in science fiction and fantasy. These beings would represent consciousness in its purest form. At first, however, they may seem to violate the laws of physics, since light always travels at the speed of light.

But in the last decade, headlines were made by physicists at Harvard University who announced that they were able to stop a beam of light dead in its tracks. These physicists apparently accomplished the impossible, slowing down a light beam to a leisurely pace until it could be placed in a bottle. Capturing a light beam in a bottle is not so fantastic if you look carefully at a glass of water. As a light beam enters the water, it slows down, bending as it enters the water at an angle. Similarly, light bends as it enters glass, making telescopes and microscopes possible. The reason for all this comes from the quantum theory.

Think of the old Pony Express, which delivered the mail in the nineteenth century in the American West. Each pony could sprint between relay stations at great speed. But the bottleneck was the delay factor at each relay station, where the mail, rider, and pony had to be exchanged. This slowed down the average velocity of the mail considerably. In the same way, in the vacuum between atoms, light still travels at c, the speed of light, which is roughly 186,282 miles per second. However, when it hits atoms, light is delayed; it is briefly absorbed and then reemitted by atoms, sending it on its way a fraction of a second later. This slight delay is responsible for light beams, on average, apparently slowing down in glass or water.

The Harvard scientists exploited this phenomenon, taking a container of gas and carefully cooling it down to near absolute zero. At these freezing temperatures, the gas atoms absorbed a light beam for longer and longer time periods before reemitting it. Thus, by increasing this delay factor, they could slow down the light beam until it came to rest. The light beam still traveled at the speed of light between the gas atoms, but it spent an increasing amount of time being absorbed by them.

This raises the possibility that a conscious being, instead of assuming control of a surrogate, may prefer to remain in the form of pure energy and roam, almost ghostlike, as pure energy.

So in the future, as laser beams are sent to the stars containing our connectomes, the beam may be transferred into a cloud of gas molecules and then contained in a bottle. This "bottle of light" is very similar to a quantum computer. Both of them have a collection of atoms vibrating in unison, in which the atoms are in phase with one another. And both of them can do complex computations that are far beyond an ordinary computer's capability. So if the problems of quantum computers can be solved, it may also give us the ability to manipulate these "bottles of light."

FASTER THAN LIGHT?

We see, then, that all these problems are ones of engineering. There is no law of physics preventing traveling on an energy beam in the next century or beyond. So this is perhaps the most convenient way of visiting the planets and stars. Instead of riding on a light beam, as the poets dreamed, we become the light beam.

To truly realize the vision expressed in Asimov's science-fiction tale, we need to ask if faster-than-light intergalactic travel is truly possible. In his short story, beings of immense power move freely between galaxies separated by millions of light-years.

Is this possible? To answer this question, we have to push the very boundaries of modern quantum physics. Ultimately, things called "wormholes" may provide a shortcut through the vastness of space and time. And beings made of pure energy rather than matter would have a decisive advantage in passing through them.

Einstein, in some sense, is like the cop on the block, stating that you cannot go faster than light, the ultimate velocity in the universe. Traveling across the Milky Way galaxy, for example, would take one hundred thousand years, even sailing on a laser beam. Although only an instant of time has passed for the traveler, the time on the home planet has progressed one hundred thousand years. And passing between galaxies involves millions to billions of light-years.

But Einstein himself left a loophole in his work. In his general theory of relativity of 1915, he showed that gravity arose from the warping of space-time. Gravity is not the "pull" of a mysterious invisible force, as Newton once thought, but actually a "push" caused by space itself bending around an object. Not only did this brilliantly explain the bending of starlight passing near stars and the expansion of the universe, it left open the possibility of the fabric of space-time stretching until it ripped.

In 1935, Einstein and his student Nathan Rosen introduced the possibility that two black-hole solutions could be joined back to back, like Siamese twins, so if you fell into one black hole, you could, in principle, pass out of the other one. (Imagine joining two funnels at their ends. Water that drains through one funnel emerges from the other.) This "wormhole," also called the Einstein-Rosen Bridge, introduced the possibility of portals or gateways between universes. Einstein himself dismissed the possibility that you could pass through a black hole, since you would be crushed in the process, but several subsequent developments have raised the possibility of faster-than-light travel through a wormhole.

First, in 1963, mathematician Roy Kerr discovered that a spinning black hole does not collapse into a single dot, as previously thought, but into a rotating ring, spinning so fast that centrifugal forces prevent it from collapsing. If you fell through the ring, then you could pass into another universe. The gravitational forces would be large, but not infinite. This would be like Alice's Looking Glass, where you could pass your hand through the mirror and enter a parallel universe. The rim of the Looking Glass would be the ring forming the black hole itself. Since Kerr's discovery, scores of other solutions of Einstein's equations have shown that you can, in principle, pass between universes without being immediately crushed. Since every black hole seen so far in space is spinning rapidly (some of them clocked at one million miles per hour), this means that these cosmic gateways could be commonplace.

In 1988, physicist Dr. Kip Thorne of Cal Tech and his colleagues showed that, with enough "negative energy," it might be possible to stabilize a black hole so that a wormhole becomes "transversable" (i.e., you can freely pass through it both ways without being crushed). Negative energy is perhaps the most exotic substance in the universe, but it actually exists and can be created (in microscopic quantities) in the laboratory.

So here is the new paradigm. First, an advanced civilization would concentrate enough positive energy at a single point, comparable to a black hole, to open up a hole through space connecting two distant points. Second, it would amass enough negative energy to keep the gateway open, so that it is stable and does not close the instant you enter it.

We can now put this idea into proper perspective. Mapping the entire human connectome should be possible late in this century. An interplanetary laser network could be established early in the next century, so that consciousness can be beamed across the solar system. No new law of physics would be required. A laser network that can go between the stars may have to wait until the century after that. But a civilization that can play with wormholes will have to be thousands of years ahead of us in technology, stretching the boundaries of known physics.

All this, then, has direct implications for whether consciousness can pass between universes. If matter comes close to a black hole, the gravity becomes so intense that your body becomes "spaghettified." The gravity pulling on your leg is greater than the gravity pulling on your head, so your body is stretched by tidal forces. In fact, as you approach the black hole, even the atoms of your body are stretched until the electrons are ripped from the nuclei, causing your atoms to disintegrate.

(To see the power of tidal forces, just look at the tides of Earth and the rings of Saturn. The gravity of the moon and sun exert a pull on Earth, causing the oceans to rise several feet during high tide. And if a moon comes too close to a giant planet like Saturn, the tidal forces will stretch the moon and eventually tear it apart. The distance at which moons get ripped apart by tidal forces is called the Roche limit. The rings of Saturn lie exactly at the Roche limit, so they might have been caused by a moon that wandered too close to the mother planet.)

Even if we enter a spinning black hole and use negative energy to stabilize it, then, the gravity fields still might be so powerful that we'd be spaghettified.

But here is where laser beams have an important advantage over matter when passing through a wormhole. Laser light is immaterial, so it cannot be stretched by tidal forces as it passes near a black hole. Instead, light becomes "blue-shifted" (i.e., it gains energy and its frequency increases). Even though the laser beam is distorted, the information stored on it is untouched. For example, a message in Morse code carried by a laser beam becomes compressed, but the information content remains unchanged. Digital information is untouched by tidal forces. So gravitational forces, which can be fatal to beings made of matter, may be harmless to beings traveling on light beams.

In this way, consciousness carried by a laser beam, because it is immaterial, has a decisive advantage over matter in passing through a wormhole.

Laser beams have another advantage over matter when passing through a wormhole. Some physicists have calculated that a microscopic wormhole, perhaps the size of an atom, might be easier to create. Matter would not be able to pass through such a tiny wormhole. But X-ray lasers, with a wavelength smaller than an atom, might possibly be able to pass through without difficulty.

Although Asimov's brilliant short story was clearly a work of fantasy, ironically a vast interstellar network of laser stations might already exist within the galaxy, yet we are so primitive that we are totally unaware of it.

To a civilization thousands of years ahead of us, the technology to digitalize their connectomes and send them to the stars would be child's play. In that case, it is conceivable that intelligent beings are already zapping their consciousness across a vast network of laser beams in the galaxy. Nothing we observe with our most advanced telescopes and satellites prepares us to detect such an intergalactic network.

Carl Sagan once lamented the possibility that we might live in a world surrounded by alien civilizations and not have the technology to realize it.

Then the next question is: What lurks in the alien mind?

If we were to encounter such an advanced civilization, what kind of consciousness might it have? One day, the destiny of the human race may rest on answering this question.

Sometimes I think that the surest sign that intelligent life exists elsewhere in the universe is that none of it has tried to contact us.

—BILL WATTERSON

Either intelligent life exists in outer space or it doesn't. Either thought is frightening.

—ARTHUR C. CLARKE

14 THE ALIEN MIND

In *War of the Worlds* by H. G. Wells, aliens from Mars attack Earth because their home planet is dying. Armed with death rays and giant walking machines, they quickly incinerate many cities and are on the verge of seizing control of Earth's major capitals. Just as the Martians are crushing all signs of resistance and our civilization is about to be reduced to rubble, they are mysteriously stopped cold in their tracks. With all their advanced science and weaponry, they failed to factor in an onslaught from the lowliest of creatures: our germs.

That single novel created an entire genre, launching a thousand movies like *Earth vs. the Flying Saucers* and *Independence Day.* Most scientists cringe, however, when they see how the aliens are described. In the movies, aliens are often depicted as creatures with some sense of human values and emotions. Even with glowing green skin and huge heads, they still look like us to a certain degree. They also tend to speak perfect English.

But, as many scientists have pointed out, we may have much more in common with a lobster or a sea slug than we do with an alien from space.

As with silicon consciousness, alien consciousness will most likely have the general features described in our space-time theory; that is, the ability to make a model of the world and then calculate how it will evolve in time to achieve a goal. But while robots can be programmed so that they emotionally bond with humans and have goals compatible with ours, alien consciousness may have neither. It's likely to have its own set of values and goals, independent of humanity. One can only speculate what these might be.

Physicist Dr. Freeman Dyson of the Institute for Advanced Study at Princeton was a consultant to the movie *2001*. When he finally saw the movie, he was delighted, not because of its dazzling special effects, but because it was the first Hollywood movie ever to present an alien consciousness, with desires, goals, and intentions totally foreign to ours. For the first time, the aliens were not simply human actors flailing about, trying to act menacing in cheesy monster costumes. Instead, alien consciousness was presented as something totally orthogonal to human experience, something entirely outside our ken.

In 2011, Stephen Hawking raised another question. The noted cosmologist made headlines when he said that we must be prepared for a possible alien attack. He said that if we ever encounter an alien civilization, it will be more advanced than ours and hence will pose a mortal threat to our very existence.

We have only to see what happened to the Aztecs when they encountered the bloodthirsty Hernán Cortés and his conquistadors to imagine what might happen with such a fateful encounter. Armed with technology that the Bronze Age Aztecs had never seen before, such as iron swords, gunpowder, and the horse, this small band of cutthroats was able to crush the ancient Aztec civilization in a matter of months in 1521.

All this raises these questions: What will alien consciousness be like? How will their thinking process and goals differ from ours? What do they want?

FIRST CONTACT IN THIS CENTURY

This is not an academic question. Given the remarkable advances in astrophysics, we may actually make contact with an alien intelligence in the coming decades. How we respond to them could determine one of the most pivotal events in human history.

Several advances are making this day possible.

First, in 2011 the Kepler satellite, for the first time in history, gave scientists a "census" of the Milky Way galaxy. After analyzing light from thousands of stars, the Kepler satellite found that one in two hundred might harbor an earthlike planet in the habitable zone. For the first time, we can therefore calculate how many stars within the Milky Way galaxy might be earthlike: about a billion. As we look at the distant stars, we have genuine reason to wonder if anyone is looking back at us.

So far, more than one thousand exoplanets have been analyzed in detail by earthbound telescopes. (Astronomers find them at the rate of about two exoplanets per week.) Unfortunately, nearly all of them are Jupiter-size planets, probably devoid of any earthlike creatures, but there are a handful of "super earths," rocky planets that are a few times larger than Earth. Already, the Kepler satellite has identified about 2,500 candidate exoplanets in space, a handful of which look very much like Earth. These planets are at just the right distance from their mother stars so that liquid oceans can exist. And liquid water is the "universal solvent" that dissolves most organic chemicals like DNA and proteins.

In 2013, NASA scientists announced their most spectacular discovery using the Kepler satellite: two exoplanets that are near twins of Earth. They are located 1,200 light-years away in the constellation Lyra. They are only 60 percent and 40 percent larger than Earth. More important, both lie within the habitable zone of their mother star, so there is a possibility that they have liquid oceans. Of all the planets analyzed so far, they are the closest to being mirror images of Earth.

Furthermore, the Hubble Space Telescope has given us an estimate of the total number of galaxies in the visible universe: one hundred billion. Therefore, we can calculate the number of earthlike planets in the visible universe: one billion times one hundred billion, or one hundred quintillion earthlike planets.

This is a truly astronomical number, so the odds of life existing in the universe are astronomically large, especially when you consider that the universe is 13.8 billion years old, and there has been plenty of time for intelligent empires to rise—and perhaps fall. In fact, it would be more miraculous if another advanced civilization did *not* exist.

SETI AND ALIEN CIVILIZATIONS

Second, radio telescope technology is becoming more sophisticated. So far, only about one thousand stars have been closely analyzed for signs of intelligent life, but in the coming decade this number could rise by a factor of one million.

Using radio telescopes to hunt for alien civilizations dates back to 1960, when astronomer Frank Drake initiated Project Ozma (after the Queen of Oz), using the twenty-five-meter radio telescope in Green Bank, West Virginia. This marked the birth of the SETI project (the Search for Extraterrestrial Intelligence). Unfortunately, no signals from aliens were picked up, but in 1971 NASA proposed Project Cyclops, which was supposed to have 1,500 radio telescopes at a cost of $10 billion.

Not surprisingly, it never went anywhere. Congress was not amused.

Funding did become available for a much more modest proposal: to send a carefully coded message in 1971 to aliens in outer space. A coded message containing 1,679 bits of information was transmitted via the giant Arecibo radio telescope in Puerto Rico toward the Globular Cluster M13, about 25,100 light-years away. It was the world's first cosmic greeting card, containing relevant information about the human race. But no reply message was received. Perhaps the aliens were not impressed with us, or possibly the speed of light got in the way. Given the large distances involved, the earliest date for a reply message would be 52,174 years from now.

Since then, some scientists have expressed misgivings about advertising our existence to aliens in space, at least until we know their intentions toward us. They disagree with the proponents of the METI Project (Messaging to Extra-Terrestrial Intelligence) who actively promote sending signals to alien civilizations in space. The reasoning behind the METI Project is that Earth already sends vast amounts of radio and TV signals into outer space, so a few more messages from the METI Project will not make much difference. But the critics of METI believe that we should not needlessly increase our chances of being discovered by potentially hostile aliens.

In 1995, astronomers turned to private sources to start the SETI Institute in Mountain View, California, to centralize research and initiate Project Phoenix, which is trying to study one thousand nearby sunlike stars in the 1,200-to-3,000-megahertz radio range. The equipment is so sensitive that it can pick up the emissions from an airport radar system two hundred light-years away. Since its founding, the SETI Institute has scanned more than one thousand stars at a cost of $5 million per year, but still no luck.

A more novel approach is the SETI@home project, initiated by astronomers at the University of California at Berkeley in 1999, which uses an informal army of millions of amateur PC owners. Anyone can join in this historic hunt. While you are sleeping at night, your screen saver crunches some of the data pouring in from the Arecibo radio telescope in Puerto Rico. So far, it has signed up 5.2 million users in 234 countries; perhaps these amateurs dream that they will be the first in human history to make contact with alien life. Like Columbus's, their names may go down in history. The SETI @home project has grown so rapidly that it is, in fact, the largest computer project of this type ever undertaken.

When I interviewed Dr. Dan Wertheimer, director of SETI@home, I asked him how they can distinguish false messages from real ones. He said something that surprised me. He told me that they sometimes deliberately "seed" the data from radio telescopes with fake signals from an imaginary intelligent civilization. If no one picks up these fake messages, then they know that there is something wrong with their software. The lesson here is that if your PC screen saver announces that it has deciphered a message from an alien civilization, please do not immediately call the police or the president of the United States. It might be a fake message.

ALIEN HUNTERS

One colleague of mine who has dedicated his life to finding intelligent life in outer space is Dr. Seth Shostak, director of the SETI Institute. With his Ph.D. in physics from the California Institute of Technology, I might have expected him to become a distinguished physics professor lecturing to eager Ph.D. students, but instead he spends his time in an entirely different fashion: asking for donations to the SETI Institute from wealthy individuals, poring over possible signals from outer space, and doing a radio show. I once asked him about the "giggle factor"—do fellow scientists giggle when he tells them that he listens to aliens from outer space? Not anymore, he claims. With all the new discoveries in astronomy, the tide has turned.

In fact, he even sticks his neck out and says flatly that we will make contact with an alien civilization in the very near future. He has gone on record as proclaiming that the 350-antenna Allen Telescope Array now being built "will trip across a signal by the year 2025."

Isn't that a bit risky, I asked him? What makes him so sure? One factor working in his favor has been the explosion in the number of radio telescopes in the last few years. Although the U.S. government does not fund his project, the SETI Institute recently hit pay dirt when it convinced Paul Allen (the Microsoft billionaire) to donate over $30 million in funds to start the Allen Telescope Array at Hat Creek, California, 290 miles north of San Francisco. It currently scans the heavens with 42 radio telescopes, and eventually will reach up to 350. (One problem, however, is the chronic lack of funding for these scientific experiments. To make up for budget cuts, the Hat Creek facility is kept alive through partial funding from the military.)

One thing, he confessed to me, makes him squirm a bit, and that is when people confuse the SETI Project with UFO hunters. The former, he claims, is based on solid physics and astronomy, using the latest in technology. The latter, however, base their theories on anecdotal hearsay evidence that may or may not be based on truth. The problem is that the mass of UFO sightings he gets in the mail are not reproducible or testable. He urges anyone who claims to have been abducted by aliens in a flying saucer to steal something—an alien pen or paperweight, for example—to prove your case. Never leave a UFO empty-handed, he told me.

He also concludes that there is no firm evidence that aliens have visited our planet. I then asked him whether he thought the U.S. government was deliberately covering up evidence of an alien encounter, as many conspiracy theorists believe. He replied, "Would they really be so efficient at covering up a big thing like this? Remember, this is the same government that runs the post office."

DRAKE'S EQUATION

When I asked Dr. Wertheimer why he is so sure that there is alien life in outer space, he replied that the numbers are in his favor. Back in 1961, astronomer Frank Drake tried to estimate the number of such intelligent civilizations by making plausible assumptions. If we start with the number one hundred billion, the number of stars in the Milky Way galaxy, then we can estimate the fraction of them that are similar to our sun. We can reduce that number further by estimating the fraction of them that have planets, the fraction of them that have earthlike planets, etc. After making a number of reasonable assumptions, we come up with an estimate of ten thousand advanced civilizations in our own Milky Way galaxy. (Carl Sagan, with a different set of estimates, came up with the number one million.)

Since then, scientists have been able to make much better estimates of the number of advanced civilizations in our galaxy. For example, we know there are more planets orbiting stars than Drake originally expected, and more earthlike planets as well. But we still face a problem. Even if we know how many earthlike twins there are in space, we still don't know how many of them support intelligent life. Even on Earth, it took 4.5 billion years before intelligent beings (us) finally arose from the swamp. For about 3.5 billion years, life-forms have existed on Earth, but only in the last one hundred thousand years or so have intelligent beings like us emerged. So even on an earthlike planet like Earth itself, the rise of truly intelligent life has been very difficult.

WHY DON'T THEY VISIT US?

But then I asked Dr. Seth Shostak of SETI this killer question: If there are so many stars in the galaxy, and so many alien civilizations, then why don't they visit us? This is the Fermi paradox, named for Enrico Fermi, the Nobel laureate who helped build the atomic bomb and unlocked the secrets of the nucleus of the atom.

Many theories have been proposed. For one, the distance between stars might be too great. It would take about seventy thousand years for our most powerful chemical rockets to reach the stars nearest to Earth. Perhaps a civilization thousands to millions of years more advanced than ours may solve this problem, but there's another possibility. Maybe they annihilated themselves in a nuclear war. As John F. Kennedy once said, "I am sorry to say there is too much point to the wisecrack that life is extinct on other planets because their scientists were more advanced than ours."

But perhaps the most logical reason is this: Imagine walking down a country road and encountering an ant hill. Do we go down to the ants and say, "I bring you trinkets. I bring you beads. I give you nuclear energy. I will create an ant paradise for you. Take me to your leader"?

Probably not.

Now imagine that workers are building an eight-lane superhighway next to the anthill. Would the ants know what frequency the workers are talking on? Would they even know what an eight-lane superhighway was? In the same fashion, any intelligent civilization that can reach Earth from the stars would be thousands of years to millions of years ahead of us, and we may have nothing to offer them. In other words, we are arrogant to believe that aliens will travel trillions upon trillions of miles just to see us.

More than likely, we are not on their radar screen. Ironically, the galaxy could be teaming with intelligent life-forms and we are so primitive that we are oblivious of them.

FIRST CONTACT

But assume for the moment that the time will come, perhaps sooner rather than later, when we make contact with an alien civilization. This moment could be a turning point in the history of humanity. So the next questions are: What do they want, and what will their consciousness be like?

In the movies and in science-fiction novels, the aliens often want to eat us, conquer us, mate with us, enslave us, or strip our planet of valuable resources. But all this is highly improbable.

Our first contact with an alien civilization will probably not begin with a flying saucer landing on the White House lawn. More likely, it will happen when some teenager, running a screen saver from the SETI@home project, announces that his or her PC has decoded signals from the Arecibo radio telescope in Puerto Rico. Or perhaps when the SETI project at Hat Creek detects a message that indicates intelligence.

Our first encounter will therefore be a one-way event. We will be able to eavesdrop on intelligent messages, but a return message may take decades or centuries to reach them.

The conversations that we hear on the radio may give us valuable insight into this alien civilization. But most of the message will likely be gossip, entertainment, music, etc., with little scientific content.

Then I asked Dr. Shostak the next key question: Will you keep it a secret once First Contact is made? After all, won't it cause mass panic, religious hysteria, chaos, and spontaneous evacuations? I was a bit surprised when he said no. They would give all the data to the governments of the world and to the people.

The next questions are: What will they be like? How do they think?

To understand alien consciousness, perhaps it is instructive to analyze another consciousness that is quite alien to us, the consciousness of animals. We live with them, yet we are totally ignorant of what goes on in their minds.

Understanding animal consciousness, in turn, may help us understand alien consciousness.

ANIMAL CONSCIOUSNESS

Do animals think? And if so, what do they think about? This question has perplexed the greatest minds in history for thousands of years. The Greek writers and historians Plutarch and Pliny both wrote about a famous question that remains unsolved even today. Over the centuries, many solutions have been given by the giants of philosophy.

A dog is traveling down a road, looking for its master, when it encounters a fork that branches in three directions. The dog first takes the left path, sniffs around, and then returns, knowing that his master has not taken that road. Then it takes the right path, sniffs, and realizes that his master has not taken this road either. But this time, the dog triumphantly takes the middle road, without sniffing.

What was going on in the dog's mind? Some of the greatest philosophers have tackled this question, to no avail. The French philosopher and essayist Michel de Montaigne wrote that the dog obviously concluded that the only possible solution was to take the middle road, a conclusion showing that dogs are capable of abstract thought.

But St. Thomas Aquinas, arguing in the thirteenth century, said the opposite—that the appearance of abstract thought is not the same thing as genuine thinking. We can be fooled by superficial appearances of intelligence, he claimed.

Centuries later, there was also a famous exchange between John Locke and George Berkeley about animal consciousness. "Beasts abstract not," proclaimed Locke flatly. To which Bishop Berkeley responded, "If the fact that brutes abstract not be made the distinguishing property of that sort of animal, I fear a great many of those that pass for man must be reckoned into their number."

Philosophers down the ages have tried to analyze this question in the same manner: by imposing human consciousness on the dog. This is the mistake of anthropomorphism, or assuming that animals think and behave like us. But perhaps the real solution might be to look at this question from the dog's point of view, which could be quite alien.

In Chapter 2, I gave a definition of consciousness in which animals were part of a continuum of consciousness. Animals can differ from us in the parameters they use to create a model of the world. Dr. David Eagleman says that psychologists call this "umwelt," or the reality perceived by other animals. He notes, "In the blind and deaf world of the tick, the important signals are temperature and the odor of butyric acid. For the black ghost knifefish, it's electrical fields. For the echo locating bat, air-compressed waves. Each organism inhabits its own umwelt, which it presumably assumes to be the entire objective reality 'out there.'"

Consider the brain of a dog, which is constantly living in a swirl of odors, by which it hunts for food or locates a mate. From these smells, the dog then constructs a mental map of what exists in its surroundings. This map of smells is totally different from the one we get from our eyes and conveys an entirely different set of information. (Recall from Chapter 1 that Dr. Penfield constructed a map of the cerebral cortex, showing the distorted self-image of the body. Now imagine a Penfield diagram of a dog's brain. Most of it would be devoted to its nose, not its fingers. Animals would have a totally different Penfield diagram. Aliens in space would likely have an even stranger Penfield diagram.)

Unfortunately, we tend to assign human consciousness to animals, even though animals may have a totally different world outlook. For example, when a dog faithfully follows its master's orders, we subconsciously assume that the dog is man's best friend because he likes us and respects us. But since the dog is descended from *Canis lupus* (the gray wolf), which hunts in packs with a rigid pecking order, more than likely the dog sees you as some sort of alpha male, or the leader of the pack. You are, in some sense, the Top Dog. (This is probably one reason why puppies are much easier to train than older dogs; it is likely easier to imprint one's presence on a puppy's brain, while more mature dogs realize that humans are not part of their pack.)

Also, when a cat enters a new room and urinates all over the carpet, we assume that the cat is angry or nervous, and we try to find out the reason why the cat is upset. But perhaps the cat is simply marking its territory with the smell of its urine to ward off other cats. So the cat is not upset at all;

it's simply warning other cats to stay out of the house, because the house belongs to it.

And if the cat purrs and rubs itself against your legs, we assume that it is grateful to you for taking care of it, that this is a sign of warmth and affection. More than likely, the cat is rubbing its hormone onto you to claim ownership of its possession (i.e., you), to ward off other cats. In the cat's viewpoint, you are a servant of some sort, trained to give it food several times a day, and rubbing its scent on you warns other cats to stay away from this servant.

As the sixteenth-century philosopher Michel de Montaigne once wrote, "When I play with my cat, how do I know that she is not playing with me rather than I with her?"

And if the cat then stalks off to be alone, it is not necessarily a sign of anger or aloofness. The cat is descended from the wildcat, which is a solitary hunter, unlike the dog. There is no alpha male to slobber over, as in the case of the dog. The proliferation of various "animal whisperer" programs on TV is probably a sign of the problems we encounter when we force human consciousness and intentions onto animals.

A bat would also have a much different consciousness, which would be dominated by sounds. Almost blind, the bat requires constant feedback from tiny squeaks it makes, which allow it to locate insects, obstacles, and other bats via sonar. The Penfield map of its brain would be quite alien to us, with a huge portion devoted to its ears. Similarly, dolphins have a different consciousness than humans, which is also based on sonar. Because dolphins have a smaller frontal cortex, it was once thought that they were not so intelligent, but the dolphin compensates for this by having a larger brain mass. If you unfold the neocortex of the dolphin brain, it would cover six magazine pages, while if you unfold the neocortex of a human, it would measure only four magazine pages. Dolphins also have very well-developed parietal and temporal cortices to analyze sonar signals in the water and are one of the few animals that can recognize themselves in a mirror, probably because of this fact.

In addition, the dolphin brain is actually structured differently from humans' because dolphin and human lineages diverged about ninety-five million years ago. Dolphins have no need for a nose, so their olfactory bulb disappears soon after birth. But thirty million years ago, their auditory cortex exploded in size because dolphins learned to use echolocation, or sonar, to find food. Like bats', their world must be one of whirling echoes and vibrations. Compared to humans, dolphins have an extra lobe in their limbic system, called the "paralimbic" region, which probably helps them forge strong social relations.

Meanwhile, dolphins also have a language that is intelligent. I once swam in a pool of dolphins for a TV special for the Science Channel. I put sonar sensors in the pool that could pick up the clicks and whistles used by dolphins to talk to one another. These signals were recorded and then analyzed by computer. There is a simple way to discern if there is an intelligence lurking among this random set of squeals and chirps. In the English language, for example, the letter *e* is the most commonly used letter of the alphabet. In fact, we can create a list of all the letters of the alphabet and how frequently they occur. No matter what book in English we analyze by computer, it will roughly obey the same list of commonly found letters of the alphabet.

Similarly, this computer program can be used to analyze the dolphins' language. Sure enough, we find a similar pattern indicating intelligence. However, as we go to other mammals, the pattern begins to break down, and it finally collapses completely as we approach lower animals with small brain sizes. Then the signals become nearly random.

INTELLIGENT BEES?

To get a sense of what alien consciousness might be like, consider the strategies adopted by nature to reproduce life on Earth. There are two basic reproductive strategies nature has taken, with profound implications for evolution and consciousness.

The first, the strategy used by mammals, is to produce a small number of young offspring and then carefully nurse each one to maturity. This is a risky strategy, because only a few progeny are produced in each generation, so it assumes that nurturing will even out the odds. This means that every life is cherished and carefully nurtured for a length of time.

But there is another, much older strategy that is used by much of the plant and animal kingdom, including insects, reptiles, and most other lifeforms on Earth. This involves creating a large number of eggs or seeds and then letting them fend for themselves. Without nurturing, most of the offspring never survive, so only a few hardy individuals will make it into the next generation. This means that the energy invested in each generation by the parents is nil, and reproduction relies on the law of averages to propagate the species.

These two strategies produce startlingly different attitudes toward life and intelligence. The first strategy treasures each and every individual. Love, nurturing, affection, and attachment are at a premium in this group. This reproductive strategy can work only if the parents invest a considerable amount of precious energy to preserve their young. The second strategy, however, does not treasure the individual at all, but rather emphasizes the survival of the species or group as a whole. To them, individuality means nothing.

Furthermore, reproductive strategy has profound implications for the evolution of intelligence. When two ants meet each other, for example, they exchange a limited amount of information using chemical scents and gestures. Although the information shared by two ants is minimal, with this information they are capable of creating elaborate tunnels and chambers necessary to build an anthill. Similarly, although honeybees communicate with one another by performing a dance, they can collectively create complex honeycombs and locate distant flower beds. So their intelligence arises not so much from the individual, but from the holistic interaction of the entire colony and from their genes.

So consider an intelligent extraterrestrial civilization based on the second strategy, such as an intelligent race of honeybees. In this society, the worker bees that fly out each day in search of pollen are expendable. Worker bees do not reproduce at all, but instead live for one purpose, to serve the hive and the queen, for which they willingly sacrifice themselves. The bonds that link mammals together mean nothing to them.

Hypothetically, this might affect the development of their space program. Since we treasure the life of every astronaut, considerable resources are devoted to bringing them back alive. Much of the cost of space travel goes into life support so the astronauts can make the return voyage home and reenter the atmosphere. But for a civilization of intelligent honeybees, each worker's life may not be worth that much, so their space program would cost considerably less. Their workers would not have to come back. Every voyage might be a one-way trip, and that would represent significant savings.

Now imagine if we were to encounter an alien from space that was actually similar to a honeybee worker. Normally, if we encounter a honeybee in the forest, chances are it will completely ignore us, unless we threaten it or the hive. It's as if we did not exist. Similarly, this worker would most likely not have the slightest interest in making contact with us or sharing its knowledge. It would go on with its primary mission and ignore us. Moreover, the values that we cherish would mean little to it.

Back in the 1970s, there were two medallions put aboard the *Pioneer 10* and *11* probes, containing crucial information about our world and society. The medallions exalted the diversity and richness of life on Earth. Scientists back then assumed that alien civilizations in space would be like us, curious and interested in making contact. But if such an alien worker bee were to find our medallion, chances are that it would mean nothing to it.

Furthermore, each worker need not be very intelligent. They need to be only intelligent enough to serve the interest of the hive. So if we were to send a message to a planet of intelligent bees, chances are that they would show little interest in sending a message back.

Even if contact could be made with such a civilization, it might be difficult communicating with them. For example, when we communicate with one another, we break ideas down into sentences, with a subject-verb structure, in order to build a narrative, often a personal story. Most of our sentences have the following structure: "I did this" or "They did that." In fact, most of our literature and conversations use storytelling, often involving experiences and adventures that we or our role models have had. This presupposes that our personal experiences are the dominant way to convey information.

However, a civilization based on intelligent honeybees may not have the least interest in personal narratives and storytelling. Being highly collective, their messages may not be personal, but matter-of-fact, containing vital information necessary for the hive rather than personal trivia and gossip that might advance an individual's social position. In fact, they might find our storytelling language to be a bit repulsive, since it puts the role of the individual before the needs of the collective.

Also, worker bees would have a totally different sense of time. Since worker bees are expendable, they might not have a long life span. They might only take on projects that are short and well defined.

However, humans live much longer, but we also have a tacit sense of time; we take on projects and occupations that we can reasonably see to the end within our lifetimes. We subconsciously pace our projects, our relations with others, and our goals to accommodate a finite life span. In other words, we live our lives in distinct phases: being single, married, raising children, and eventually retiring. Often without being conscious of it, we assume that we will live and eventually die within a finite time frame.

But imagine beings that can live for thousands of years, or are perhaps immortal. Their priorities, their goals, and their ambitions would be completely different. They could take on projects that would normally require scores of human lifetimes. Interstellar travel is often dismissed as pure science fiction because, as we have seen, the time it takes for a conventional rocket to reach nearby stars is roughly seventy thousand years. For us, this is prohibitively long. But for an alien life-form, that time may be totally irrelevant. For example, they might be able to hibernate, slow down their metabolism, or simply live for an indefinite amount of time.

WHAT DO THEY LOOK LIKE?

Our first translations of these alien messages will probably give us some insight into the aliens' culture and way of life. For example, it is likely that the aliens will have evolved from predators and hence still share some of their characteristics. (In general, predators on Earth are smarter than prey. Hunters like tigers, lions, cats, and dogs use their cunning to stalk, ambush, and hide, all of which require intelligence. All these predators have eyes on the front of the face, for stereo vision as they focus their attention. Prey, which have eyes to the sides of the face to spot a predator, have only to run. That is why we say "sly as a fox" and "dumb bunny.") The alien life-forms may have outgrown many of the predator instincts of their distant ancestors, but it is likely that they will still have some of a predator's consciousness (i.e., territoriality, expansion, and violence when necessary).

If we examine the human race, we see that there were at least three basic ingredients that set the stage for our becoming intelligent:

1. the opposable thumb, which gives us the ability to manipulate and reshape our environment via tools
2. stereo eyes or the 3-D eyes of a hunter
3. language, which allows us to accumulate knowledge, culture, and wisdom across generations

When we compare these three ingredients with the traits found in the animal kingdom, we see that very few animals meet these criteria for intelligence. Cats and dogs, for example, do not have grasping ability or a complex language. Octopi have sophisticated tentacles, but they don't see well and don't have a complex language.

There may be variations of these three criteria. Instead of an opposable thumb, an alien might have claws or tentacles. (The only prerequisite is that they should be able to manipulate their environment with tools created by these appendages.) Instead of having two eyes, they may have many more, like insects. Or they may have sensors that detect sound or UV light rather than visible light. More than likely, they will have the stereo eyes of a hunter, because predators generally have a higher level of intelligence than prey. Also, instead of a language based on sounds, they may communicate via different forms of vibrations. (The only requirement is that they exchange information among themselves to create a culture spanning many generations.)

But beyond these three criteria, anything goes.

Next, the aliens may have a consciousness colored by their environment. Astronomers now realize that the most plentiful habitat for life in the universe may not be earthlike planets, where they can bask in the warm sunlight of the mother star, but on icy-cold satellites orbiting Jupiter-size planets billions of miles from the star. It is widely believed that Europa, an ice-covered moon of Jupiter, has a liquid ocean beneath the icy surface, heated by tidal forces. Because Europa tumbles as it orbits Jupiter, it is squeezed in different directions by the huge gravitational pull of Jupiter, which causes friction deep inside the moon. This generates heat, forming volcanoes and ocean vents that melt the ice and create liquid oceans. It is estimated that the oceans of Europa are quite deep, and that their volume may be many times the volume of the oceans of Earth. Since 50 percent of all stars in the heavens may have Jupiter-size planets (a hundred times more plentiful than earthlike planets), the most plentiful form of life may be on the icy moons of gas giants like Jupiter.

Therefore, when we encounter our first alien civilization in space, more than likely it will have an aquatic origin. (Also, it is likely that they will have migrated from the ocean and learned to live on the icy surface of their moon away from the water, for several reasons. First, any species that lives perpetually under the ice will have a quite limited view of the universe. They will never develop astronomy or a space program if they think that the universe is just the ocean underneath the ice cover. Second, because water short-circuits electrical components, they will never develop radio or TV if they stay underwater. If this civilization is to advance, it must master electronics, which cannot exist in the oceans. So, most likely, these aliens will have learned to leave the oceans and survive on the land, as we did.)

But what happens if this life-form evolves to create a space-faring civilization, capable of reaching Earth? Will they still be biological organisms like us, or will they be post-biological?

THE POST-BIOLOGICAL ERA

One person who has spent considerable time thinking about these questions is my colleague Dr. Paul Davies of Arizona State University, near Phoenix. When I interviewed him, he told me that we have to expand our own horizon to contemplate what a civilization that is thousands or more years ahead of us may look like.

Given the dangers of space travel, he believes that such beings will have abandoned their biological form, much like the bodiless minds we considered in the previous chapter. He writes, "My conclusion is a startling one. I think it very likely—in fact inevitable—that biological intelligence is only a transitory phenomenon, a fleeting phase in the evolution of intelligence in the universe. If we ever encounter extraterrestrial intelligence, I believe it is overwhelmingly likely to be post-biological in nature, a conclusion that has obvious and far-reaching ramifications for SETI."

In fact, if the aliens are thousands of years ahead of us, chances are that they have abandoned their biological bodies eons ago to create the most efficient computational body: a planet whose entire surface is completely covered with computers. Dr. Davies says, "It isn't hard to envision the entire surface of a planet being covered with a single integrated processing system. . . . Ray Bradbury has coined the term 'Matrioshka brains' for these awesome entities."

So to Dr. Davies, alien consciousness may lose the concept of "self" and be absorbed into the collective World Wide Web of Minds, which blankets the entire surface of the planet. Dr. Davies adds, "A powerful computer network with no sense of self would have an enormous advantage over human intelligence because it could redesign 'itself,' fearlessly make changes, merge with whole systems, and grow. 'Feeling personal' about it would be a distinct impediment to progress."

So in the name of efficiency and increased computational ability, he envisions members of this advanced civilization giving up their identity and being absorbed into a collective consciousness.

Dr. Davies acknowledges that critics of his idea may find this concept rather repulsive. It appears as if this alien species is sacrificing individuality and creativity to the greater good of the collective or the hive. This is not inevitable, he cautions, but it is the most efficient option for civilization.

Dr. Davies also has a conjecture that he admits is rather depressing. When I asked him why these civilizations don't visit us, he gave me a strange answer. He said that any civilization that advanced would also have developed virtual realities far more interesting and challenging than reality. The virtual reality of today would be a children's toy compared to the virtual reality of a civilization thousands of years more advanced than us.

This means that perhaps their finest minds might have decided to play out imaginary lives in different virtual worlds. It's a discouraging thought, he admitted, but certainly a possibility. In fact, it might even be a warning for us as we perfect virtual reality.

WHAT DO THEY WANT?

In the movie *The Matrix*, the machines take over and put humans into pods, where they exploit us as batteries to energize themselves. That is why they keep us alive. But since a single electrical plant produces more power than the bodies of millions of humans, any alien looking for an energy source would quickly see there is no need for human batteries. (This seems to be lost on the machine overlords in the Matrix, but hopefully aliens would see reason.)

Another possibility is that they might want to eat us. This was explored in an episode of *The Twilight Zone,* in which aliens land on Earth and promise us the benefits of their advanced technology. They even ask for volunteers to visit their beautiful home planet. The aliens accidentally leave behind a book, called *To Serve Man,* which scientists anxiously try to decipher in order to discover what wonders the aliens will share with us. Instead, the scientists find out that the book is actually a cookbook. (But since we will be made of entirely different DNA and proteins from theirs, we could be difficult for their digestive tracts to process.)

Another possibility is that the aliens will want to strip Earth of resources and valuable minerals. There may be some truth to this argument, but if the aliens are advanced enough to travel effortlessly from the stars, then there are plenty of uninhabited planets to plunder for resources, without having to worry about restive natives. From their point of view, it would be a waste of time to try to colonize an inhabited planet when there are easy alternatives.

So if the aliens do not want to enslave us or plunder our resources, then what danger do they pose? Think of deer in a forest. Whom should they fear the most—the ferocious hunter armed with a shotgun, or the mild-mannered developer armed with a blueprint? Although the hunter may scare the deer, only a few deer are threatened by him. More dangerous to the deer is the developer, because the deer are not even on his radar screen. The developer may not even think about the deer at all, concentrating instead on developing the forest into usable property. In view of this, what would an invasion actually look like?

In Hollywood movies, there is one glaring flaw: the aliens are only a century or so ahead of us, so we can usually devise a secret weapon or exploit a simple weakness in their armor to fight them off, as in *Earth vs. the Flying Saucers*. But as SETI director Dr. Seth Shostak once told me, a battle with an advanced alien civilization will be like a battle between Bambi and Godzilla.

In reality, the aliens might be millennia to millions of years ahead of us in their weaponry. So, for the most part, there will be little we can do to defend ourselves. But perhaps we can learn from the barbarians who defeated the greatest military empire of its time, the Roman Empire.

The Romans were masters of engineering, able to create weapons that could flatten barbarian villages and roads to supply distant military outposts of a vast empire. The barbarians, who were barely emerging from a nomadic existence, had little chance when encountering the juggernaut of the Roman Imperial Army.

But history records that as the empire expanded, it was spread too thin, with too many battles to fight, too many treaties bogging it down, and not enough of an economy to support all this, especially with a gradual decline in population. Moreover, the empire, always short on recruits, had to enlist young barbarian soldiers and promote them to leadership positions. Inevitably, the superior technology of the empire began to filter down to the barbarians as well. In time, the barbarians began to master the very military technologies that at first had conquered them.

Toward the end, the empire, weakened by internal palace intrigues, severe crop shortages, civil wars, and an overstretched army, faced barbarians who were able to fight the Roman Imperial Army to a standstill. The sacking of Rome in A.D. 410 and 455 paved the way for the empire's ultimate fall in A.D. 476.

In the same way, it is likely that earthlings will initially offer no real threat to an alien invasion, but over time earthlings could learn the weak points of the alien army, its power supplies, its command centers, and most of all its weaponry. In order to control the human population, the aliens will have to recruit collaborators and promote them. This will result in a diffusion of their technology to the humans.

Then a ragtag army of earthlings might be able to mount a counterattack. In Eastern military strategy, like the classic teachings of Sun Tzu in *The Art of War,* there is a way to defeat even a superior army. You first allow it to enter your territory. Once it has entered unfamiliar land and its ranks are diffused, you can counterattack where they are weakest.

Another technique is to use the enemy's strength against it. In judo, the principal strategy is to turn the momentum of the attacker to your advantage. You let the enemy attack, and then trip them or throw them off guard, exploiting the enemy's own mass and energy. The bigger they are, the harder they fall. In the same way, perhaps the only way to fight a superior alien army is to allow it to invade your territory, learn its weaponry and military secrets, and turn those very weapons and secrets against it.

So a superior alien army cannot be defeated head-on. But it will withdraw if it cannot win and the cost of a stalemate is too high. Success means depriving the enemy of a victory.

But more than likely, I believe the aliens will be benevolent and, for the most part, ignore us. We simply have nothing to offer them. If they visit us, then it will be mainly out of curiosity or for reconnaissance. (Since curiosity was an essential feature in our becoming intelligent, it is likely that any alien species will be curious, and hence want to analyze us, but not necessarily to make contact.)

MEETING AN ALIEN ASTRONAUT

Unlike in the movies, we will probably not meet the flesh-and-blood alien creatures themselves. It would simply be too dangerous and unnecessary. In the same way that we sent the Mars Rover to explore, aliens will more than likely send organic/mechanical surrogates or avatars instead, which can better handle the stresses of interstellar travel. In this way, the "aliens" we meet on the White House lawn may look nothing like their masters back on the home planet. Instead, the masters will project their consciousness into space through proxies.

More than likely, though, they will send a robotic probe to our moon, which is geologically stable, with no erosion. These probes are self-replicating; that is, they will create a factory and manufacture, say, a thousand copies of themselves. (These are called von Neumann probes, after mathematician John von Neumann, who laid the foundation for digital computers. Von Neumann was the first mathematician to seriously consider the problem of machines that could reproduce themselves.) These second-generation probes are then launched to other star systems, where each one in turn creates a thousand more third-generation probes, making a total of a million. Then these probes fan out and create more factories, making a billion probes. Starting with just one probe, we have one thousand, then a million, then a billion. Within five generations, we have a quadrillion probes. Soon we have a gigantic sphere, expanding at near light speed, containing trillions upon trillions of probes, colonizing the entire galaxy within a few hundred thousand years.

Dr. Davies takes this idea of self-replicating von Neumann probes so seriously that he has actually applied for funding to search the surface of the moon for evidence of a previous alien visitation. He wishes to scan the moon for radio emissions or radiation anomalies that would indicate evidence of an alien visitation, perhaps millions of years ago. He wrote a paper with Dr. Robert Wagner in the scientific journal *Acta Astronautica* calling for a close examination of the photos from the Lunar Reconnaissance Orbiter down to a resolution of about 1.5 feet.

They wrote, "Although there is only a tiny probability that alien technology would have left traces on the moon in the form of an artifact or surface modification of lunar features, this location has the virtue of being close," and also traces of an alien technology would remain preserved over long periods of time. Since there is no erosion on the moon, treadmarks left by aliens would still be visible (in the same way that footprints left by our astronauts in the 1970s could, in principle, last for billions of years).

One problem is that the von Neumann probe might be very small. Nanoprobes use molecular machines and MEMs, and hence it might be only as big as a bread box, he said to me, or even smaller. (In fact, if such a probe landed on Earth in someone's backyard, the owner might not even notice.)

This method, however, represents the most efficient way of colonizing the galaxy, using the exponential growth of self-replicating von Neumann probes. (This is also the way in which a virus infects our body. Starting with a handful of viruses, they land on our cells, hijack the reproductive machinery, and convert our cells into factories to create more viruses. Within two weeks, a single virus can infect trillions of cells, and we eventually sneeze.)

If this scenario is correct, it means that our own moon is the most likely place for an alien visitation. This is also the basis of the movie *2001: A Space Odyssey,* which even today represents the most plausible encounter with an extraterrestrial civilization. In the movie, a probe was placed on our moon millions of years ago, mainly to observe the evolution of life on Earth. At times, it interferes in our evolution and gives us an added boost. This information is then sent to Jupiter, which is a relay station, before heading to the home planet of this ancient alien civilization.

From the point of view of this advanced civilization, which can simultaneously scan billions of star systems, we can see that they have a considerable choice in what planetary systems to colonize. Given the sheer enormity of the galaxy, they can collect data and then best choose which planets or moons would yield the best resources. From their perspective, they might not find Earth very appealing.

The empires of the future will be empires of the mind.

—WINSTON CHURCHILL

If we continue to develop our technology without wisdom or prudence, our servant may prove to be our executioner.

—GENERAL OMAR BRADLEY

15 CONCLUDING REMARKS

In 2000, a raging controversy erupted in the scientific community. One of the founders of Sun Computers, Bill Joy, wrote an inflammatory article denouncing the mortal threat we face from advanced technology. In an article in *Wired* magazine with the provocative title "The Future Does Not Need Us," he wrote, "Our most powerful 21st century technologies—robotics, genetic engineering, and nanotech—are threatening to make humans an endangered species." That incendiary article questioned the very morality of hundreds of dedicated scientists toiling in their labs on the cutting edge of science. He challenged the very core of their research, stating that the benefits of these technologies were vastly overshadowed by the enormous threats they posed to humanity.

He described a macabre dystopia in which all our technologies conspire to destroy civilization. Three of our key creations will eventually turn on us, he warned:

- One day, bioengineered germs may escape from the laboratory and wreak havoc on the world. Since you cannot recapture these life-forms, they might proliferate wildly and unleash a fatal plague on the planet worse than those of the Middle Ages. Biotechnology may even alter human evolution, creating "several separate and unequal species . . . that would threaten the notion of equality that is the very cornerstone of our democracy."

- One day, nanobots may go berserk and spew out unlimited quantities of "gray goo," which will blanket Earth, smothering all life. Since these nanobots "digest" ordinary matter and create new forms of matter, malfunctioning nanobots could run amok and digest much of Earth. "Gray goo would surely be a depressing ending to our human adventure on Earth, far worse than mere fire or ice, and one that could stem from a simple laboratory accident. Oops," he wrote.
- One day, the robots will take over and replace humanity. They will become so intelligent that they will simply push humanity aside. We will be left as an evolutionary footnote. "The robots would in no sense be our children. . . . On this path our humanity may well be lost," he wrote.

Joy claimed that the dangers unleashed by these three technologies dwarfed the dangers posed by the atomic bomb in the 1940s. Back then, Einstein warned of the power of nuclear technology to destroy civilization: "It has become appallingly obvious that our technology has exceeded our humanity." But the atomic bomb was built by a huge government program that could be tightly regulated, while these technologies are being developed by private companies that are lightly regulated, if at all, Joy pointed out.

Sure, he conceded, these technologies may alleviate some suffering in the short term. But in the long term, the benefits are overwhelmed by the fact that they may unleash a scientific Armageddon that may doom the human race.

Joy even accused scientists of being selfish and naïve as they try to create a better society. He wrote, "A traditional utopia is a good society and a good life. A good life involves other people. This techno utopia is all about 'I don't get diseases; I don't die; I get to have better eyesight and be smarter' and all this. If you described this to Socrates or Plato, they would laugh at you."

He concluded by stating, "I think it is no exaggeration to say we are on the cusp of the further perfection of extreme evil, an evil whose possibility spreads well beyond that which weapons of mass destruction bequeathed to the nation-states. . . ."

The conclusion to all this? "Something like extinction," he warned.

As expected, the article sparked a firestorm of controversy.

That article was written over a decade ago. In terms of high technology, that is a lifetime. It is now possible to view certain of its predictions with some hindsight. Looking back at the article and putting his warnings into perspective, we can easily see that Bill Joy exaggerated many of the threats coming from these technologies, but he also spurred scientists to face up to the ethical, moral, and societal consequences of their work, which is always a good thing.

And his article opened up a discussion about who we are. In unraveling the molecular, genetic, and neural secrets of the brain, haven't we in some sense dehumanized humanity, reducing it to a bucket of atoms and neurons? If we completely map every neuron of the brain and trace every neural pathway, doesn't that remove the mystery and magic of who we are?

A RESPONSE TO BILL JOY

In retrospect, the threats from robotics and nanotechnology are more distant than Bill Joy thought, and I would argue that with enough warning, we can take a variety of countermeasures, such as banning certain avenues of research if they lead to uncontrollable robots, placing chips in them to shut them off if they become dangerous, and creating fail-safe devices to immobilize all of them in an emergency.

More immediate is the threat from biotechnology, where there is the realistic danger of biogerms that might escape the laboratory. In fact, Ray Kurzweil and Bill Joy jointly wrote an article criticizing the publication of the complete genome of the 1918 Spanish flu virus, one of the most lethal germs in modern history, which killed more people than World War I. Scientists were able to reassemble the long-dead virus by examining the corpses and blood of its victims and sequencing its genes, and then they published it on the web.

Safeguards already exist against the release of such a dangerous virus, but steps must be taken to further strengthen them and add new layers of security. In particular, if a new virus suddenly erupts in some distant place on Earth, scientists must strengthen rapid-response teams that can isolate the virus in the wild, sequence its genes, and then quickly prepare a vaccine to prevent its spread.

IMPLICATIONS FOR THE FUTURE OF THE MIND

This debate also has a direct impact on the future of the mind. At present, neuroscience is still rather primitive. Scientists can read and videotape simple thoughts from the living brain, record a few memories, connect the brain to mechanical arms, enable locked-in patients to control machines around them, silence specific regions of the brain via magnetism, and identify the regions of the brain that malfunction in mental illness.

In the coming decades, however, the power of neuroscience may become explosive. Current research is on the threshold of new scientific discoveries that will likely leave us breathless. One day, we might routinely control objects around us with the power of the mind, download memories, cure mental illness, enhance our intelligence, understand the brain neuron by neuron, create backup copies of the brain, and communicate with one another telepathically. The world of the future will be the world of the mind.

Bill Joy did not dispute the potential of this technology to relieve human suffering and pain. But what made him look on it with horror was the prospect of enhanced individuals who might cause the human species to split apart. In the article, he painted a dismal dystopia in which only a tiny elite have their intelligence and mental processes enhanced, while the masses of people live in ignorance and poverty. He worried that the human race would fission in two, or perhaps cease to be human at all.

But as we have pointed out, almost all technologies when they are first introduced are expensive and hence exclusively for the well-off. Because of mass production, the falling cost of computers, competition, and cheaper shipping, technologies inevitably filter down to the poor as well. This was also the trajectory taken by phonographs, radio, TV, PCs, laptops, and cell phones.

Far from creating a world of haves and have-nots, science has been the engine of prosperity. Of all the tools that humanity has harnessed since the dawn of time, by far the most powerful and productive has been science. The incredible wealth we see all around us is directly due to science. To appreciate how technology reduces, rather than accentuates, societal fault lines, consider the lives of our ancestors around 1900. Life expectancy in the United States back then was forty-nine years. Many children died in infancy.

Communicating with a neighbor involved yelling out the window. The mail was delivered by horse, if it came at all. Medicine was largely snake oil. The only treatments that actually worked were amputations (without anesthetics) and morphine to deaden the pain. Food rotted within days. Plumbing was nonexistent. Disease was a constant threat. And the economy could support only a handful of the rich and a tiny middle class.

Technology has changed everything. We no longer have to hunt for our food; we simply go to the supermarket. We no longer have to carry back-breaking supplies but instead simply get into our cars. (In fact, the main threat we face from technology, one that has killed millions of people, is not murderous robots or mad nanobots run amok—it's our indulgent lifestyle, which has created near-epidemic levels of diabetes, obesity, heart disease, cancer, etc. And this problem is self-inflicted.)

We also see this on the global level. In the last few decades the world has witnessed hundreds of millions of people being lifted out of grinding poverty for the first time in history. If we view the bigger picture, we see that a significant fraction of the human race has left the punishing lifestyle of sustenance farming and entered the ranks of the middle class.

It took several hundred years for Western nations to industrialize, yet China and India are doing it within a few decades, all due to the spread of high technology. With wireless technology and the Internet, these nations can leapfrog past other, more developed nations that have laboriously wired their cities. While the West struggles with an aging, decaying urban infrastructure, developing nations are building entire cities with sparkling, state-of-the-art technology.

(When I was a graduate student getting my Ph.D., my counterparts in China and India would have to wait several months to a year for scientific journals to come in the mail. Plus, they had almost no direct contact with scientists and engineers in the West, because few if any could afford to travel here. This vastly impeded the flow of technology, which moved at a glacial pace for these nations. Today, however, scientists can read one another's papers as soon as they are posted on the Internet, and can electronically collaborate with other scientists around the world. This has vastly accelerated the flow of information. And with this technology comes progress and prosperity.)

Furthermore, it's not clear that having some form of enhanced intelligence will cause a catastrophic splitting of the human race, even if many people are unable to afford this procedure. For the most part, being able to solve complex mathematical equations or have perfect recall does not guarantee a higher income, respect from your peers, or more popularity with the opposite sex, which are the incentives that motivate most people. The Caveman Principle trumps having a brain boost.

As Dr. Michael Gazzaniga notes, "The idea that we are messin' with our innards is disturbing to many. And just what would we do with expanded intelligence? Are we going to use it for solving problems, or will it just allow us to have longer Christmas card lists . . . ?"

But as we discussed in Chapter 5, unemployed workers may benefit from this technology, drastically reducing the time required to master new technologies and skills. This might not only reduce the problems associated with unemployment, it could also have an impact on the world economy, making it more efficient and responsive to change.

WISDOM AND DEMOCRATIC DEBATE

In responding to Joy's article, some critics pointed out that the debate is not about a struggle between scientists and nature, as portrayed in the article. The debate is actually between three parties: scientists, nature, and society.

Computer scientists Drs. John Brown and Paul Duguid responded to the article by stating, "Technologies—such as gunpowder, the printing press, the railroad, the telegraph, and the Internet—can change society in profound ways. But on the other hand, social systems—in the form of governments, the courts, formal and informal organizations, social movements, professional networks, local communities, market institutions, and so forth—shape, moderate, and redirect the raw power of technologies."

The point is to analyze them in terms of society, and ultimately it is up to us to adopt a new vision of the future that incorporates all the best ideas.

To me, the ultimate source of wisdom in this respect comes from vigorous democratic debate. In the coming decades, the public will be asked to vote on a number of crucial scientific issues. Technology cannot be debated in a vacuum.

PHILOSOPHICAL QUESTIONS

Lastly, some critics have claimed that the march of science has gone too far in unveiling the secrets of the mind, an unveiling that has become dehumanizing and degrading. Why bother to rejoice at discovering something new, learning a new skill, or enjoying a leisurely vacation when it can all be reduced to a few neurotransmitters activating a few neural circuits?

In other words, just as astronomy has reduced us to insignificant pieces of cosmic dust floating in an uncaring universe, neuroscience has reduced us to electrical signals circulating within neural circuits. But is this really true?

We began our discussion by highlighting the two greatest mysteries in all of science: the mind and the universe. Not only do they have a common history and narrative, they also share a similar philosophy and perhaps even destiny. Science, with all its power to peer into the heart of black holes and land on distant planets, has given birth to two overarching philosophies about the mind and the universe: the Copernican Principle and the Anthropic Principle. Both are consistent with everything known about science, but they are diametrical opposites.

The first great philosophy, the Copernican Principle, was born with the discovery of the telescope more than four hundred years ago. It states that there is no privileged position for humanity. Such a deceptively simple idea has overthrown thousands of years of cherished myths and entrenched philosophies.

Ever since the biblical tale of Adam and Eve being exiled from the Garden of Eden for biting into the Apple of Knowledge, there has been a series of humiliating dethronements. First, the telescope of Galileo clearly showed that Earth was not the center of the solar system—the sun was. This picture was then overthrown when it was realized that the solar system was just a speck in the Milky Way galaxy circulating about thirty thousand light-years from the center. Then in the 1920s, Edwin Hubble discovered there was a multitude of galaxies. The universe suddenly got billions of times bigger. Now the Hubble Space Telescope can reveal the presence of up to one hundred billion galaxies in the visible universe. Our own Milky Way galaxy has been reduced to a pinpoint in a much larger cosmic arena.

More recent cosmological theories further downgrade the position of humanity in the universe. The inflationary universe theory states that our visible universe, with its one hundred billion galaxies, is just a pinprick on a much larger, inflated universe that is so big that most light has not had time to reach us yet from distant regions. There are vast reaches of space that we cannot see with our telescopes and will never be able to visit because we cannot go faster than light. And if string theory (my specialty) is correct, it means that even the entire universe coexists with other universes in eleven-dimensional hyperspace. So even three-dimensional space is not the final word. The true arena for physical phenomena is the multiverse of universes, full of floating bubble universes.

The science-fiction writer Douglas Adams tried to summarize the sense of being constantly overthrown by inventing the Total Perspective Vortex in *The Hitchhiker's Guide to the Galaxy*. It was designed to drive any sane person insane. When you enter the chamber, all you see is a gigantic map of the entire universe. And on the map there is a tiny, almost invisible arrow that says, "You are here."

So on one hand, the Copernican Principle indicates that we are just insignificant cosmic debris drifting aimlessly among the stars. But on the other hand, all the latest cosmological data are consistent with yet another theory, which gives us the opposite philosophy: the Anthropic Principle.

This theory states that the universe is compatible with life. Again, this deceptively simple statement has profound implications. On one hand, it is impossible to dispute that life exists in the universe. But it's clear that the forces of the universe must be calibrated to a remarkable degree to make life possible. As physicist Freeman Dyson once said, "The universe seemed to know that we were coming."

For example, if the nuclear force were just a bit stronger, the sun would have burned out billions of years ago, too soon to allow DNA to get off the ground. If the nuclear force were a bit weaker, then the sun would never have ignited to begin with, and we still would not be here.

Likewise, if gravity were stronger, the universe would have collapsed into a Big Crunch billions of years ago, and we would all be roasted to death. If it were a bit weaker, then the universe would have expanded so fast it would have reached the Big Freeze, so we would all have frozen to death.

This fine-tuning extends to every atom of the body. Physics says that we are made of star dust, that the atoms we see all around us were forged in the heat of a star. We are literally children of the stars.

But the nuclear reactions that burned hydrogen to create the higher elements of our body are very complex and could have been derailed at any number of points. Then it would have been impossible to create the higher elements of our bodies, and the atoms of DNA and life would not exist.

In other words, life is precious and a miracle.

There are so many parameters that have to be fine-tuned that some claim this is not a coincidence. The weak form of the Anthropic Principle implies that the existence of life forces the physical parameters of the universe to be defined in a very precise way. The strong form of the Anthropic Principle goes even further, stating that God or some designer had to create a universe "just right" to make life possible.

PHILOSOPHY AND NEUROSCIENCE

The debate between the Copernican Principle and the Anthropic Principle also resonates in neuroscience. For example, some claim that humans can be reduced to atoms, molecules, and neurons, and hence there is no distinguished place for humanity in the universe.

Dr. David Eagleman writes, "The *you* that all your friends know and love cannot exist unless the transistors and screws of our brain are in place. If you don't believe this, step into any neurology ward in any hospital. Damage to even small parts of the brain can lead to the loss of shockingly specific abilities; the ability to name animals, or to hear music, or to manage risky behavior, or to distinguish colors, or to arbitrate simple decisions."

It seems that the brain cannot function without all its "transistors and screws." He concludes, "Our reality depends on what our biology is up to."

So on one hand, our place in the universe seems to be diminished if we can be reduced, like robots, to (biological) nuts and bolts. We are just wetware, running software called the mind, nothing more or less. Our thoughts, desires, hopes, and aspirations can be reduced to electrical impulses circulating in some region of the prefrontal cortex. That is the Copernican Principle applied to the mind.

But the Anthropic Principle can also be applied to the mind, and we then reach the opposite conclusion. It simply says that conditions of the universe make consciousness possible, even though it is extraordinarily difficult to create the mind out of random events. The great Victorian biologist Thomas

Huxley said, "How it is that anything so remarkable as a state of consciousness comes about as a result of irritating nervous tissue, is just as unaccountable as the appearance of the Djinn, when Aladdin rubbed his lamp."

Furthermore, most astronomers believe that although one day we may find life on other planets, it will most likely be microbial life, which ruled our oceans for billions of years. Instead of seeing great cities and empires, we might only find oceans of drifting microorganisms.

When I interviewed the late Harvard biologist Stephen Jay Gould about this, he explained to me his thinking as follows. If we were to somehow create a twin of Earth as it was 4.5 billion years ago, would it turn out the same way 4.5 billion years later? Most likely not. There is a large probability that DNA and life would never have gotten off the ground, and an even larger probability that intelligent life with consciousness would never have risen from the swamp.

Gould wrote, "*Homo sapiens* is one small twig [on the tree of life]. . . . Yet our twig, for better or worse, has developed the most extraordinary new quality in all the history of multicellular life since the Cambrian explosion (500 million years ago). We have invented consciousness with all its sequelae from Hamlet to Hiroshima."

In fact, in the history of Earth, there are many times when intelligent life was almost extinguished. In addition to the mass extinctions that wiped out the dinosaurs and most life on Earth, humans have faced additional near extinctions. For example, humans are all genetically related to one another to a considerable degree, much closer than two typical animals of the same species. Although humans may look diverse from the outside, our genes and internal chemistry tell a different story. In fact, any two humans are so closely related genetically that we can actually do the math and calculate when a "genetic Eve" or "genetic Adam" gave birth to the entire human race. Moreover, we can calculate how many of us there were in the past.

The numbers are remarkable. Genetics shows that there were only a few hundred to a few thousand humans alive about seventy to one hundred thousand years ago and that they gave birth to the entire human race. (One theory holds that the titanic explosion of the Toba volcano in Indonesia about seventy thousand years ago caused temperatures to drop so dramatically that most of the human race perished, leaving only a handful to populate Earth.) From that small band of humans came the adventurers and explorers who would eventually colonize the entire planet.

Repeatedly in the history of Earth, intelligent life might have come to a dead end. It is a miracle we survived. We can also conclude that although life may exist on other planets, conscious life may exist on only a tiny fraction of them. So we should treasure the consciousness that is found on Earth. It is the highest form of complexity known in the universe, and probably also the rarest.

Sometimes, when contemplating the future destiny of the human race, I have to come to grips with the distinct possibility of its self-destruction. Although volcanic eruptions and earthquakes could spell doom for the human race, our worst fears may be realized through man-made disasters, such as nuclear wars or bioengineered germs. If so, then perhaps the only conscious life-form in this sector of the Milky Way galaxy might be extinguished. This, I feel, would be a tragedy not just for us, but for the universe as well. We take for granted that we are conscious, but we don't understand the long, tortuous sequence of biological events that have transpired to make this possible. Psychologist Steven Pinker writes, "I would argue that nothing gives life more purpose than the realization that every moment of consciousness is a precious and fragile gift."

THE MIRACLE OF CONSCIOUSNESS

Lastly, there is the criticism of science that says to understand something is to remove its mystery and magic. Science, by lifting the veil concealing the secrets of the mind, is also making it more ordinary and mundane. However, the more I learn about the sheer complexity of the brain, the more amazed I am that something that sits on our shoulders is the most sophisticated object we know about in the universe. As Dr. David Eagleman says, "What a perplexing masterpiece the brain is, and how lucky we are to be in a generation that has the technology and the will to turn our attention to it. It is the most wondrous thing we have discovered in the universe, and it is us." Instead of diminishing the sense of wonder, learning about the brain only increases it.

More than two thousand years ago, Socrates said, "To know thyself is the beginning of wisdom." We are on a long journey to complete his wishes.

APPENDIX

QUANTUM CONSCIOUSNESS?

In spite of all the miraculous advances in brain scans and high technology, some people claim that we will never understand the secret of consciousness, since consciousness is beyond our puny technology. In fact, in their view consciousness is more fundamental than atoms, molecules, and neurons and determines the nature of reality itself. To them, consciousness is the fundamental entity out of which the material world is created. And to prove their point, they refer to one of the greatest paradoxes in all of science, which challenges our very definition of reality: the Schrödinger's Cat paradox. Even today, there is no universal consensus on the question, with Nobel laureates taking divergent stances. What is at stake is nothing less than the nature of reality and thought.

The Schrödinger's Cat paradox cuts to the very foundation of quantum mechanics, a field that makes lasers, MRI scans, radio and TV, modern electronics, the GPS, and telecommunications possible, upon which the world economy depends. Many of quantum theory's predictions have been tested to an accuracy of one part in one hundred billion.

I have spent my entire professional career working on the quantum theory. Yet I realize that it has feet of clay. It's an unsettling feeling knowing my life's work is based on a theory whose very foundation is based on a paradox.

This debate was sparked by Austrian physicist Erwin Schrödinger, who was one of the founding fathers of the quantum theory. He was trying to explain the strange behavior of electrons, which seemed to exhibit both wave and particle properties. How can an electron, a point particle, have two divergent behaviors? Sometimes electrons acted like a particle, creating well-defined tracks in a cloud chamber. Other times, electrons acted like a wave, passing through tiny holes and creating wavelike interference patterns, like those on the surface of a pond.

In 1925, Schrödinger put forward his celebrated wave equation, which bears his name and is one of the most important equations ever written. It was an instant sensation, and won him the Nobel Prize in 1933. The Schrödinger equation accurately described the wavelike behavior of electrons and, when applied to the hydrogen atom, explained its strange properties. Miraculously, it could also be applied to any atom and explain most of the features of the periodic table of elements. It seemed as if all chemistry (and hence all biology) were nothing but solutions of this wave equation. Some physicists even claimed that the entire universe, including all the stars, planets, and even us, was nothing but a solution of this equation.

But then physicists began to ask a problematic question that resonates even today: If the electron is described by a wave equation, then what is waving?

In 1927, Werner Heisenberg proposed a new principle that split the physics community down the middle. Heisenberg's celebrated uncertainty principle states that you cannot know both the location and the momentum of an electron with certainty. This uncertainty was not a function of how crude your instruments were but was inherent in physics itself. Even God or some celestial being could not know the precise location and momentum of an electron.

So the wave function of Schrödinger actually described the probability of finding the electron. Scientists had spent thousands of years painfully trying to eliminate chance and probabilities in their work, and now Heisenberg was allowing it in through the back door.

The new philosophy can be summed up as follows: the electron is a point particle, but the probability of finding it is given by a wave. And this wave obeys Schrödinger's equation and gives rise to the uncertainty principle.

The physics community cracked in half. On one side, we had physicists like Niels Bohr, Werner Heisenberg, and most atomic physicists eagerly adopting this new formulation. Almost daily, they were announcing new breakthroughs in understanding the properties of matter. Nobel Prizes were being handed out to quantum physicists like Oscars. Quantum mechanics was becoming a cookbook. You did not need to be a master physicist to make stellar contributions—you just followed the recipes given by quantum mechanics and you would make stunning breakthroughs.

On the other side, we had aging Nobel laureates like Albert Einstein, Erwin Schrödinger, and Louis de Broglie who were raising philosophical objections. Schrödinger, whose work helped start this whole process, grumbled that if he had known that his equation would introduce probability into physics, he would never have created it in the first place.

Physicists embarked on an eighty-year debate that continues even today. On one hand, Einstein would proclaim that "God does not play dice with the world." Niels Bohr, on the other hand, reportedly replied, "Stop telling God what to do."

In 1935, to demolish the quantum physicists once and for all, Schrödinger proposed his celebrated cat problem. Place a cat in a sealed box, with a container of poison gas. In the box, there is a lump of uranium. The uranium atom is unstable and emits particles that can be detected by a Geiger counter. The counter triggers a hammer, which falls and breaks the glass, releasing the gas, which can kill the cat.

How do you describe the cat? A quantum physicist would say that the uranium atom is described by a wave, which can either decay or not decay. Therefore you have to add the two waves together. If the uranium fires, then the cat dies, so that is described by one wave. If the uranium does not fire, then the cat lives, and that is also described by a wave. To describe the cat, you therefore have to add the wave of a dead cat to the wave of a live cat.

This means that the cat is neither dead nor alive! The cat is in a netherworld, between life and death, the sum of the wave describing a dead cat with the wave of a live cat.

This is the crux of the problem, which has reverberated in the halls of physics for almost a century. So how do you resolve this paradox? There are at least three ways (and hundreds of variations on these three).

The first is the original Copenhagen interpretation proposed by Bohr and Heisenberg, the one that is quoted in textbooks around the world. (It is the one that I start with when I teach quantum mechanics.) It says that to determine the state of the cat, you must open the box and make a measurement. The cat's wave (which was the sum of a dead cat and a live cat) now "collapses" into a single wave, so the cat is now known to be alive (or dead). Thus, observation determines the existence and state of the cat. The measurement process is thus responsible for two waves magically dissolving into a single wave.

Einstein hated this. For centuries, scientists have battled something called "solipsism" or "subjective idealism," which claims that objects cannot exist unless there is someone there to observe them. Only the mind is real—the material world exists only as ideas in the mind. Thus, say the solipsists (such as Bishop George Berkeley), if a tree falls in the forest but no one is there to observe it, perhaps the tree never fell. Einstein, who thought all this was pure nonsense, promoted an opposing theory called "objective reality," which says simply that the universe exists in a unique, definite state independent of any human observation. It is the commonsense view of most people.

Objective reality goes back to Isaac Newton. In this scenario, the atom and subatomic particles are like tiny steel balls, which exist at definite points in space and time. There is no ambiguity or chance in locating the position of these balls, whose motions can be determined by using the laws of motion. Objective reality was spectacularly successful in describing the motions of planets, stars, and galaxies. Using relativity, this idea can also describe black holes and the expanding universe. But there is one place where it fails miserably, and that is inside the atom.

Classical physicists like Newton and Einstein thought that objective reality finally banished solipsism from physics. Walter Lippmann, the columnist, summed it up when he wrote, "The radical novelty of modern science lies precisely in the rejection of the belief . . . that the forces which move the stars and atoms are contingent upon the preferences of the human heart."

But quantum mechanics allowed a new form of solipsism back into physics. In this picture, before it is observed, a tree can exist in any possible state (e.g., sapling, burned, sawdust, toothpicks, decayed). But when you look at it, the wave suddenly collapses and it looks like a tree. The original solipsists talked about trees that either fell or didn't. The new quantum solipsists were introducing *all* possible states of a tree.

This was too much for Einstein. He would ask guests at his house, "Does the moon exist because a mouse looks at it?" To a quantum physicist, in some sense the answer might be yes.

Einstein and his colleagues would challenge Bohr by asking: How can the quantum microworld (with cats being dead and alive simultaneously) coexist with the commonsense world we see around us? The answer was that there is a "wall" that separates our world from the atomic world. On one side of the wall, common sense rules. On the other side of the wall, the quantum theory rules. You can move the wall if you want and the results are still the same.

This interpretation, no matter how strange, has been taught for eighty years by quantum physicists. More recently, there have been some doubts cast on the Copenhagen interpretation. Today we have nanotechnology, with which we can manipulate individual atoms at will. On a scanning tunneling microscope screen, atoms appear to be fuzzy tennis balls. (For BBC-TV, I had a chance to fly out to IBM's Almaden Lab in San Jose, California, and actually push individual atoms around with a tiny probe. It is now possible to play with atoms, which were once thought to be so small they could never be seen.)

As we've discussed, the Age of Silicon is slowly coming to an end, and some believe that molecular transistors will replace silicon transistors. If so, then the paradoxes of the quantum theory may lie at the very heart of every computer of the future. The world economy may eventually rest on these paradoxes.

COSMIC CONSCIOUSNESS AND MULTIPLE UNIVERSES

There are two alternate interpretations of the cat paradox, which take us to the strangest realms in all science: the realm of God and multiple universes.

In 1967, the second resolution to the cat problem was formulated by Nobel laureate Eugene Wigner, whose work was pivotal in laying the foundation of quantum mechanics and also building the atomic bomb. He said that only a conscious person can make an observation that collapses the wave function. But who is to say that this person exists? You cannot separate the observer from the observed, so maybe this person is also dead and alive. In other words, there has to be a new wave function that includes both the cat and the observer. To make sure that the observer is alive, you need a second observer to watch the first observer. This second observer is called "Wigner's friend," and is necessary to watch the first observer so that all waves collapse. But how do we know that the second observer is alive? The second observer has to be included in a still-larger wave function to make sure he is alive, but this can be continued indefinitely. Since you need an infinite number of "friends" to collapse the previous wave function to make sure they are alive, you need some form of "cosmic consciousness," or God.

Wigner concluded: "It was not possible to formulate the laws (of quantum theory) in a fully consistent way without reference to consciousness." Toward the end of his life, he even became interested in the Vedanta philosophy of Hinduism.

In this approach, God or some eternal consciousness watches over all of us, collapsing our wave functions so that we can say we are alive. This interpretation yields the same physical results as the Copenhagen interpretation, so this theory cannot be disproven. But the implication is that consciousness is the fundamental entity in the universe, more fundamental than atoms. The material world may come and go, but consciousness remains as the defining element, which means that consciousness, in some sense, creates reality. The very existence of the atoms we see around us is based on our ability to see and touch them.

(At this point, it's important to note that some people think that because consciousness determines existence, then consciousness can therefore control existence, perhaps by meditation. They think that we can create reality according to our wishes. This thinking, as attractive as it might sound, goes against quantum mechanics. In quantum physics, consciousness makes observations and therefore determines the state of reality, but consciousness cannot choose ahead of time which state of reality actually exists. Quantum mechanics allows you only to determine the chance of finding one state, but we cannot bend reality to our wishes. For example, in gambling, it is possible to mathematically calculate the chances of getting a royal flush. However, this does not mean that you can somehow control the cards to get the royal flush. You cannot pick and choose universes, just as we have no control over whether the cat is dead or alive.)

MULTIPLE UNIVERSES

The third way to resolve the paradox is the Everett, or many-worlds, interpretation, which was proposed in 1957 by Hugh Everett. It is the strangest theory of all. It says that the universe is constantly splitting apart into a multiverse of universes. In one universe, we have a dead cat. In another universe, we have a live cat. This approach can be summarized as follows: wave functions never collapse, they just split. The Everett many-worlds theory differs from the Copenhagen interpretation only in that it drops the final assumption: the collapse of the wave function. In some sense, it is the simplest formulation of quantum mechanics, but also the most disturbing.

There are profound consequences to this third approach. It means that all possible universes might exist, even ones that are bizarre and seemingly impossible. (However, the more bizarre the universe, the more unlikely it is.)

This means people who have died in our universe are still alive in another universe. And these dead people insist that their universe is the correct one, and that our universe (in which they are dead) is fake. But if these "ghosts" of dead people are still alive somewhere, then why can't we meet them? Why can't we touch these parallel worlds? (As strange as it may seem, in this picture Elvis is still alive in one of these universes.)

What's more, some of these universes may be dead, without any life, but others may look exactly like ours, except for one key difference. For example, the collision of a single cosmic ray is a tiny quantum event. But what happens if this cosmic ray goes through Adolf Hitler's mother, and the infant Hitler dies in a miscarriage? Then a tiny quantum event, the collision of a single cosmic ray, causes the universe to split in half. In one universe, World War II never happened, and sixty million people did not have to die. In the other universe, we've had the ravages of World War II. These two universes grow to be quite far apart, yet they are initially separated by one tiny quantum event.

This phenomenon was explored by science-fiction writer Philip K. Dick in his novel *The Man in the High Tower,* where a parallel universe opens up because of a single event: a bullet is fired at Franklin Roosevelt, who is killed by an assassin. This pivotal event means that the United States is not prepared for World War II, and the Nazis and Japanese are victorious and eventually partition the United States in half.

But whether the bullet fires or misfires depends, in turn, on whether a microscopic spark is set off in the gunpowder, which itself depends on complex molecular reactions involving the motions of electrons. So perhaps quantum fluctuations in the gunpowder may determine whether the gun fires or misfires, which in turn determines whether the Allies or the Nazis emerge victorious during World War II.

So there is no "wall" separating the quantum world and the macroworld. The bizarre features of the quantum theory can creep into our "common-sense" world. These wave functions never collapse—they keep splitting endlessly into parallel realities. The creation of alternative universes never stops. The paradoxes of the microworld (i.e., being dead and alive simultaneously, being in two places at the same time, disappearing and reappearing somewhere else) now enter into our world as well.

But if the wave function is continually splitting apart, creating entirely new universes in the process, then why can't we visit them?

Nobel laureate Steven Weinberg compares this to listening to the radio in your living room. There are hundreds of radio waves simultaneously filling up your room from all over the world, but your radio dial is tuned to only one frequency. In other words, your radio has "decohered" from all the other stations. (Coherence is when all waves vibrate in perfect unison, as in a laser beam. Decoherence is when these waves begin to fall out of phase, so they no longer vibrate in unison.) These other frequencies all exist, but your radio cannot pick them up because they are not vibrating at the same frequency that we are anymore. They have decoupled; that is, they have decohered from us.

In the same way, the wave function of the dead and alive cat have decohered as time goes on. The implications are rather staggering. In your living room, you coexist with the waves of dinosaurs, pirates, aliens from space, and monsters. Yet you are blissfully unaware that you are sharing the same space as these strange denizens of quantum space, because your atoms are no longer vibrating in unison with them. These parallel universes do not exist in some distant never-never land. They exist in your living room.

Entering one of these parallel worlds is called "quantum jumping" or "sliding" and is a favorite gimmick of science fiction. To enter a parallel universe, we need to take a quantum jump into it. (There was even a TV series called *Sliders* where people slide back and forth between parallel universes. The series began when a young boy read a book. That book is actually my book *Hyperspace,* but I take no responsibility for the physics behind that series.)

Actually, it's not so simple to jump between universes. One problem we sometimes give our Ph.D. students is to calculate the probability that you will jump through a brick wall and wind up on the other side. The result is sobering. You would have to wait longer than the lifetime of the universe to experience jumping or sliding through a brick wall.

LOOKING IN THE MIRROR

When I look at myself in a mirror, I don't really see myself as I truly am. First, I see myself about a billionth of a second ago, since that is the time that it takes a light beam to leave my face, hit a mirror, and enter my eyes. Second, the image I see is really an average over billions and billions of wave functions. This average certainly does resemble my image, but it is not exact. Surrounding me are multiple images of myself oozing in all directions. I am continually surrounded by alternate universes, forever branching into different worlds, but the probability of sliding between them is so tiny that Newtonian mechanics seems to be correct.

At this point, some people ask this question: Why don't scientists simply do an experiment to determine which interpretation is valid? If we run an experiment with an electron, all three interpretations will yield the same result. All three are therefore serious, viable interpretations of quantum mechanics, with the same underlying quantum theory. What is different is how we explain the results.

Hundreds of years in the future, physicists and philosophers may still be debating this question, with no resolution, because all three interpretations yield the same physical results. But perhaps there is one way in which this philosophical debate touches on the brain, and that is the question of free will, which in turn affects the moral foundation of human society.

FREE WILL

Our entire civilization is based on the concept of free will, which impacts on the notions of reward, punishment, and personal responsibility. But does free will really exist? Or is it a clever way of keeping society together although it violates scientific principles? The controversy goes to the very heart of quantum mechanics itself.

It is safe to say that more and more neuroscientists are gradually coming to the conclusion that free will does not exist, at least not in the usual sense. If certain bizarre behaviors can be linked to precise defects in the brain, then a person is not scientifically responsible for the crimes he might commit. He might be too dangerous to be left walking the streets and must be locked up in an institution of some sort, but punishing someone for having a stroke or tumor in the brain is misguided, they say. What that person needs is medical and psychological help. Perhaps the brain damage can be treated (e.g., by removing a tumor), and the person can become a productive member of society.

For example, when I interviewed Dr. Simon Baron-Cohen, a psychologist at Cambridge University, he told me that many (but not all) pathological killers have a brain anomaly. Their brain scans show that they lack empathy when seeing someone else in pain, and in fact they might even take pleasure in watching this suffering (in these individuals, the amygdala and the nucleus accumbens, the pleasure center, light up when they view videos of people experiencing pain).

The conclusion some might draw from this is that these people are not truly responsible for their heinous acts, although they should still be removed from society. They need help, not punishment, because of a problem with their brain. In a sense, they may not be acting with free will when they commit their crimes.

An experiment done by Dr. Benjamin Libet in 1985 casts doubt on the very existence of free will. Let's say that you are asking subjects to watch a clock and then to note precisely when they decide to move a finger. Using EEG scans, one can detect exactly when the brain makes this decision. When you compare the two times, you will find a mismatch. The EEG scans show that the brain has actually made the decision about three hundred milliseconds before the person becomes aware of it.

This means that, in some sense, free will is a fake. Decisions are made ahead of time by the brain, without the input of consciousness, and then later the brain tries to cover this up (as it's wont to do) by claiming that the decision was conscious. Dr. Michael Sweeney concludes, "Libet's findings suggested that the brain knows what a person will decide *before* the person does. . . . The world must reassess not only the idea of movements divided between voluntary and involuntary, but also the very idea of free will."

All this seems to indicate that free will, the cornerstone of society, is a fiction, an illusion created by our left brain. So are we masters of our fate, or just pawns in a swindle perpetuated by the brain?

There are several ways to approach this sticky question. Free will goes against a philosophy called determinism, which simply says that all future events are determined by physical laws. According to Newton himself, the universe was some sort of clock, ticking away since the beginning of time, obeying the laws of motion. Hence all events are predictable.

The question is: Are we part of this clock? Are all our actions also determined? These questions have philosophical and theological implications. For example, most religions adhere to some form of determinism and predestination. Since God is omnipotent, omniscient, and omnipresent, He knows the future, and hence the future is determined ahead of time. He knows even before you are born whether you will go to Heaven or Hell.

The Catholic Church split in half on this precise question during the Protestant revolution. According to Catholic doctrine at that time, one could change one's ultimate fate with an indulgence, usually by making generous financial donations to the Church. In other words, determinism could be altered by the size of your wallet. Martin Luther specifically singled out the corruption of the Church over indulgences when he tacked his 95 Theses on the door of a church in 1517, triggering the Protestant Reformation. This was one of the key reasons why the Church split down the middle, causing casualties in the millions and laying waste to entire regions of Europe.

But after 1925, uncertainty was introduced into physics via quantum mechanics. Suddenly everything became uncertain; all you could calculate was probabilities. In this sense, perhaps free will does exist, and it's a manifestation of quantum mechanics. So some claim that the quantum theory reestablishes the concept of free will. The determinists have fought back, however, claiming that quantum effects are extremely small (at the level of atoms), too small to account for the free will of large human beings.

The situation today is actually rather muddled. Perhaps the question "Does free will exist?" is like the question "What is life?" The discovery of DNA has rendered that question about life obsolete. We now realize that the question has many layers and complexities. Perhaps the same applies to free will, and there are many types.

If so, the very definition of "free will" becomes ambiguous. For example, one way to define free will is to ask whether behavior can be predicted. If free will exists, then behavior cannot be determined ahead of time. Let's say you watch a movie, for example. The plot is completely determined, with no free will whatsoever. So the movie is completely predictable. But our world cannot be like a movie, for two reasons. The first is the quantum theory, as we have seen. The movie represents only one possible timeline. The second reason is chaos theory. Although classical physics says that all of the motions of atoms are completely determined and predictable, in practice it is impossible to predict their motions because there are so many atoms involved. The slightest disturbance of a single atom can have a ripple effect, which can cascade down to create enormous disturbances.

Think of the weather. In principle, if you knew the behavior of every atom in the air, you could predict the weather a century from now if you had a big enough computer. But in practice, this is impossible. After just a few hours, the weather becomes so turbulent and complex that any computer simulation is rendered useless.

This creates what is called the "butterfly effect," which means that even the beat of butterfly wings can cause tiny ripples in the atmosphere, which grow and in turn can escalate into a thunderstorm. So if even the flapping of butterfly wings can create thunderstorms, the hope of accurately predicting the weather is far-fetched.

Let's go back to the thought experiment described to me by Stephen Jay Gould. He asked me to imagine Earth 4.5 billion years ago, when it was born. Now imagine you could somehow create an identical copy of Earth, and let it evolve. Would we still be here on this different Earth 4.5 billion years later?

One could easily imagine, due to quantum effects or the chaotic nature of the weather and oceans, that humanity would never evolve into precisely the same creatures on this version of Earth. So ultimately, it seems a combination of uncertainty and chaos makes a perfectly deterministic world impossible.

THE QUANTUM BRAIN

This debate also affects the reverse engineering of the brain. If you can successfully reverse engineer a brain made of transistors, this success implies that the brain is deterministic and predictable. Ask it any question and it repeats the exact same answer. Computers are deterministic in this way, since they always give the same answer for any question.

So it seems we have a problem. On one hand, quantum mechanics and chaos theory claim that the universe is not predictable, and therefore, free will seems to exist. But a reverse-engineered brain, made of transistors, would by definition be predictable. Since the reverse-engineered brain is theoretically identical to a living brain, then the human brain is also deterministic and there is no free will. Clearly, this contradicts the first statement.

A minority of scientists claim that you cannot authentically reverse engineer the brain, or ever create a true thinking machine, because of the quantum theory. The brain, they argue, is a quantum device, not just a collection of transistors. Hence this project is doomed to fail. In this camp is Oxford physicist Dr. Roger Penrose, an authority on Einstein's theory of relativity, who claims that it is quantum processes that may account for the consciousness of the human brain. Penrose starts by saying that mathematician Kurt Gödel has proven that arithmetic is incomplete; that is, that there are true statements in arithmetic that cannot be proven using the axioms of arithmetic. Similarly, not only is mathematics incomplete, but so is physics. He concludes by stating that the brain is basically a quantum mechanical device and there are problems that no machine can solve because of Gödel's incompleteness theorem. Humans, however, can make sense of these conundrums using intuition.

Similarly, the reverse-engineered brain, no matter how complex, is still a collection of transistors and wires. In such a deterministic system, you can accurately predict its future behavior because the laws of motion are well known. In a quantum system, however, the system is inherently unpredictable. All you can calculate are the chances that something will occur, because of the uncertainty principle.

If it turns out that the reverse-engineered brain cannot reproduce human behavior, then scientists may be forced to admit that there are unpredictable forces at work (i.e., quantum effects inside the brain). Dr. Penrose argues that inside the neuron there are tiny structures, called microtubules, where quantum processes dominate.

At present, there is no consensus on this problem. Judging from the reaction to Penrose's idea when it was first proposed, it would be safe to say that most of the scientific community is skeptical of his approach. Science, however, is never conducted as a popularity contest, but instead advances through testable, reproducible, and falsifiable theories.

For my own part, I believe transistors cannot truly model all the behaviors of neurons, which carry out both analog and digital calculations. We know that neurons are messy. They can leak, misfire, age, die, and are sensitive to the environment. To me, this suggests that a collection of transistors can only approximately model the behavior of neurons. For example, we saw earlier, in discussing the physics of the brain, that if the axon of the neuron becomes thinner, then it begins to leak and also does not carry out chemical reactions that well. Some of this leakage and these misfires will be due to quantum effects. As you try to imagine neurons that are thinner, denser, and faster, quantum effects become more obvious. This means that even for normal neurons there are problems of leakage and instabilities, and these problems exist both classically and quantum mechanically.

In conclusion, a reverse-engineered robot will give a good but not perfect approximation of the human brain. Unlike Penrose, I think it is possible to create a deterministic robot out of transistors that gives the appearance of consciousness, but without any free will. It will pass the Turing test. But I think there will be differences between such a robot and humans due to these tiny quantum effects.

Ultimately, I think free will probably does exist, but it is not the free will envisioned by rugged individualists who claim they are complete masters of their fate. The brain is influenced by thousands of unconscious factors that predispose us to make certain choices ahead of time, even if we think we made them ourselves. This does not necessarily mean that we are actors in a film that can be rewound anytime. The end of the movie hasn't been written yet, so strict determinism is destroyed by a subtle combination of quantum effects and chaos theory. In the end, we are still masters of our destiny.

NOTES

INTRODUCTION

1 **You may have to travel**: To see this, define "complex" in terms of the total amount of information that can be stored. The closest rival to the brain might be the information contained within our DNA. There are three billion base pairs in our DNA, each one containing one of four nucleic acids, labeled A,T,C,G. Therefore the total amount of information we can store in our DNA is four raised to the three-billionth power. But the brain can store much more information among its one hundred billion neurons, which can either fire or not fire. Hence, there are two raised to the one-hundred-billionth power possible initial states of the human brain. But while DNA is static, the states of the brain change every few milliseconds. A simple thought may contain one hundred generations of neural firings. Hence, there are two raised to one hundred billion, all raised to the hundredth power, possible thoughts contained in one hundred generations. But our brains are continually firing, day and night, ceaselessly computing. Therefore the total number of thoughts possible within N generations is two raised to the one-hundred-billionth power, all raised to the Nth power, which is truly astronomical. Therefore the amount of information that we can store in our brains far exceeds the information stored within our DNA by a wide margin. In fact, it is the largest amount of information that we can store in our solar system, and even possibly in our sector of the Milky Way galaxy.

5 **"The most valuable insights":** Boleyn-Fitzgerald, p. 89.

5 **"All of these questions that philosophers":** Boleyn-Fitzgerald, p. 137.

CHAPTER 1: UNLOCKING THE MIND

15 **He was semiconscious for weeks:** See Sweeney, pp. 207–8.
15 **Dr. John Harlow, the doctor who treated:** Carter, p. 24.
15 **In the year A.D. 43, records show:** Horstman, p. 87.
16 **"It was like . . . standing in the doorway":** Carter, p. 28.
30 **The Transparent Brain:** *New York Times,* April 10, 2013, p. 1.
34 **"Emotions are not feelings at all":** Carter, p. 83.
34 **the mind is more like a "society of minds":** Interview with Dr. Minsky for the BBC-TV series *Visions of the Future,* February 2007. Also, interview for *Science Fantastic* national radio broadcast, November 2009.
35 **consciousness was like a storm raging:** Interview with Dr. Pinker in September 2003 for *Exploration* national radio broadcast.
35 **"the intuitive feeling we have":** Pinker, "The Riddle of Knowing You're Here," in *Your Brain: A User's Guide* (New York: Time Inc. Specials, 2011).
35 **Consciousness turns out to consist of:** Boleyn-Fitzgerald, p. 111.
38 **"indeed a conscious system in its own right":** Carter, p. 52.
38 **I asked him how experiments:** Interview with Dr. Michael Gazzaniga in September 2012 for *Science Fantastic* national radio broadcast.
39 **"The possible implications of this":** Carter, p. 53.
39 **"If that person dies, what happens?":** Boleyn-Fitzgerald, p. 119.
39 **a young king who inherits:** Interview with Dr. David Eagleman in May 2012 for *Science Fantastic* national radio broadcast.
40 **"people named Denise or Dennis":** Eagleman, p. 63.
40 **"at least 15% of human females":** Eagleman, p. 43.

CHAPTER 2: CONSCIOUSNESS—A PHYSICIST'S VIEWPOINT

41 **"We cannot see ultraviolet light":** Pinker, *How the Mind Works,* pp. 561–65.
42 **"Everybody knows what consciousness is":** *Biological Bulletin* 215, no. 3 (December 2008): 216.
45 **We will do so in the notes:** Level II consciousness can be counted by listing the total number of distinct feedback loops when an animal interacts with members of its species. As a rough guess, Level II consciousness can be approximated by multiplying the number of others in an animal's pack or tribe, multiplied by the total number of distinct emotions or gestures it uses to communicate with others. There are caveats to this ranking, however, since this is just a first guess.

For example, animals like the wildcat are social, but they are also solitary hunters, so it appears as if the number of animals in its pack is one. But that is true only when it is hunting. When it is time to reproduce, wildcats engage in complex mating rituals, so its Level II consciousness must take this into account.

Furthermore, when female wildcats give birth to litters of kittens, which have to be nursed and fed, the number of social interactions increases as a consequence. So even for solitary hunters, the number of members of its species that it interacts with is not one, and the total number of distinct feedback loops can be quite large.

Also, if the number of wolves in the pack decreases, then it appears as if its Level II number decreases correspondingly. To account for this, we have to introduce the concept of an average Level II number that is common for the entire species, as well as a specific Level II consciousness for an individual animal.

The average Level II number for a given species does not change if the pack gets smaller, because it is common for the entire species, but the individual Level II number (because it measures individual mental activity and consciousness) does change.

When applied to humans, the average Level II number must take into account the Dunbar number, which is 150, and represents roughly the number of people in our social grouping that we can keep track of. So the Level II number for humans as a species would be the total number of distinct emotions and gestures we use to communicate, multiplied by the Dunbar number of 150. (Individuals can have different levels of Level II consciousness, since their circle of friends and the ways they interact with them can vary considerably.)

We should also note that certain Level I organisms (like insects and reptiles) can exhibit social behaviors. Ants, when they bump into one another, exchange information via chemical scents, and bees dance to communicate the location of flower beds. Reptiles even have a primitive limbic system. But in the main, they do not exhibit emotions.

46 **"The difference between man":** Gazzaniga, p. 27.
46 **"The greatest achievement of the human brain":** Gilbert, p. 5.
46 **"area 10 (the internal granular layer IV)":** Gazzaniga, p. 20.
48 **The male gets confused, because it wants:** Eagleman, p. 144.
55 **"I predict that mirror neurons":** Brockman, p. xiii.
58 **Biologist Carl Zimmer writes:** Bloom, p. 51.
58 **"Most of the time we daydream":** Bloom, p. 51.
59 **I asked one person who may:** Interview with Dr. Michael Gazzaniga in September 2012 for *Science Fantastic* national radio broadcast.
59 **"It is the left hemisphere":** Gazzaniga, p. 85.

CHAPTER 3: TELEPATHY—A PENNY FOR YOUR THOUGHTS

64 **Indeed, in a recent "Next 5 in 5":** http://www.ibm.com/5in5.
65 **I had the pleasure of touring:** Interview with Dr. Gallant on July 11, 2012, at the University of California, Berkeley. Also, interview with Dr. Gallant on *Science Fantastic* for national radio, July 2012.
65 **"This is a major leap forward":** Berkeleyan Newsletter, September 22, 2011, http://newscenter.bekrely.edu/20211/09/22/brain-movies.htm.
67 **"If you take 200 voxels":** Brockman, p. 236.
67 **Dr. Brian Pasley and his colleagues:** Visit to Dr. Pasley's laboratory on July 11, 2012, at the University of California, Berkeley.
68 **Similar results were obtained:** The Brain Institute, University of Utah, Salt Lake City, http://brain.utah.edu.
69 **This could have applications for artists:** http://io9 /543338/a-device-that-lets-io9 .com/543338/a-device-that-lets-ou-type-with-your-mind.
69 **According to their officials:** http://news.discovery.com/tech/type-with-your -mind-110309.html.
71 **being explored by Dr. David Poeppel:** *Discover Magazine Presents the Brain,* Spring 2012, p. 43.
72 **In 1993 in Germany:** *Scientific American,* November 2008, p. 68.
73 **The only justification for its existence:** Garreau, pp. 23–24.
74 **I once had lunch with:** Symposium on the future of science sponsored by the Sci-

ence Fiction Channel at the Chabot Pace and Science Center, Oakland, California, in May 2004.

74 **On another occasion:** Conference in Anaheim, California, April 2009.

74 **He says, "Imagine if soldiers":** Garreau, p. 22.

75 **"What he is doing is spending":** Ibid., p. 19.

75 **When I asked Dr. Nishimoto:** Visit to Dr. Gallant's laboratory at the University of California, Berkeley, on July 11, 2012.

78 **"There are ethical concerns":** http://www.nbcnews.com/health/words-from-brain-waves-may-let-scientists-read-your-mind-1C6435988.

CHAPTER 4: TELEKINESIS: MIND CONTROLLING MATTER

81 **"I would love to have":** *New York Times,* May 17, 2012, p. A17 and http://www.msnbc.mns.com/id/47447302/ns/health-health_care/t/parallyzed-woman-gets-robotic-arm.html.

81 **"We have taken a tiny sensor":** Interview with Dr. John Donoghue in November 2009 for *Science Fantastic* national radio broadcast.

82 **In the United States alone, more than two hundred thousand:** Centers for Disease Control and Prevention, Washington, D.C. http://www.cdc.gov/traumaticbraininjury/scifacts.html.

82 **When the monkey wanted to move:** http://deptwww.physio.northwestern.edu/faculty/Miller.htm; http://www.northwestern.edu/newscenter/stories/2012/04/miller-paralyzed-technology.html.

83 **"We are eavesdropping on the natural":** http://www.northwestern.edu/newscenter/stories/2012/04/miller-paralyzed-technology.html.

84 **More than 1,300 service members:** http://www.darpa.mil/Our_Work/DSO/Programs/Revolutionizing_Prosthetics.aspx. CBS *60 Minutes,* broadcast on December 30, 2012.

84 **"They thought we were crazy":** Ibid.

84 **she appeared on *60 Minutes*:** Ibid.

85 **"There's going to be a whole ecosystem":** *Wall Street Journal,* May 29, 2012.

85 **But perhaps the most novel applications:** Interview with Dr. Nicolelis in April 2011 for *Science Fantastic* national radio broadcast.

85 **Smart Hands and Mind Melds:** *New York Times,* March 13, 2013, http://nytimes.com/2013/03/01/science/new-research-suggests-two-rat-brains-can-be-linked. See also *Huffington Post,* February 28, 2013, http://huffingtonpost.com/2013/2/28mind-melds-brain-communication.

87 **In 2013, the next important step:** *USA Today,* August 8, 2013, p. 1D.

91 **About ten years ago:** Interview with Dr. Nicolelis in April 2011.

91 **"so there's nothing sticking out":** For a full discussion of the exoskeleton, see Nicolelis, pp. 303–7.

95 **The Honda Corporation has:** http://www.asimo.honda.com. Also, interview with the creators of ASIMO in April 2007 for the BBC-TV series *Visions of the Future.*

96 **Eventually, you get the hang:** http://discovermagazine.com/2007/may/review-test-driving-the-future.

96 **Then, by thinking, the patient:** *Discover,* December 9, 2011, http://discovermagazine.com/2011/dec/09-mind-over-motor-controlling-robots-with-your-thoughts.

96 **"We will likely be able to operate":** Nicolelis, p. 315.

100 **I saw a demonstration of this:** Interview with the scientists at Carnegie Mellon in August 2010 for the Discovery/Science Channel TV series *Sci Fi Science.*

CHAPTER 5: MEMORIES AND THOUGHTS MADE TO ORDER

105 **"It has all come together":** Wade, p. 89.
107 **So far, scientists have identified:** Ibid., p. 91.
107 **For instance, Dr. Antonio Damasio:** Damasio, pp. 130–53.
107 **One fragment of memory might:** Wade, p. 232.
108 **"If you can't do it":** http://www.newscientist.com/article/dn3488.
109 **"Turn the switch on":** http://www.eurekalert.org/pub_releases/2011-06/uosc-rmr06211.php.
109 **"Using implantables to enhance competency":** http://hplusmagazine.com/2009/03/18/artificial-hipppocampus.
109 **Not surprisingly, with so much at stake:** http://articles.washingtonpost.com/2013-07-12/national/40863765_1_brain-cells-mice-new-memories.
113 **If encoding the memory:** This brings up the question of whether carrier pigeons, migratory birds, whales, etc., have a long-term memory, given that they can migrate over hundreds to thousands of miles in search of feeding and breeding grounds. Science knows little about this question. But it is believed that their long-term memory is based on locating certain landmarks along the way, rather than recalling elaborate memories of past events. In other words, they do not use memory of past events to help them simulate the future. Their long-term memory consists of just a series of markers. Apparently, only in humans are long-term memories used to help simulate the future.
113 **"The purpose of memory is":** Michael Lemonick, "Your Brain: A User's Guide," *Time,* December 2011, p. 78.
113 **"You might look at it":** http://sciencedaily.com/videos/2007/0210-brain_scans_of_the_future.htm.
114 **their study proves a "tentative answer":** http://www.sciencedaily.com/videos/2007/0710.
114 **"The whole idea is that the device":** *New York Times,* September 12, 2012, p. A18.
115 **"It will likely take us":** http://www.tgdaily.com/general-sciences-features/58736-artificial-cerebellum-restores-rats.
116 **There are 5.3 million Americans:** Alzheimer's Foundation of America, http://www.alzfdn.org.
118 **"This adds to the notion":** ScienceDaily.com, October 2009, http://www.sciencedaily.com/releases/2009/10/091019122647.htm.
118 **"We can never turn it into":** Ibid.
119 **"This implies these flies have":** Wade, p. 113.
119 **This effect is not just restricted:** Ibid.
119 **"We can now give you":** Ibid., p. 114.
120 **Basically, the more CREB proteins:** Bloom, p. 244.
123 **"Propranolol sits on that nerve cell":** SATI e-News, June 28, 2007, http://www.mysati.com/enews/June2007/ptsd.htm.
123 **Its report concluded:** Boleyn-Fitzgerald, p. 104.
123 **"Our breakups, our relationships":** Ibid.
123 **"should we deprive them of morphine":** Ibid., p. 105.

124 **"If further work confirms this view":** Ibid., p. 106.

127 **"Each of these perennial records":** Nicolelis, p. 318.

128 **"Forgetting is the most beneficial process":** *New Scientist,* March 12, 2003, http://www.newscientists.com/article/dn3488.

CHAPTER 6: EINSTEIN'S BRAIN AND ENHANCING INTELLIGENCE

131 **"got caught up in the moment":** http://abcnews.go.com/blogs/headlines/2012/03/einsteins-brainarrives-in-on-afterdd.

133 **"I have always maintained that":** Gould, p. 109.

134 **"The human brain remains 'plastic'":** www.sciencedaily.com/releases/2011/12/111208257120.htm.

135 **"The emerging picture from such studies":** Gladwell, p. 40.

136 **Five years later, Terman started:** See C. K. Holahan and R.R. Sears, *The Gifted Group in Later Maturity* (Stanford, CA: Stanford University Press, 1995).

137 **"Your grades in school":** Boleyn-Fitzgerald, p. 48.

137 **"Tests don't measure motivation":** Sweeney, p. 26.

138 **The pilots who scored highest:** Bloom, p. 12.

138 **"The left hemisphere is responsible":** Ibid., p. 15.

142 **Dr. Darold Treffert, a Wisconsin physician:** http://www.daroldtreffert.com.

142 **It took him just forty-five seconds:** Tammet, p. 4.

142 **I had the pleasure of interviewing:** Interview with Mr. Daniel Tammet in October 2007 for *Science Fantastic* national radio broadcast.

144 **"Our study confirms":** *Science Daily,* March 2012, http://www.sciencedaily.com/releases/2012/03/120322100313.htm.

144 **Kim Peek's brain:** AP wire story, November 8, 2004, http://www.Space.com.

145 **In 1998, Dr. Bruce Miller:** *Neurology* 51 (October 1998): pp. 978–82. See also http://www.wisconsinmedicalsociety.org/savant_syndrome/savant-articles/acquired_savant.

145 **In addition to the savants:** Sweeney, p. 252.

146 **This idea has actually been tried:** Center of the Mind, Sydney, Australia, http://www.centerofthemind.com.

146 **In another experiment, Dr. R. L. Young:** R. L. Young, M. C. Ridding, and T. L. Morrell, "Switching Skills on by Turning Off Part of the Brain," *Neurocase* 10 (2004): 215, 222.

146 **"When applied to the prefrontal lobes":** Sweeney, p. 311.

147 **Until recently, it was thought:** *Science Daily,* May 2012, http://www.sciencedaily.com/releases/2012/05/120509180113.htm.

148 **"Savants have a high capacity":** Ibid.

150 **In 2007, a breakthrough occurred:** Sweeney, p. 294.

150 **"Stem cell research and regenerative medicine":** Sweeney, p. 295.

150 **Scientists have focused on a few genes:** Katherine S. Pollard, "What Makes Us Different," *Scientific American Special Collectors Edition* (Winter 2013): 31–35.

151 **"I jumped at the opportunity":** Ibid.

152 **"With my mentor David Haussler":** Ibid.

155 **One such gene was discovered:** *TG Daily,* November 15, 2012. http://www.tgdaily.com/general-sciences-features/67503-new-found-gene-separates-man-from-apes.

157 **Many theories have been proposed:** See, for example, Gazzaniga, *Human: The Science Behind What Makes Us Unique.*

159 **"For the first few hundred million years":** Gilbert, p. 15.
161 **"Cortical gray matter neurons are working":** Douglas Fox, "The Limits of Intelligence," *Scientific American,* July 2011, p. 43.
161 **"You might call it the mother":** Ibid., p. 42.

CHAPTER 7: IN YOUR DREAMS

171 **He followed this up with one thousand:** C. Hall and R. Van de Castle, *The Content Analysis of Dreams* (New York: Appleton-Century-Crofts, 1966).
174 **When I interviewed him, he told me:** Interview with Dr. Allan Hobson in July 2012 for *Science Fantastic* national radio broadcast.
175 **Studies have shown that it is possible:** Wade, p. 229.
176 **ATR chief scientist Yukiyasu Kamitani:** *New Scientist,* December 12, 2008, http://www.newscientistcom/article/dn16267-mindreading-software-coudl-record-your-dreams.html.
176 **When I visited the laboratory:** Visit to Dr. Gallant's laboratory on July 11, 2012.
178 **"Our dreams are therefore not":** *Science Daily,* October 28, 2011, http://www.sciencedaily.com/releases/2011/111028113626.htm.
179 **Already, prototypes of Internet contact lenses:** See the work of Dr. Babak Parviz, http://www.wearable-technologies.com/262.

CHAPTER 8: CAN THE MIND BE CONTROLLED?

181 **A raging bull is released:** Miguel Nicolelis, *Beyond Boundaries* (New York: Henry Holt, 2011), pp. 228–32.
184 **The cold war hysteria eventually reached:** "Project MKUltra, the CIA's Program of Research into Behavioral Modification. Joint Hearings Before the Select Committee on Human Resources, U.S. Senate, 95th Congress, First Session," Government Printing Office, August 8, 1977, Washington, D.C., http://www.nytimes.com/packages/pdf/national/13inmate_ProjectMKULTRA.pdf; "CIA Says It Found More Secret Papers on Behavior Control," *New York Times,* September 3, 1977; "Government Mind Control Records of MKUltra and Bluebird/Artichoke," http://wanttoknow.info/mindcontrol.shtml; "The Select Committee to Study Governmental Operations with Respect to Intelligence Activities, Foreign and Military Intelligence." The Church Committee Report No. 94-755, 94th Congress, 2nd Session, p. 392, Government Printing Office, Washington, D.C., 1976; "Project MKUltra, the CIA's Program of Research in Behavior Modification," http://scribd.com/doc/75512716/Project-MKUltra-The-CIA-s-Program-of-Research-in-Behavior-Modification.
185 **"great potential for development":** Rose, p. 292.
185 **"neuro-scientific impossibility":** Ibid., p. 293.
186 **"It is probably significant that":** "Hypnosis in Intelligence," Black Vault Freedom of Information Act Archive, 2008, http://documents.theblackvault.com/documents/mindcontrol/hypnosisinintelligence.pdf.
188 **To see how widespread this problem:** Boleyn-Fitzgerald, p. 57.
189 **Drugs like LSD:** Sweeney, p. 200.
189 **"This is the first time we've shown":** Boleyn-Fitzgerald, p. 58.
191 **"If you want to turn off":** http://www.nytimes.com/2011/05/17/science/17 optics.html.

191 **"By feeding information from sensors":** *New York Times,* March 17, 2011, http://nytimes.com/2011/05/17/science/17optics.html.

CHAPTER 9: ALTERED STATES OF CONSCIOUSNESS

196 **"Some fraction of history's prophets":** Eagleman, p. 207.
197 **"Sometimes it's a personal God":** Boleyn-Fitzgerald, p. 122.
197 **"'Finally, I see what it is'":** Ramachandran, p. 280.
197 **"During the three minute bursts":** David Biello, *Scientific American,* p. 41, www.sciammind.com.
198 **To test these ideas:** Ibid., p. 42.
199 **"Although atheists might argue":** Ibid., p. 45.
199 **"If you are an atheist":** Ibid., p. 44.
200 **One theory holds that Parkinson's:** Sweeney, p. 166.
201 **"Neurons wired for the sensation":** Ibid., p. 90.
203 **"The brain's gonna do what":** Ibid., p. 165.
204 **"Brain scans have led researchers":** Ibid., p. 208.
204 **"If left unchecked, the left hemisphere":** Ramachandran, p. 267.
205 **Underactivity in this area:** Carter, pp. 100–103.
208 **Ten percent of them, in turn:** Baker, pp. 46–53.
208 **"Depression 1.0 was psychotherapy":** Ibid., p. 3.
210 **One to three percent of DBS patients:** Carter, p. 98.
211 **"The calcium channels findings suggest":** *New York Times,* February 26, 2013, http://www.nytimes.com/2013/03/01/health/study-finds-genetic-risk-factors-shared-by-5.
211 **"What we have identified here":** Ibid.

CHAPTER 10: THE ARTIFICIAL MIND AND SILICON CONSCIOUSNESS

216 **"Machines will be capable":** Crevier, p. 109.
216 **"within a generation . . . the problem":** Ibid.
217 **"It's as though a group of people":** Kaku, p. 79.
217 **"I would pay a lot for a robot":** Brockman, p. 2.
219 **However, I met privately with:** Interview with the creators of ASIMO during a visit to Honda's laboratory in Nagoya, Japan, in April 2007 for the BBC-TV series *Visions of the Future.*
221 **he used to marvel at the mosquito:** Interview with Dr. Rodney Brooks in April 2002 for *Exploration* national radio broadcast.
227 **I have had the pleasure of visiting:** Visit to MIT Media Laboratory for the Discovery/Science Channel TV series *Sci Fi Science,* April 13, 2010.
227 **"That is why Breazeal decided":** Moss, p. 168.
227 **At Waseda University:** Gazzaniga, p. 352.
229 **Their goal is to integrate:** Ibid., p. 252.
229 **Meet Nao:** *Guardian*, August 9, 2010, http://www.guardian.co.uk/technology/2010/aug/09/nao-robot-develop-emotions.htm.
230 **"It's hard to predict the future":** http://cosmomagazine.com/news/4177/reverse-engineering-brain.

230 **Neuroscientists like Dr. Antonio Damasio:** Damasio, pp. 108–29.
240 **"In mathematics, you don't understand":** Kurzweil, p. 248.
240 **"There could not be an objective test":** Pinker, "The Riddle of Knowing You're Here," *Your Brain: A User's Guide,* Winter, 2011, p. 19.
241 **At Meiji University:** Gazzaniga, p. 352.
241 **"To our knowledge, this is the first":** Kurzweil.net, August 24, 2012, http://www.kurzweilai.net/robot-learns-self-awareness. See also *Yale Daily News,* September 25, 2012, http://yaledailynews.blog/2012/09/25/first-self-aware-robot-created.
245 **When I interviewed Dr. Hans Moravec:** Interview with Dr. Hans Moravec in November 1998 for *Exploration* national radio broadcast.
245 **"Unleashed from the plodding pace":** Sweeney, p. 316.
247 **When I interviewed Dr. Rodney Brooks:** Interview with Dr. Brooks in April 2002 for *Exploration* national radio broadcast.
248 **"We don't like to give up":** TEDTalks, http://www.ted.com/talks/lang/en/rodney_brooks_on_robots.html.
249 **Similarly, at the University of Southern California:** http://phys.org/news205059692.html.

CHAPTER 11: REVERSE ENGINEERING THE BRAIN

250 **Almost simultaneously, the European Commission:** http://actu.epfl.ch/news/the-human-brain-project-wins-top-european-science.
256 **"It's essential for us to understand":** http://ted.com/talks/henry_markram_supercomputing_the_brain's_secrets.html.
256 **"There's not a single neurological disease":** Kushner, p. 19.
257 **"I think we're far from playing God":** Ibid., p. 2.
260 **"In a hundred years, I'd like":** Sally Adee, "Reverse Engineering the Brain," *IEEE Spectrum,* http://spectrum.ieee.org/biomedical/ethics/reverse-engineering-the-brain.
261 **"Researchers have conjectured":** http://cnn.com/2012/01/tech/innovation/brain-map-connectome/index.html.
261 **"In the seventeenth century":** http://www.ted.com/talks/lang.en/sebastian_seung.html.
262 **"The Allen Human Brain Atlas provides":** http://ts-si.org/neuroscience/29735-allen-human-brain-atlas-updates-with-comprehensive).
264 **According to Dr. V. S. Ramachandran:** TED Talks, January 2010, http://www.ted.com.

CHAPTER 12: THE FUTURE: MIND BEYOND MATTER

267 **5.8 percent claimed they had an out-of-body:** Nelson, p. 137.
268 **"I see myself lying in bed":** Ibid., p. 140.
269 **Notably, temporary loss of blood:** *National Geographic News,* April 8, 2010, http://news.nationalgeographic.com/news/2010/04/100408-near-death-experiences-blood-carbon.htm; Nelson, p. 126
270 **Dr. Thomas Lempert, neurologist:** Nelson, p. 126.
270 **The U.S. Air Force, for example:** Ibid., p. 128.

271 **We once spoke at a conference:** Dubai, United Arab Emirates, November 2012. Interviewed in February 2003 for *Exploration* national radio broadcast. Interviewed in October 2012 for *Science Fantastic* national radio broadcast.

271 **By 2055, $1,000 of computing power:** Bloom, p. 191.

272 **For example, Bill Gates, cofounder:** Sweeney, p. 298.

272 **"People who predict a very utopian future":** Carter, p. 298.

273 **He told me that the San Diego Zoo:** Interview with Dr. Robert Lanza in September 2009 for *Exploration* national radio broadcast.

275 **"Should we ridicule the modern seekers":** Sebastian Seung, TEDTalks, http://www.ted.com/talks/lang/en/sebastian_seung.html.

276 **In 2008, BBC-TV aired:** http://www.bbc.com.uk/sn/tvradio/programmes/horizon/broadband/tx/isolation.

280 **we will be able to reverse engineer:** Interview with Dr. Moravec in November 1998 for *Exploration* national radio broadcast.

282 **On the other side was Eric Drexler:** See a series of letter in *Chemical and Engineering News* from 2003 to 2004.

283 **"I'm not planning to die":** Garreau, p. 128.

CHAPTER 13: THE MIND AS PURE ENERGY

284 **"Wormholes, extra dimensions":** Sir Martin Rees, *Our Final Hour* (New York: Perseus Books, 2003), p. 182.

CHAPTER 14: THE ALIEN MIND

297 **So far, more than one thousand:** Kepler Web Page, http://kepler.nasa.gov.

297 **In 2013, NASA scientists announced:** Ibid.

299 **how they can distinguish false messages:** Interview with Dr. Wertheimer in June 1999 for *Exploration* national radio broadcast.

299 **I once asked him about the giggle factor:** Interview with Dr. Seth Shostak in May 2012 for *Science Fantastic* national radio broadcast.

299 **He has gone on record:** Ibid.

300 **"Remember, this is the same government":** Davies, p. 22.

303 **The Greek writers:** Sagan, p. 221.

303 **But St. Thomas Aquinas:** Ibid.

303 **We can be fooled:** Ibid.

303 **"If the fact that brutes abstract":** Ibid., p. 113.

304 **"In the blind and deaf world":** Eagleman, p. 77.

311 **we have to expand our own horizon:** Interview with Dr. Paul Davies in April 2012 for *Science Fantastic* national radio broadcast.

311 **"My conclusion is a startling one":** Davies, p. 159.

315 **"Although there is only a tiny probability":** *Discovery News,* December 27, 2011, http://news.discovery.com/space/seti-to-scour-the-moon-for-alien-tech-111227.htm.

CHAPTER 15: CONCLUDING REMARKS

317 **In an article in *Wired*:** *Wired,* April 2000, http://www.wired.com/wired/archive/8.04/joy.html.

318 **"several separate and unequal species":** Garreau, p. 139.

318 **"This techno utopia is all about":** Ibid., p. 180.

322 **"The idea that we are messin'":** Ibid., p. 353.

322 **"Technologies—such as gunpowder":** Ibid., p. 182.

325 **"The *you* that all your friends know":** Eagleman, p. 205.

325 **"Our reality depends on what":** Ibid., p. 208.

326 **"How it is that anything so remarkable":** Pinker, p. 132.

326 **somehow create a twin of the Earth:** Interview with Dr. Stephen Jay Gould in November 1996 for *Exploration* national radio broadcast.

326 **"*Homo sapiens* is one small twig":** Pinker, p. 133.

327 **"nothing gives life more purpose":** Pinker, "The Riddle of Knowing You're Here," *Time: Your Brain: A User's Guide* (Winter 2011), p. 19.

327 **"What a perplexing masterpiece":** Eagleman, p. 224.

APPENDIX: QUANTUM CONSCIOUSNESS?

338 **many (but not all) pathological killers:** Interview with Dr. Simon Baron-Cohen in July 2005 for *Exploration* national radio broadcast.

338 **Dr. Michael Sweeney concludes, "Libet's findings":** Sweeney, p. 150.

SUGGESTED READING

Baker, Sherry. "Helen Mayberg." *Discover Magazine Presents the Brain.* Waukesha, WI: Kalmbach Publishing Co., Fall 2012.

Bloom, Floyd. *Best of the Brain from Scientific American: Mind, Matter, and Tomorrow's Brain.* New York: Dana Press, 2007.

Boleyn-Fitzgerald, Miriam. *Pictures of the Mind: What the New Neuroscience Tells Us About Who We Are.* Upper Saddle River, N.J.: Pearson Education, 2010.

Brockman, John, ed. *The Mind: Leading Scientists Explore the Brain, Memory, Personality, and Happiness.* New York: Harper Perennial, 2011.

Calvin, William H. *A Brief History of the Mind.* New York: Oxford University Press, 2004.

Carter, Rita. *Mapping the Mind.* Berkeley: University of California Press, 2010.

Crevier, Daniel. *AI: The Tumultuous History of the Search for Artificial Intelligence.* New York: Basic Books, 1993.

Crick, Francis. *The Astonishing Hypothesis: The Science Search for the Soul.* New York: Touchstone, 1994.

Damasio, Antonio. *Self Comes to Mind: Constructing the Conscious Brain.* New York: Pantheon Books, 2010.

Davies, Paul. *The Eerie Silence: Renewing Our Search for Alien Intelligence.* New York: Houghton Mifflin Harcourt, 2010.

Dennet, Daniel C. *Breaking the Spell: Religion as a Natural Phenomenon.* New York: Viking, 2006.

———. *Conscious Explained.* New York: Back Bay Books, 1991.

DeSalle, Rob, and Ian Tattersall. *The Brain: Big Bangs, Behaviors, and Beliefs.* New Haven, CT: Yale University Press, 2012.

Eagleman, David. *Incognito: The Secret Lives of the Brain.* New York: Pantheon Books, 2011.

Fox, Douglas. "The Limits of Intelligence," *Scientific American,* July 2011.

Garreau, Joel. *Radical Evolution: The Promise and Peril of Enhancing Our Minds, Our Bodies—and What It Means to Be Human.* New York: Random House, 2005.

Gazzaniga, Michael S. *Human: The Science Behind What Makes Us Unique.* New York: HarperCollins, 2008.

Gilbert, Daniel. *Stumbling on Happiness.* New York: Alfred A. Knopf, 2006.

Gladwell, Malcolm. *Outliers: The Story of Success.* New York: Back Bay Books, 2008.

Gould, Stephen Jay. *The Mismeasure of Man.* New York: W. W. Norton, 1996.

Horstman, Judith. *The Scientific American Brave New Brain.* San Francisco: John Wiley and Sons, 2010.

Kaku, Michio. *Physics of the Future.* New York: Doubleday, 2009.

Kurzweil, Ray. *How to Create a Mind: The Secret of Human Thought Revealed.* New York: Viking Books, 2012.

Kushner, David. "The Man Who Builds Brains." *Discover Magazine Presents the Brain.* Waukesha, WI: Kalmbach Publishing Co., Fall 2001.

Moravec, Hans. *Mind Children: The Future of Robot and Human Intelligence.* Cambridge, MA: Harvard University Press, 1988.

Moss, Frank. *The Sorcerers and Their Apprentices: How the Digital Magicians of the MIT Media Lab Are Creating the Innovative Technologies That Will Transform Our Lives.* New York: Crown Business, 2011.

Nelson, Kevin. *The Spiritual Doorway in the Brain.* New York: Dutton, 2011.

Nicolelis, Miguel. *Beyond Boundaries: The New Neuroscience of Connecting Brains with Machines—and How It Will Change Our Lives.* New York: Henry Holt and Co., 2011.

Pinker, Steven. *How the Mind Works.* New York: W. W. Norton, 2009.

———. *The Stuff of Thought: Language as a Window into Human Nature.* New York: Viking, 2007.

———. "The Riddle of Knowing You're Here." In *Your Brain: A User's Guide.* New York: Time Inc. Specials, 2011.

Piore, Adam. "The Thought Helmet: The U.S. Army Wants to Train Soldiers to Communicate Just by Thinking." *The Brain, Discover Magazine Special,* Spring 2012.

Purves, Dale, et al., eds. *Neuroscience.* Sunderland, MA: Sinauer Associates, 2001.

Ramachandran, V. S. *The Tell-Tale Brain: A Neuroscientist's Quest for What Makes Us Human.* New York: W. W. Norton, 2011.

Rose, Steven. *The Future of the Brain: The Promise and Perils of Tomorrow's Neuroscience.* Oxford, UK: Oxford University Press, 2005.

Sagan, Carl. *The Dragons of Eden: Speculations on the Evolution of Human Intelligence.* New York: Ballantine Books, 1977.

Sweeney, Michael S. *Brain: The Complete Mind: How It Develops, How It Works, and How to Keep It Sharp.* Washington, D.C.: National Geographic, 2009.

Tammet, Daniel. *Born on a Blue Day: Inside the Extraordinary Mind of an Autistic Savant.* New York: Free Press, 2006.

Wade, Nicholas, ed. *The Science Times Book of the Brain.* New York: New York Times Books, 1998.

ILLUSTRATION CREDITS

Page 16: Jeffrey L. Ward

Page 17: Jeffrey L. Ward

Page 19: Jeffrey L. Ward

Page 21: Jeffrey L. Ward

Page 25 (top): AP Photo / David Duprey

Page 25 (bottom): Tom Barrick, Chris Clark / Science Source

Page 28: Jeffrey L. Ward

Page 53: Jeffrey L. Ward

Page 54: Jeffrey L. Ward

Page 56: Jeffrey L. Ward

Page 92: The Laboratory of Dr. Miguel Nicolelis, Duke University

Page 106: Jeffrey L. Ward

Page 228 (top): MIT Media Lab, Personal Robots Group

Page 228 (bottom): MIT Media Lab, Personal Robots Group, Mikey Siegel

INDEX

Page numbers in *italics* refer to illustrations. Page numbers beginning with 343 refer to endnotes.

ALSO BY

Michio Kaku

PHYSICS OF THE FUTURE

How Science Will Shape Human Destiny and Our Daily Lives by the Year 2100

Renowned theoretical physicist Michio Kaku details the developments in computer technology, artificial intelligence, medicine, space travel, and more, that are poised to happen over the next hundred years. He also considers how these inventions will affect the world economy, addressing the key questions: Who will have jobs? Which nations will prosper? In *Physics of the Future*, Kaku forecasts a century of earthshaking advances in technology that could make even the last centuries' leaps and bounds seem insignificant.

Physics

PARALLEL WORLDS

A Journey Through Creation, Higher Dimensions, and the Future of the Cosmos

In this thrilling journey into the mysteries of our cosmos, bestselling author Michio Kaku takes us on a dizzying ride to explore black holes and time machines, multidimensional space and, most tantalizing of all, the possibility that parallel universes may lay alongside our own. Kaku skillfully guides us through the latest innovations in string theory and its latest iteration, M-theory, which posits that our universe may be just one in an endless multiverse, a singular bubble floating in a sea of infinite bubble universes. If M-theory is proven correct, we may perhaps finally find answer to the question, "What happened before the big bang?" This is an exciting and unforgettable introduction into the new cutting-edge theories of physics and cosmology from one of the preeminent voices in the field.

Physics

PHYSICS OF THE IMPOSSIBLE

A Scientific Exploration into the World of Phasers, Force Fields, Teleportation, and Time Travel

Inspired by the fantastic worlds of *Star Trek*, *Star Wars*, and *Back to the Future*, renowned theoretical physicist and bestselling author Michio Kaku takes an informed, serious, and often surprising look at what our current understanding of the universe's physical laws may permit in the near and distant future. Entertaining, informative, and imaginative, *Physics of the Impossible* probes the very limits of human ingenuity and scientific possibility.

Physics

VISIONS

How Science Will Revolutionize the 21st Century

In *Visions*, physicist and author Michio Kaku examines the great scientific revolutions that have dramatically reshaped the twentieth century—the quantum mechanics, biogenetics, and artificial intelligence—and shows how they will change and alter science and the way we live. Science, for all its breathtaking change, evolves slowly; we can accurately predict, asserts Kaku, what the direction of science will be, based on the paths that are being forged today. A thrilling, unique narrative that brings together the thinking of many of the world's most accomplished scientists to explore the world of the future, *Visions* is science writing at its best.

Science